APPLIED
VECTOR
ANALYSIS

HARCOURT BRACE JOVANOVICH COLLEGE OUTLINE SERIES

APPLIED VECTOR ANALYSIS

Hwei P. Hsu

Fairleigh Dickinson University

Books for Professionals
Harcourt Brace Jovanovich, Publishers
San Diego   New York   London

Requests for permission to make copies of any part of the work should be mailed to:

Permissions
Harcourt Brace Jovanovich, Publishers
Orlando, Florida 32887

Printed in the United States of America

Library of Congress Cataloging in Publication Data

Hsu, Hwei P. (Hwei Piao), 1930-
    Applied vector analysis.

    (College outline series) (Books for professionals)
    Includes index.
    1. Vector analysis.   I. Title.   II. Series.
III. Series: Books for professionals
QA433.H77 1984   515'.63   84-592
ISBN 0-15-601697-4

First edition

B   C   D   E

PREFACE

In recent years vector analysis has become an essential and integral requirement for engineers, physicists, and other scientists, as well as for mathematicians. Practically, the methods of vector analysis provide a concise and precise means of mathematically analyzing physical and geometric phenomena, while a study of the theory aids in the development of an intuitive understanding of physical and geometric ideas.

Although intended as a supplement to all standard textbooks for formal courses in vector analysis, this Outline may be used as a self-contained textbook because of its unique combination of rigor and informality in the treatment of its subject matter. Moreover, hundreds of applications to elementary geometry, mechanics, electromagnetic theory, and fluid mechanics make this Outline a valuable companion text for courses in the numerous fields that employ vector methods.

The first half of the Outline is designed for any student who has had an eight-semester-hour course in elementary calculus, or the equivalent. However, the second half of the Outline, whose intent is primarily practical, assumes a basic familiarity with advanced calculus and applied mathematics. The first three chapters develop vector algebra and its application to elementary geometry. In Chapter 1, vectors are defined and vector algebra is treated without the introduction of a coordinate system. An analytic approach to vector algebra is offered in Chapter 2, and applications to geometry are treated in Chapter 3. Vector differentiation and gradient, divergence, and curl are discussed in Chapters 4 and 5. Vector integration and the divergence theorem, Stokes' theorem, and other integral theorems are treated in Chapters 6 and 7. Chapter 8 introduces curvilinear coordinates. The last three chapters, Chapters 9 through 11, cover the practical applications of vectors to mechanics, fluid mechanics, and electromagnetic theory.

I am grateful to my daughter Diana for helping with the typing and to my wife Daisy, whose understanding and constant supportiveness were necessary factors in the completion of this work. I also wish to thank Prof. John Van Iwaarden of Hope College, Holland, Michigan, for his careful review of the manuscript and his constructive suggestions.

Hwei P. Hsu

CONTENTS

1 ALGEBRA OF VECTORS

THIS CHAPTER IS ABOUT

- ☑ **Scalars and Vectors**
- ☑ **Equality of Vectors**
- ☑ **Multiplication of Vectors by Scalars**
- ☑ **Addition and Subtraction of Vectors**
- ☑ **Scalar or Dot Product**
- ☑ **Vector or Cross Product**
- ☑ **Scalar Triple Product**
- ☑ **Vector Triple Product**
- ☑ **Basis**

1-1. Scalars and Vectors

A. Definition of a scalar

A scalar is a quantity that is characterized solely by *magnitude*, such as mass, time, or temperature. The value of a scalar is an ordinary number.

B. Definition and characteristics of vectors

Definition: A vector is a quantity that is characterized by both *magnitude* and *direction*, such as displacement, force, momentum, or velocity. In most printed material, as in this text, a vector is represented by boldface type, such as **A**.

    Geometric representation: Graphically, a vector **A** is represented by a *directed line segment* $\overrightarrow{PQ}$, as illustrated in Figure 1-1. The vector **A** has a direction from P to Q. The point P is called the *initial point*, and the point Q is called the *terminal point* of **A**. The length $|\overrightarrow{PQ}|$ of the line segment is the magnitude of **A**. We can denote the magnitude of **A** by either A or $|A|$.

    When the initial point of a vector is fixed, it is called a **fixed** or **localized vector**. If the initial point is not fixed, it is called a **free** or **nonlocalized vector**. (You can assume that all vectors are free vectors in this text unless we tell you that they are fixed.)

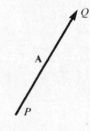

Figure 1-1
Directed line segment representation of a vector.

1-2. Equality of Vectors

Two free vectors **A** and **B** are equal, that is,

         A = B   (1.1)

when they have the same magnitude and direction, as shown in Figure 1-2.

    This definition of equality does not imply that two equal vectors coincide in space. Neither does the equality of vectors (1.1) apply to fixed vectors because only one vector can have a given magnitude, direction, and initial point.

    As a direct consequence of the equality of vectors (1.1), we can obtain the *commutative* property (1.2)

$$\text{If}\quad \mathbf{A} = \mathbf{B}, \quad \text{then}\quad \mathbf{B} = \mathbf{A} \qquad (1.2)$$

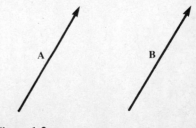

Figure 1-2
Equal vectors.

and the *transitive* property (1.3)

$$\text{If} \quad \mathbf{A} = \mathbf{B} \quad \text{and} \quad \mathbf{B} = \mathbf{C}, \quad \text{then} \quad \mathbf{A} = \mathbf{C} \tag{1.3}$$

1-3. Multiplication of Vectors by Scalars

A. Definition of scalar multiplication

Let $\mathbf{A}$ be any vector, and m any scalar. Then the vector $m\mathbf{A}$ (shown in Figure 1-3) is defined as follows:

SCALAR MULTIPLICATION

(1) The magnitude of $m\mathbf{A}$ is $|m||\mathbf{A}|$; that is, $|m\mathbf{A}| = |m||\mathbf{A}|$.

(2) If $m > 0$, the direction of $m\mathbf{A}$ is the same as that of $\mathbf{A}$.

(3) If $m < 0$, the direction of $m\mathbf{A}$ is opposite to that of $\mathbf{A}$.

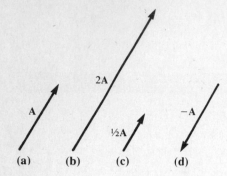

Figure 1-3
Multiplication of a vector by scalars:
(a) The vector **A**; (b) *m***A** for *m* = 2;
(c) *m* = 1/2; (d) *m* = −1.

B. Types of scalar multiplication

1. The zero vector

If $m = 0$, we obtain the zero vector $\mathbf{0}$; that is,

ZERO VECTOR
$$(0)\mathbf{A} = \mathbf{0} \quad \text{where } \mathbf{A} \text{ is any vector} \tag{1.4}$$

So the zero vector $\mathbf{0}$ is a vector that has magnitude 0 and any direction.

2. The negative of a vector

If $m = -1$, we obtain the negative of vector $\mathbf{A}$, denoted by $-\mathbf{A}$; that is,

NEGATIVE VECTOR
$$(-1)\mathbf{A} = -\mathbf{A} \tag{1.5}$$

So a negative vector $-\mathbf{A}$ is a vector whose magnitude is that of $\mathbf{A}$, but whose direction is opposite to that of $\mathbf{A}$.

3. The unit vector

If $\mathbf{A} \neq \mathbf{0}$ and $m = 1/|\mathbf{A}| = 1/A$, we obtain a unit vector

UNIT VECTOR
$$\mathbf{e}_A = \frac{1}{|\mathbf{A}|}\mathbf{A} \tag{1.6}$$

So a unit vector $\mathbf{e}_A$ is a vector whose magnitude is $|\mathbf{e}_A| = 1$ and whose direction is the same as that of $\mathbf{A}$. By using the definition of the unit vector (1.6) and scalar multiplication, we can see that the vector $\mathbf{A}$ may be represented by the product of its magnitude and the unit vector $\mathbf{e}_A$; that is,

$$\mathbf{A} = A\mathbf{e}_A \tag{1.7}$$

C. Properties of scalar multiplication

Multiplication of a vector by a scalar has the following properties:

PROPERTIES OF SCALAR MULTIPLICATION
$$|m\mathbf{A}| = |m||\mathbf{A}| \tag{1.8}$$
$$m(n\mathbf{A}) = (mn)\mathbf{A} \quad \text{for any scalars } m, n \tag{1.9}$$
$$n(m\mathbf{A}) = m(n\mathbf{A}) \quad \text{for any scalars } m, n \tag{1.10}$$

The magnitude of a vector has the following properties:

$$|\mathbf{A}| > 0 \quad \text{for any } \mathbf{A} \neq \mathbf{0} \tag{1.11}$$
$$|\mathbf{A}| = 0 \quad \text{if and only if } \mathbf{A} = \mathbf{0} \tag{1.12}$$

D. Condition for parallelism

Geometrically, the result of multiplying a given vector by a scalar is a vector parallel to the given vector. Thus, two nonzero vectors $\mathbf{A}$ and $\mathbf{B}$ are parallel (designated $\mathbf{A} \parallel \mathbf{B}$) if and only if there exists a scalar m such that

CONDITION FOR
PARALLELISM
$$\mathbf{B} = m\mathbf{A} \qquad (1.13)$$

Since the zero vector **0** has an arbitrary direction, **0** is defined as parallel to any vector **A** and **A** is defined as parallel to **0**. Using this convention, we can extend the condition for parallelism:

$\mathbf{A} \parallel \mathbf{B}$ if and only if $\mathbf{B} = m\mathbf{A}$ or $\mathbf{A} = n\mathbf{B}$ for any scalars m, n

1-4. Addition and Subtraction of Vectors

A. Definitions of vector addition and subtraction

1. Addition

Given two vectors **A** and **B**, the **sum (A + B)**, or **resultant**

VECTOR ADDITION
$$\mathbf{C} = \mathbf{A} + \mathbf{B} \qquad (1.14)$$

is a unique vector **C**, which can be determined geometrically. (See Figure 1-4.) If the initial point of **B** is placed at the terminal point of **A**, the resultant **C** is the vector whose initial point is at the initial point of **A** and whose terminal point is at the terminal point of **B**. This geometric method of vector addition is known as the **triangle rule**.

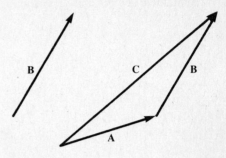

Figure 1-4
Addition of vectors.

2. Subtraction

Given two vectors **A** and **B**, the **difference (A − B)** is the sum (**C**) of **A** and (−**B**); that is,

VECTOR SUBTRACTION $\quad \mathbf{C} = \mathbf{A} - \mathbf{B} = \mathbf{A} + (-\mathbf{B}) \qquad (1.15)$

We can see the difference **A − B** in Figure 1-5a.

Geometrically, if vectors **A** and **B** have a common initial point, then **A − B** is the vector that goes from the terminal point of **B** to the terminal point of **A**. See Figure 1-5b for an illustration of the geometric results of vector subtraction.

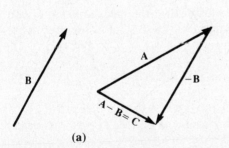

(a)

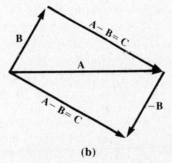

(b)

Figure 1-5
Subtraction of vectors.

B. Properties of vector addition

Vector addition has the following properties:

PROPERTIES
OF VECTOR
ADDITION

$\mathbf{A} + \mathbf{B} = \mathbf{B} + \mathbf{A}$	[Commutative law]	(1.16)
$\mathbf{A} + (\mathbf{B} + \mathbf{C}) = (\mathbf{A} + \mathbf{B}) + \mathbf{C}$	[Associative law]	(1.17)
$m(\mathbf{A} + \mathbf{B}) = m\mathbf{A} + m\mathbf{B}$	[Distributive law]	(1.18)
$(m + n)\mathbf{A} = m\mathbf{A} + n\mathbf{A}$	[Scalar distributive law]	(1.19)
$\mathbf{A} + \mathbf{0} = \mathbf{A}$		(1.20)
$\mathbf{A} + (-\mathbf{A}) = \mathbf{0} \quad$ or $\quad \mathbf{A} - \mathbf{A} = \mathbf{0}$		(1.21)

These properties can be verfied geometrically.

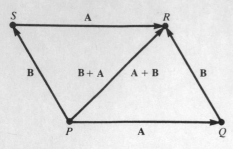

Figure 1-6
Commutative law of vector addition.

EXAMPLE 1-1: Verify the commutative law (1.16).

Proof: Let **A** and **B** be two vectors. Using the triangle method, add **A** + **B** and **B** + **A**, constructing the triangles PQR and PSR, as shown in Figure 1-6. Then

$$\mathbf{A} + \mathbf{B} = \overrightarrow{PQ} + \overrightarrow{QR} = \overrightarrow{PR}$$
$$\mathbf{B} + \mathbf{A} = \overrightarrow{PS} + \overrightarrow{SR} = \overrightarrow{PR}$$

Hence **A** + **B** = **B** + **A**.

EXAMPLE 1-2: Verify the associative law (1.17).

Proof: Construct a polygon $PQRS$ having the vectors **A**, **B**, and **C** as consecutive sides (as shown in Figure 1-7).

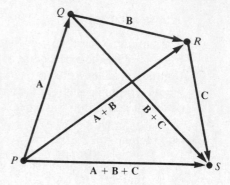

Figure 1-7
Associative law of vector addition.

Then

$$\mathbf{A} + (\mathbf{B} + \mathbf{C}) = \mathbf{A} + \overrightarrow{QS} = \overrightarrow{PQ} + \overrightarrow{QS} = \overrightarrow{PS}$$
$$(\mathbf{A} + \mathbf{B}) + \mathbf{C} = \overrightarrow{PR} + \mathbf{C} = \overrightarrow{PR} + \overrightarrow{RS} = \overrightarrow{PS}$$

Hence **A** + (**B** + **C**) = (**A** + **B**) + **C**.

1-5. Scalar or Dot Product

A. Geometric definition of the scalar product

The **scalar** or **dot product** (sometimes also called the **inner product**) of two vectors **A** and **B** is denoted by **A** · **B** (read "**A** *dot* **B**") and is defined as the scalar given by

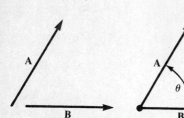

Figure 1-8
Angle between two vectors.

SCALAR PRODUCT (GEOMETRIC)
$$\mathbf{A} \cdot \mathbf{B} = |\mathbf{A}||\mathbf{B}|\cos\theta = AB\cos\theta \qquad (1.22)$$

where θ is the angle between **A** and **B** and $0 \leqslant \theta \leqslant \pi$. (See Figure 1-8.)

From the definition of the scalar product (1.22), the angle θ between **A** and **B** can be expressed as

$$\cos\theta = \frac{\mathbf{A} \cdot \mathbf{B}}{|\mathbf{A}||\mathbf{B}|} = \frac{\mathbf{A} \cdot \mathbf{B}}{AB} \qquad (1.23)$$

or

$$\theta = \cos^{-1}\left(\frac{\mathbf{A} \cdot \mathbf{B}}{AB}\right) \qquad (1.24)$$

provided **A** ≠ **0** and **B** ≠ **0**.

Two vectors **A** and **B** are perpendicular, or orthogonal, to each other (denoted by **A** ⊥ **B**) if the angle θ between them is a right angle (i.e., $\theta = \pi/2$ radians = 90°). Since the vector **0** has arbitrary direction, it is considered perpendicular to any vector **A**, and conversely, **A** is considered perpendicular to **0**.

EXAMPLE 1-3: Show that **A** and **B** are perpendicular to each other if and only if $\mathbf{A} \cdot \mathbf{B} = 0$.

Solution: If $\mathbf{A} \cdot \mathbf{B} = AB \cos \theta = 0$, we can conclude that $A = |\mathbf{A}| = 0$ or $B = |\mathbf{B}| = 0$ or $\cos \theta = 0$. Since the zero vector **0** has arbitrary direction, it is defined to be perpendicular to every vector. So if $\mathbf{A} = \mathbf{0}$ or $\mathbf{B} = \mathbf{0}$, then $\mathbf{A} \perp \mathbf{B}$. When $\mathbf{A} \neq \mathbf{0}$, $\mathbf{B} \neq \mathbf{0}$, and $\cos \theta = 0$, then $\theta = \pi/2$, so again $\mathbf{A} \perp \mathbf{B}$.

Conversely, if $\mathbf{A} \perp \mathbf{B}$ and $\mathbf{A} \neq \mathbf{0}$ and $\mathbf{B} \neq \mathbf{0}$, the angle between **A** and **B** is $\pi/2$. Thus

$$\mathbf{A} \cdot \mathbf{B} = |\mathbf{A}||\mathbf{B}| \cos \frac{\pi}{2} = 0$$

B. Projection and component of a vector

1. The **projection** of a vector **A** onto **B**, denoted by $\text{proj}_B \mathbf{A}$, is a *vector*

PROJECTION OF A VECTOR

$$\text{proj}_B \mathbf{A} = (A \cos \theta)\mathbf{e}_B \qquad (1.25)$$

where θ is the angle between **A** and **B**, and $\mathbf{e}_B = (1/B)\mathbf{B}$ is a unit vector in the direction of **B** (See Figure 1-9).

2. The **component** of vector **A** along nonzero vector **B**, denoted by $\text{comp}_B \mathbf{A}$, is a *scalar*

COMPONENT OF A VECTOR

$$\text{comp}_B \mathbf{A} = A \cos \theta = \mathbf{A} \cdot \mathbf{e}_B \qquad (1.26a)$$

where θ is the angle between **A** and **B** (Figure 1-9). Similarly, the component of a vector **B** along a nonzero vector **A**, denoted by $\text{comp}_A \mathbf{B}$, is given by

$$\text{comp}_A \mathbf{B} = B \cos \theta = \mathbf{B} \cdot \mathbf{e}_A \qquad (1.26b)$$

Using (1.26b), the scalar product $\mathbf{A} \cdot \mathbf{B}$ defined in (1.22) can now be expressed as

$$\mathbf{A} \cdot \mathbf{B} = AB \cos \theta = A(B \cos \theta) = A \, \text{comp}_A \mathbf{B} \qquad (1.27a)$$

or

$$\mathbf{A} \cdot \mathbf{B} = AB \cos \theta = B(A \cos \theta) = B \, \text{comp}_B \mathbf{A} \qquad (1.27b)$$

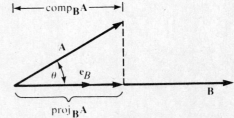

Figure 1-9
Component ($\text{comp}_B \mathbf{A} = A \cos \theta$) and projection ($\text{proj}_B \mathbf{A} = (A \cos \theta)\mathbf{e}_B$) of a vector.

C. Properties of the scalar product

The scalar product has the following properties:

PROPERTIES OF THE SCALAR PRODUCT

$$\mathbf{A} \cdot \mathbf{B} = \mathbf{B} \cdot \mathbf{A} \qquad \text{[Commutative law]} \quad (1.28)$$

$$\mathbf{A} \cdot (\mathbf{B} + \mathbf{C}) = \mathbf{A} \cdot \mathbf{B} + \mathbf{A} \cdot \mathbf{C} \qquad \text{[Distributive law]} \quad (1.29)$$

$$(m\mathbf{A}) \cdot \mathbf{B} = \mathbf{A} \cdot (m\mathbf{B}) = m(\mathbf{A} \cdot \mathbf{B}) \qquad (1.30)$$

$$\mathbf{A} \cdot \mathbf{A} = |\mathbf{A}|^2 = A^2 \quad \text{for every } \mathbf{A} \qquad (1.31a)$$

$$\mathbf{A} \cdot \mathbf{A} = 0 \quad \text{if and only if } \mathbf{A} = \mathbf{0} \qquad (1.31b)$$

where m is an arbitrary scalar.

EXAMPLE 1-4: Verify the commutative law of the scalar product (1.28).

Proof: From the definition of the scalar product,

$$\mathbf{A} \cdot \mathbf{B} = AB \cos \theta = BA \cos \theta = \mathbf{B} \cdot \mathbf{A}$$

EXAMPLE 1-5: Verify the distributive law of the scalar product (1.29).

Proof: From Figure 1-10, we have

$$PR = PQ + QR$$

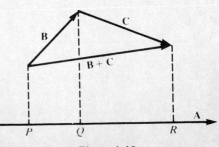

Figure 1-10

From the definition of a vector component (1.26) and Figure 1-9, we have $PQ = \text{comp}_A \mathbf{B}$, $QR = \text{comp}_A \mathbf{C}$, and $PR = \text{comp}_A(\mathbf{B} + \mathbf{C})$. So

$$\text{comp}_A(\mathbf{B} + \mathbf{C}) = \text{comp}_A \mathbf{B} + \text{comp}_A \mathbf{C} \tag{1.32}$$

Knowing that $\mathbf{e}_A$ is a unit vector in the direction of $\mathbf{A}$, we can use the definition (1.26) of a component to rewrite (1.32) as

$$(\mathbf{B} + \mathbf{C}) \cdot \mathbf{e}_A = \mathbf{B} \cdot \mathbf{e}_A + \mathbf{C} \cdot \mathbf{e}_A \tag{1.33}$$

Multiplying both sides of (1.33) by A and using property (1.30), we get

$$(\mathbf{B} + \mathbf{C}) \cdot A\mathbf{e}_A = \mathbf{B} \cdot A\mathbf{e}_A + \mathbf{C} \cdot A\mathbf{e}_A \tag{1.34a}$$

Then, because $\mathbf{A} = A\mathbf{e}_A$, we can rewrite (1.34a) as

$$(\mathbf{B} + \mathbf{C}) \cdot \mathbf{A} = \mathbf{B} \cdot \mathbf{A} + \mathbf{C} \cdot \mathbf{A} \tag{1.34b}$$

Then using the commutative law of the scalar product (1.28), we obtain the distributive law of the scalar product (1.29):

$$\mathbf{A} \cdot (\mathbf{B} + \mathbf{C}) = \mathbf{A} \cdot \mathbf{B} + \mathbf{A} \cdot \mathbf{C}$$

EXAMPLE 1-6: Verify the property $\mathbf{A} \cdot \mathbf{A} = |\mathbf{A}|^2 = A^2$ (1.31a).

Proof: Set $\mathbf{A} = \mathbf{B}$ in the definition of the scalar product (1.22). Then we obtain

$$\mathbf{A} \cdot \mathbf{A} = |\mathbf{A}||\mathbf{A}|\cos 0 = |\mathbf{A}|^2 = A^2$$

since the angle θ between $\mathbf{A}$ and itself is 0 and $\cos 0 = 1$.

1-6. Vector or Cross Product

A. Geometric definition of the vector product

The **vector product**, or **cross product** (sometimes also called the **outer product**) of two vectors $\mathbf{A}$ and $\mathbf{B}$ is a vector denoted by $\mathbf{A} \times \mathbf{B}$ (read "$\mathbf{A}$ *cross* $\mathbf{B}$") defined by

VECTOR PRODUCT (GEOMETRIC)
$$\mathbf{A} \times \mathbf{B} = |\mathbf{A}||\mathbf{B}|\sin\theta\,\mathbf{u} = (AB\sin\theta)\mathbf{u} \tag{1.35}$$

where θ is the angle between $\mathbf{A}$ and $\mathbf{B}$, $0 \leqslant \theta \leqslant \pi$, and $\mathbf{u}$ is a unit vector such that $\mathbf{u} \perp \mathbf{A}$ and $\mathbf{u} \perp \mathbf{B}$. The direction of $\mathbf{u}$ is that of the advance of a right-hand screw as $\mathbf{A}$ rotates toward $\mathbf{B}$ through the angle θ. (See Figure 1-11.)

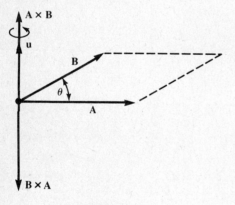

Figure 1-11
Vector product.

EXAMPLE 1-7: Show that the area S of a parallelogram with vectors $\mathbf{A}$ and $\mathbf{B}$ forming adjacent sides is given by

$$S = |\mathbf{A} \times \mathbf{B}| \tag{1.36}$$

Solution: Figure 1-12 shows the parallelogram. If h is the perpendicular distance from the terminal point of $\mathbf{B}$ to the vector $\mathbf{A}$, then $S = Ah$. But $h = B\sin\theta$. Hence

$$S = AB\sin\theta$$

Using the definition (1.35) of the vector product, we obtain

$$|\mathbf{A} \times \mathbf{B}| = |(AB\sin\theta)\mathbf{u}| = (AB\sin\theta)|\mathbf{u}| = AB\sin\theta$$

since $|\mathbf{u}| = 1$. So

$$S = |\mathbf{A} \times \mathbf{B}|$$

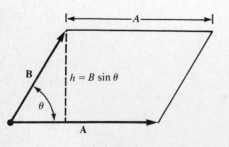

Figure 1-12

EXAMPLE 1-8: If $\mathbf{B}'$ and $\mathbf{C}'$ are the respective projections of $\mathbf{B}$ and $\mathbf{C}$ on a plane perpendicular to a vector $\mathbf{A}$, then show that

$$\mathbf{A} \times \mathbf{B} = \mathbf{A} \times \mathbf{B}' \tag{1.37}$$

$$\mathbf{A} \times (\mathbf{B} + \mathbf{C}) = \mathbf{A} \times (\mathbf{B}' + \mathbf{C}') \tag{1.38}$$

Solution: As Figure 1-13 shows, **A**, **B**, and **B′** are coplanar (i.e., lie in the same plane). By the definition of the vector product, it is clear that **A** × **B** and **A** × **B′** have the same direction.

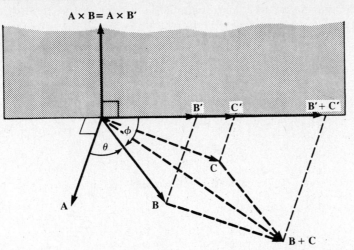

Figure 1-13

Now

$$|\mathbf{A} \times \mathbf{B}| = AB \sin \theta$$

But $B' = B \cos \phi = B \cos(90° - \theta) = B \sin \theta$, so

$$|\mathbf{A} \times \mathbf{B}'| = AB' \sin 90° = AB' = AB \sin \theta$$

Therefore

$$|\mathbf{A} \times \mathbf{B}| = |\mathbf{A} \times \mathbf{B}'|$$

Since **A** × **B** and **A** × **B′** have the same direction and the same magnitude, they are equal and (1.37) is proved.

Next, from the component equation $comp_\mathbf{A}(\mathbf{B} + \mathbf{C}) = comp_\mathbf{A}\mathbf{B} + comp_\mathbf{A}\mathbf{C}$ (1.32), we note that the projection of **B** + **C** on the plane is **B′** + **C′**. Hence by (1.37) we have

$$\mathbf{A} \times (\mathbf{B} + \mathbf{C}) = \mathbf{A} \times (\mathbf{B}' + \mathbf{C}')$$

B. Properties of the vector product

The vector product has the following properties:

PROPERTIES OF THE VECTOR PRODUCT

$$\mathbf{A} \times \mathbf{B} = -\mathbf{B} \times \mathbf{A} \qquad \text{[Anticommutative or skew-symmetry law]} \quad (1.39)$$

$$\mathbf{A} \times (\mathbf{B} + \mathbf{C}) = \mathbf{A} \times \mathbf{B} + \mathbf{A} \times \mathbf{C} \quad \text{[Distributive law]} \qquad (1.40)$$

$$(m\mathbf{A}) \times \mathbf{B} = \mathbf{A} \times (m\mathbf{B}) = m\mathbf{A} \times \mathbf{B} \qquad (1.41)$$

$$\mathbf{A} \times \mathbf{A} = 0 \qquad (1.42)$$

$$\mathbf{A} \times 0 = 0 \quad \text{for any } \mathbf{A} \qquad (1.43)$$

EXAMPLE 1-9: Verify the anticommutative law (1.39) of the vector product.

Proof: Let θ be the angle between **A** and **B**, with $0 \leqslant \theta \leqslant \pi$. Then

$$|\mathbf{B} \times \mathbf{A}| = |\mathbf{B}||\mathbf{A}| \sin \theta = BA \sin \theta = AB \sin \theta = |\mathbf{A} \times \mathbf{B}| \qquad (1.44)$$

But by the right-hand rule (Figure 1-11), the direction of **B** × **A** is opposite to that of **A** × **B**. Hence we have **B** × **A** = −**A** × **B** or **A** × **B** = −**B** × **A** (1.39).

EXAMPLE 1-10: Verify the distributive law (1.40) of the vector product.

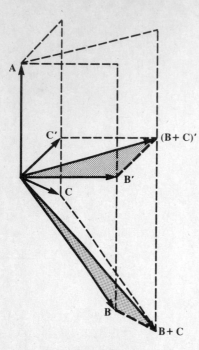

Figure 1-14

Proof: Consider a triangle whose sides are **B, C,** and **B + C.** Projecting this triangle onto a plane that is perpendicular to **A** and contains its initial point, we get a triangle with sides **B′, C′,** and **(B + C)′,** as shown in Figure 1-14.

Now imagine rotating this triangle 90° counterclockwise about **A** as an axis. This rotation gives us another triangle with sides **B″, C″,** and **(B + C)″.** But

$$(\mathbf{B} + \mathbf{C})'' = \mathbf{B}'' + \mathbf{C}''$$

Hence by the distributive law of vector addition (1.18) we have

$$A(\mathbf{B} + \mathbf{C})'' = A(\mathbf{B}'' + \mathbf{C}'') = A\mathbf{B}'' + A\mathbf{C}'' \tag{1.45}$$

Now using the definition of the vector product (1.35), we can show that

$$\mathbf{A} \times \mathbf{B}' = A\mathbf{B}''$$
$$\mathbf{A} \times \mathbf{C}' = A\mathbf{C}''$$
$$\mathbf{A} \times (\mathbf{B} + \mathbf{C})' = A(\mathbf{B} + \mathbf{C})''$$

So from (1.45), we have

$$\mathbf{A} \times (\mathbf{B} + \mathbf{C})' = \mathbf{A} \times \mathbf{B}' + \mathbf{A} \times \mathbf{C}'$$

Since $(\mathbf{B} + \mathbf{C})' = \mathbf{B}' + \mathbf{C}'$,

$$\mathbf{A} \times (\mathbf{B}' + \mathbf{C}') = \mathbf{A} \times \mathbf{B}' + \mathbf{A} \times \mathbf{C}' \tag{1.46}$$

Hence by (1.37) and (1.38) we obtain the distributive law of the vector product (1.40):

$$\mathbf{A} \times (\mathbf{B} + \mathbf{C}) = \mathbf{A} \times \mathbf{B} + \mathbf{A} \times \mathbf{C}$$

EXAMPLE 1-11: Verify the property (1.42): $\mathbf{A} \times \mathbf{A} = \mathbf{0}$.

Proof: From the anticommutative law (1.39), we can see that $\mathbf{A} \times \mathbf{A} = -\mathbf{A} \times \mathbf{A}$ when $\mathbf{A} = \mathbf{B}$. Thus $\mathbf{A} \times \mathbf{A} = \mathbf{0}$.

Note: This is consistent with the definition of the vector product (1.35), since $\theta = 0$ implies $\sin \theta = 0$ and hence $|\mathbf{A} \times \mathbf{A}| = A^2 \sin 0 = 0$.

EXAMPLE 1-12: Show the following identity:

$$(\mathbf{A} + \mathbf{B}) \times \mathbf{C} = \mathbf{A} \times \mathbf{C} + \mathbf{B} \times \mathbf{C} \tag{1.47}$$

Solution: Using (1.39) and (1.40), we have

$$(\mathbf{A} + \mathbf{B}) \times \mathbf{C} = -\mathbf{C} \times (\mathbf{A} + \mathbf{B})$$
$$= -(\mathbf{C} \times \mathbf{A} + \mathbf{C} \times \mathbf{B})$$
$$= -\mathbf{C} \times \mathbf{A} - \mathbf{C} \times \mathbf{B}$$
$$= \mathbf{A} \times \mathbf{C} + \mathbf{B} \times \mathbf{C}$$

1-7. Scalar Triple Product

A. Definition of the scalar triple product

The **scalar triple product** of three vectors **A, B,** and **C** is a scalar $\mathbf{A} \cdot (\mathbf{B} \times \mathbf{C})$, or simply

SCALAR TRIPLE PRODUCT $\mathbf{A} \cdot \mathbf{B} \times \mathbf{C}$ (1.48)

Note: The expression $(\mathbf{A} \cdot \mathbf{B}) \times \mathbf{C}$ is undefined, since it indicates the vector product of a scalar and a vector. Thus we can omit the parentheses in $\mathbf{A} \cdot (\mathbf{B} \times \mathbf{C})$.

EXAMPLE 1-13: Give a geometric interpretation of $\mathbf{A} \cdot \mathbf{B} \times \mathbf{C}$.

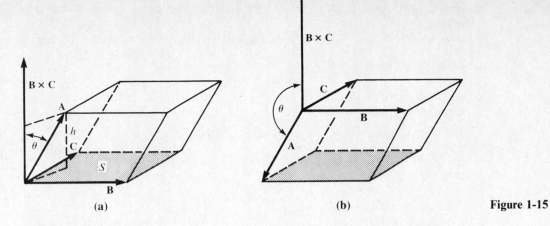

(a)   (b)   **Figure 1-15**

Solution: Figure 1-15 shows a parallelepiped whose sides are **A**, **B**, and **C**. From condition (1.36), the area S of which the vectors **B** and **C** are adjacent edges is

$$S = |\mathbf{B} \times \mathbf{C}|$$

If h is the altitude of the parallelepiped, then

$$h = |\mathbf{A}||\cos\theta|$$

where θ is the angle between **A** and **B** $\times$ **C**. So the volume of the parallelepiped is

$$V = hS = |\mathbf{A}||\mathbf{B} \times \mathbf{C}||\cos\theta| = |\mathbf{A} \cdot \mathbf{B} \times \mathbf{C}| \qquad (1.49)$$

You can see from (1.49) that when $0 < \theta < \pi/2$ (as in Figure 1-15a), $\mathbf{A} \cdot \mathbf{B} \times \mathbf{C} > 0$, and when $\pi/2 < \theta < \pi$ (as in Figure 1-15b), $\mathbf{A} \cdot \mathbf{B} \times \mathbf{C} < 0$. Hence **A**, **B**, and **C** form a **positive triple** if and only if $\mathbf{A} \cdot \mathbf{B} \times \mathbf{C} > 0$.

B. Condition for coplanarity

Vectors **A**, **B**, and **C** are coplanar if and only if

CONDITION FOR COPLANARITY $\mathbf{A} \cdot \mathbf{B} \times \mathbf{C} = 0$ $\qquad\qquad\qquad$ (1.50)

EXAMPLE 1-14: Show geometrically that the condition for coplanarity is given by (1.50).

Proof: If **A**, **B**, and **C** are coplanar, the altitude of the parallelepiped that they form is zero. Hence from (1.49)

$$\mathbf{A} \cdot \mathbf{B} \times \mathbf{C} = 0$$

Conversely, if $\mathbf{A} \cdot \mathbf{B} \times \mathbf{C} = 0$, the altitude of the parallelepiped is zero, and that means that **A**, **B**, and **C** must be coplanar.

EXAMPLE 1-15: If any two vectors in a scalar triple product are equal, show that the product is zero; that is,

$$\mathbf{A} \cdot \mathbf{A} \times \mathbf{C} = 0, \qquad \mathbf{C} \cdot \mathbf{B} \times \mathbf{C} = 0, \qquad \mathbf{A} \cdot \mathbf{B} \times \mathbf{B} = 0 \qquad (1.51)$$

Solution: From the definition of a vector product (1.35), $\mathbf{A} \times \mathbf{C}$ is perpendicular to **A**. Hence their dot product is zero; that is, $\mathbf{A} \cdot \mathbf{A} \times \mathbf{C} = 0$.

Similarly, since $\mathbf{B} \times \mathbf{C}$ is perpendicular to **C**, their scalar product is zero; that is, $\mathbf{C} \cdot (\mathbf{B} \times \mathbf{C}) = 0$.

From (1.42) we have $\mathbf{B} \times \mathbf{B} = \mathbf{0}$, so $\mathbf{A} \cdot \mathbf{B} \times \mathbf{B} = \mathbf{A} \cdot \mathbf{0} = 0$.

Alternate Solution: If any two of the vectors **A**, **B**, and **C** are equal, then **A**, **B**, and **C** are coplanar. Hence (1.51) follows from the condition for coplanarity (1.50).

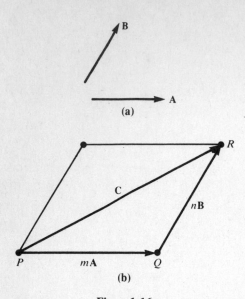

Figure 1-16

EXAMPLE 1-16: If **A**, **B**, and **C** are coplanar and we assume that **A** and **B** are not parallel, show that **C** can be expressed as a linear combination of **A** and **B**; that is,

$$C = mA + nB \tag{1.52}$$

where m and n are two uniquely determined scalars.

Solution: Since **A** and **B** are not parallel, there exists a parallelogram whose diagonal is **C** and whose edges are parallel to **A** and **B** (see Figure 1-16). Then

$$C = \overrightarrow{PQ} + \overrightarrow{QR}$$

Now $\overrightarrow{PQ} \parallel \mathbf{A}$ and $\overrightarrow{QR} \parallel \mathbf{B}$, which imply that there exist scalars m and n such that $\overrightarrow{PQ} = m\mathbf{A}$ and $\overrightarrow{QR} = n\mathbf{B}$. Hence

$$C = mA + nB$$

To show that m and n are uniquely determined, assume that there exist m' and n' different from m and n such that

$$C = m'A + n'B$$

Then by subtraction

$$(m - m')\mathbf{A} + (n - n')\mathbf{B} = 0$$

or

$$(m - m')\mathbf{A} = (n' - n)\mathbf{B}$$

But **A** and **B** are nonparallel—hence nonzero—vectors. So it is clearly required that $m - m' = 0$ and $n' - n = 0$. Hence $m = m'$ and $n = n'$. This contradicts our assumption. Thus the scalar multiples m and n are unique.

C. Fundamental identity for the scalar triple product

For any vectors **A**, **B**, and **C**,

SCALAR TRIPLE PRODUCT FUNDAMENTAL IDENTITY
$$\mathbf{A} \cdot \mathbf{B} \times \mathbf{C} = \mathbf{A} \times \mathbf{B} \cdot \mathbf{C} \tag{1.53}$$

EXAMPLE 1-17: Verify the fundamental identity (1.53) for the scalar triple product.

Proof: Since by (1.51) the scalar triple product is zero whenever two of its vectors are equal, we can construct

$$(\mathbf{A} + \mathbf{C}) \cdot (\mathbf{A} + \mathbf{C}) \times \mathbf{B} = 0 \tag{1.54}$$

From (1.47) we have

$$(\mathbf{A} + \mathbf{C}) \times \mathbf{B} = \mathbf{A} \times \mathbf{B} + \mathbf{C} \times \mathbf{B} \tag{1.55}$$

Substituting (1.55) into (1.54) and using the distributive law (1.29), we obtain

$$(\mathbf{A} + \mathbf{C}) \cdot [(\mathbf{A} \times \mathbf{B}) + (\mathbf{C} \times \mathbf{B})]$$
$$= \mathbf{A} \cdot (\mathbf{A} \times \mathbf{B}) + \mathbf{C} \cdot (\mathbf{A} \times \mathbf{B}) + \mathbf{A} \cdot (\mathbf{C} \times \mathbf{B}) + \mathbf{C} \cdot (\mathbf{C} \times \mathbf{B}) = 0 \tag{1.56}$$

Again, by (1.51) we have

$$\mathbf{A} \cdot (\mathbf{A} \times \mathbf{B}) = \mathbf{C} \cdot (\mathbf{C} \times \mathbf{B}) = 0$$

Thus (1.56) reduces to

$$\mathbf{C} \cdot (\mathbf{A} \times \mathbf{B}) + \mathbf{A} \cdot (\mathbf{C} \times \mathbf{B}) = 0$$

By transposing and using the anticommutative property (1.39),

$$\mathbf{C} \cdot (\mathbf{A} \times \mathbf{B}) = -\mathbf{A} \cdot (\mathbf{C} \times \mathbf{B}) = \mathbf{A} \cdot [-(\mathbf{C} \times \mathbf{B})] = \mathbf{A} \cdot (\mathbf{B} \times \mathbf{C})$$

But by the commutative property of the scalar product (1.28)

$$\mathbf{C} \cdot (\mathbf{A} \times \mathbf{B}) = (\mathbf{A} \times \mathbf{B}) \cdot \mathbf{C}$$

So we have proved

$$(\mathbf{A} \times \mathbf{B}) \cdot \mathbf{C} = \mathbf{A} \cdot (\mathbf{B} \times \mathbf{C})$$

The fundamental identity (1.53) says that in a scalar triple product, dot and cross can be interchanged without changing its value. In other words, the position of the dot and cross in the scalar triple product is immaterial. Therefore $\mathbf{A} \cdot \mathbf{B} \times \mathbf{C}$ is often denoted by

$$\mathbf{A} \cdot \mathbf{B} \times \mathbf{C} = [\mathbf{ABC}] \qquad (1.57)$$

1-8. Vector Triple Product

A. Definition of the vector triple product

The **vector triple product** of three vectors **A**, **B**, and **C** is the vector

VECTOR TRIPLE PRODUCT $\qquad \mathbf{A} \times (\mathbf{B} \times \mathbf{C}) \qquad (1.58)$

Here, we need the parentheses because an expression like $\mathbf{A} \times \mathbf{B} \times \mathbf{C}$ is ambiguous, the result depending on whether we form $\mathbf{A} \times \mathbf{B}$ first or $\mathbf{B} \times \mathbf{C}$ first. In general,

$$\mathbf{A} \times (\mathbf{B} \times \mathbf{C}) \neq (\mathbf{A} \times \mathbf{B}) \times \mathbf{C}$$

B. Fundamental identities for the vector triple product

VECTOR TRIPLE PRODUCT $\qquad \mathbf{A} \times (\mathbf{B} \times \mathbf{C}) = (\mathbf{A} \cdot \mathbf{C})\mathbf{B} - (\mathbf{A} \cdot \mathbf{B})\mathbf{C} \qquad (1.59)$
FUNDAMENTAL IDENTITIES $\qquad (\mathbf{A} \times \mathbf{B}) \times \mathbf{C} = (\mathbf{A} \cdot \mathbf{C})\mathbf{B} - (\mathbf{B} \cdot \mathbf{C})\mathbf{A} \qquad (1.60)$

The fundamental identities (1.59) and (1.60) are easily remembered by the "middle-term rule":

The vector product is equal to the middle vector whose coefficient is the scalar product of the remaining vectors minus the other vector in the parentheses whose coefficient is the scalar product of the remaining vectors.

From the identities (1.59) and (1.60) it is clear that

$$\mathbf{A} \times (\mathbf{B} \times \mathbf{C}) \neq (\mathbf{A} \times \mathbf{B}) \times \mathbf{C}$$

EXAMPLE 1-18: Prove the fundamental identity (1.59) for the vector triple product.

Proof: If any one of the vectors **A**, **B**, or **C** is **0**, then (1.59) is trivially true. If **B** and **C** are parallel, it follows that either $\mathbf{B} = m\mathbf{C}$ or $\mathbf{C} = n\mathbf{B}$ for some scalars m and n. Hence both sides of (1.59) are again **0**. So we can assume that **A**, **B**, and **C** are nonzero vectors and that **B** and **C** are not parallel.

By (1.51) we can construct

$$\mathbf{B} \times \mathbf{C} \cdot \mathbf{A} \times (\mathbf{B} \times \mathbf{C}) = 0$$

So, by the condition for coplanarity (1.50), the vectors **B**, **C**, and $\mathbf{A} \times (\mathbf{B} \times \mathbf{C})$ are in the same plane. Hence by (1.52) there exist scalars m and n such that

$$\mathbf{A} \times (\mathbf{B} \times \mathbf{C}) = m\mathbf{B} - n\mathbf{C} \qquad (1.61)$$

We dot both sides with **A** to obtain

$$\mathbf{A} \cdot [\mathbf{A} \times (\mathbf{B} \times \mathbf{C})] = m(\mathbf{A} \cdot \mathbf{B}) - n(\mathbf{A} \cdot \mathbf{C}) \qquad (1.62a)$$

But by (1.51) the left side is zero:

$$m(\mathbf{A} \cdot \mathbf{B}) - n(\mathbf{A} \cdot \mathbf{C}) = 0 \qquad (1.62b)$$

Now if $(\mathbf{A} \cdot \mathbf{B}) = 0$, then $\mathbf{A}$ is perpendicular to $\mathbf{B}$. But from the left side of (1.62a), we see that $\mathbf{A}$ is perpendicular to $\mathbf{A} \times (\mathbf{B} \times \mathbf{C})$, so that $\mathbf{A}$ is perpendicular to the plane of $\mathbf{B}$ and $\mathbf{C}$. Hence $\mathbf{A}$ is parallel to $\mathbf{B} \times \mathbf{C}$; that is, $\mathbf{A} \cdot \mathbf{C} = 0$ and $\mathbf{A} \times (\mathbf{B} \times \mathbf{C}) = \mathbf{0}$, and identity (1.59) is trivially true.

An identical argument holds under an assumption that $\mathbf{A} \cdot \mathbf{C} = 0$.

Now we assume that $\mathbf{A} \cdot \mathbf{B} \neq 0$ and $\mathbf{A} \cdot \mathbf{C} \neq 0$. Then (1.62b) yields

$$\frac{n}{\mathbf{A} \cdot \mathbf{B}} = \frac{m}{\mathbf{A} \cdot \mathbf{C}} = \lambda$$

and $m = \lambda(\mathbf{A} \cdot \mathbf{C})$ and $n = \lambda(\mathbf{A} \cdot \mathbf{B})$. Thus (1.61) becomes

$$\mathbf{A} \times (\mathbf{B} \times \mathbf{C}) = \lambda[(\mathbf{A} \cdot \mathbf{C})\mathbf{B} - (\mathbf{A} \cdot \mathbf{B})\mathbf{C}] \tag{1.63}$$

To determine λ, we proceed as follows: Consider a vector $\mathbf{D}$ such that $\mathbf{B}, \mathbf{C},$ and $\mathbf{D}$ are coplanar and $\mathbf{D}$ is perpendicular to $\mathbf{C}$, that is, $\mathbf{C} \cdot \mathbf{D} = 0$, and $\mathbf{D}, \mathbf{C},$ and $\mathbf{B} \times \mathbf{C}$ form a positive triple (see Figure 1-17.) We dot both sides of (1.63) with $\mathbf{D}$ to obtain

$$[\mathbf{A} \times (\mathbf{B} \times \mathbf{C})] \cdot \mathbf{D} = \lambda[(\mathbf{A} \cdot \mathbf{C})(\mathbf{B} \cdot \mathbf{D}) - (\mathbf{A} \cdot \mathbf{B})(\mathbf{C} \cdot \mathbf{D})]$$

Since $\mathbf{C} \cdot \mathbf{D} = 0$, using the fundamental identity for the scalar triple product (1.53), we have

$$\mathbf{A} \times [(\mathbf{B} \times \mathbf{C}) \cdot \mathbf{D}] = \mathbf{A} \cdot [(\mathbf{B} \times \mathbf{C}) \times \mathbf{D}] = \lambda(\mathbf{A} \cdot \mathbf{C})(\mathbf{B} \cdot \mathbf{D}) \tag{1.64}$$

Now, by (1.50) and the definition of a vector product (1.35), $(\mathbf{B} \times \mathbf{C}) \times \mathbf{D}$ is coplanar with $\mathbf{B}$ and $\mathbf{C}$ and perpendicular to $\mathbf{D}$. Thus it is parallel to $\mathbf{C}$. Since $\mathbf{D}, \mathbf{C},$ and $\mathbf{B} \times \mathbf{C}$ form a positive triple, $(\mathbf{B} \times \mathbf{C}) \times \mathbf{D}$ is in the same direction as $\mathbf{C}$. Thus there exists a scalar $k > 0$ such that

$$(\mathbf{B} \times \mathbf{C}) \times \mathbf{D} = k\mathbf{C} \tag{1.65}$$

Now from Figure 1-17,

$$|(\mathbf{B} \times \mathbf{C}) \times \mathbf{D}| = BCD \sin \theta = |k\mathbf{C}| = kC$$

Hence we obtain

$$k = BD \sin \theta = BD \cos(90^\circ - \theta) = \mathbf{B} \cdot \mathbf{D}$$

Thus (1.65) reduces to

$$(\mathbf{B} \times \mathbf{C}) \times \mathbf{D} = (\mathbf{B} \cdot \mathbf{D})\mathbf{C} \tag{1.66}$$

Substituting (1.66) into (1.64) yields

$$\mathbf{A} \cdot [(\mathbf{B} \times \mathbf{C}) \times \mathbf{D}] = \mathbf{A} \cdot [(\mathbf{B} \cdot \mathbf{D})\mathbf{C}]$$
$$= (\mathbf{B} \cdot \mathbf{D})(\mathbf{A} \cdot \mathbf{C}) = (\mathbf{A} \cdot \mathbf{C})(\mathbf{B} \cdot \mathbf{D}) = \lambda(\mathbf{A} \cdot \mathbf{C})(\mathbf{B} \cdot \mathbf{D})$$

or $\lambda = 1$. Hence (1.63) becomes (1.59):

$$\mathbf{A} \times (\mathbf{B} \times \mathbf{C}) = (\mathbf{A} \cdot \mathbf{C})\mathbf{B} - (\mathbf{A} \cdot \mathbf{B})\mathbf{C}$$

EXAMPLE 1-19: Verify the second of the fundamental identities (1.60),

$$(\mathbf{A} \times \mathbf{B}) \times \mathbf{C} = (\mathbf{A} \cdot \mathbf{C})\mathbf{B} - (\mathbf{B} \cdot \mathbf{C})\mathbf{A}$$

Proof: From the anticommutative law of a vector product (1.39),

$$(\mathbf{A} \times \mathbf{B}) \times \mathbf{C} = -\mathbf{C} \times (\mathbf{A} \times \mathbf{B})$$

Replacing $\mathbf{A}, \mathbf{B},$ and $\mathbf{C}$ in fundamental identity (1.59) with $\mathbf{C}, \mathbf{A},$ and $\mathbf{B}$, respectively, we have

$$\mathbf{C} \times (\mathbf{A} \times \mathbf{B}) = (\mathbf{C} \cdot \mathbf{B})\mathbf{A} - (\mathbf{C} \cdot \mathbf{A})\mathbf{B}$$

Hence

$$(\mathbf{A} \times \mathbf{B}) \times \mathbf{C} = -[(\mathbf{C} \cdot \mathbf{B})\mathbf{A} - (\mathbf{C} \cdot \mathbf{A})\mathbf{B}] = (\mathbf{A} \cdot \mathbf{C})\mathbf{B} - (\mathbf{B} \cdot \mathbf{C})\mathbf{A}$$

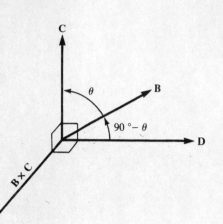

Figure 1-17

1-9. Basis

A. Linear dependence

A vector that can be obtained as a sum

LINEAR COMBINATION $\qquad m_1\mathbf{A}_1 + m_2\mathbf{A}_2 + \cdots + m_n\mathbf{A}_n$ $\qquad$ (1.67)

is called a **linear combination** of the vectors $\mathbf{A}_1, \mathbf{A}_2, \ldots, \mathbf{A}_n$. (The $m_1, m_2, \ldots, m_n$ are arbitrary scalars.)

If we have

LINEAR DEPENDENCE $\qquad m_1\mathbf{A}_1 + m_2\mathbf{A}_2 + \cdots + m_n\mathbf{A}_n = \mathbf{0}$ $\qquad$ (1.68)

where at least one of the coefficients $m_1, m_2, \ldots, m_n$ is nonzero, then the set of vectors $\mathbf{A}_1, \mathbf{A}_2, \ldots, \mathbf{A}_n$ is said to be **linearly dependent**. If a set of vectors is *not* linearly dependent, we say that it is **linearly independent**.

Note: In order to prove that a set of vectors $\mathbf{A}_1, \mathbf{A}_2, \ldots, \mathbf{A}_n$ is linearly independent, we have to show that if (1.68) holds, then $m_1 = m_2 = \cdots = m_n = 0$.

B. Orthonormal basis

If every vector $\mathbf{A}$ in three-dimensional space can be expressed as a linear combination of nonzero vectors $\mathbf{u}_1, \mathbf{u}_2$, and $\mathbf{u}_3$, we say that the three vectors $\mathbf{u}_1, \mathbf{u}_2$, and $\mathbf{u}_3$ *span* the space. In addition, if these three vectors are linearly independent, they constitute a **basis**.

A set of three nonzero vectors $\mathbf{u}_1, \mathbf{u}_2, \mathbf{u}_3$, is called an **orthogonal basis** if and only if they are mutually orthogonal; that is, each is perpendicular to the other two, or in general

ORTHOGONAL BASIS $\qquad \mathbf{u}_m \cdot \mathbf{u}_n = 0 \quad$ for all $m \neq n$ $\qquad$ (1.69)

It is called an **orthonormal basis** if and only if

ORTHONORMAL BASIS $\qquad \mathbf{u}_m \cdot \mathbf{u}_n = \delta_{mn} = \begin{cases} 1 & \text{if } m = n \\ 0 & \text{if } m \neq n \end{cases}$ $\qquad$ (1.70)

where δ_{mn} is the Kronecker delta.

Note: An orthonormal basis is an orthogonal basis. Also, each vector $\mathbf{u}_1, \mathbf{u}_2$, and $\mathbf{u}_3$ of the orthonormal basis is a unit vector.

EXAMPLE 1-20: Let $\mathbf{u}_1, \mathbf{u}_2$, and $\mathbf{u}_3$ form an orthonormal basis. Show that this set of vectors is linearly independent.

Solution: To prove that an orthonormal set is linearly independent, it is sufficient, from (1.68), to show that if

$$\lambda_1\mathbf{u}_1 + \lambda_2\mathbf{u}_2 + \lambda_3\mathbf{u}_3 = \mathbf{0} \qquad (1.71)$$

then

$$\lambda_1 = \lambda_2 = \lambda_3 = 0$$

Taking the dot product of both sides of (1.71) with $\mathbf{u}_1$, we obtain

$$(\lambda_1\mathbf{u}_1 + \lambda_2\mathbf{u}_2 + \lambda_3\mathbf{u}_3) \cdot \mathbf{u}_1 = \lambda_1\mathbf{u}_1 \cdot \mathbf{u}_1 + \lambda_2\mathbf{u}_2 \cdot \mathbf{u}_1 + \lambda_3\mathbf{u}_3 \cdot \mathbf{u}_1 = \mathbf{0} \cdot \mathbf{u}_1$$

and consequently

$$\lambda_1\mathbf{u}_1 \cdot \mathbf{u}_1 + \lambda_2\mathbf{u}_2 \cdot \mathbf{u}_1 + \lambda_3\mathbf{u}_3 \cdot \mathbf{u}_1 = 0$$

From the definition of an orthonormal basis (1.70), we know that

$$\mathbf{u}_1 \cdot \mathbf{u}_1 = 1, \qquad \mathbf{u}_2 \cdot \mathbf{u}_1 = \mathbf{u}_3 \cdot \mathbf{u}_1 = 0 \qquad (1.72)$$

Hence we obtain $\lambda_1 = 0$.

Similarly, we dot (1.71) with $\mathbf{u}_2$ and $\mathbf{u}_3$, respectively, to obtain $\lambda_2 = \lambda_3 = 0$. Hence the orthonormal set $\mathbf{u}_1, \mathbf{u}_2, \mathbf{u}_3$ is linearly independent.

EXAMPLE 1-21: Let $\mathbf{u}_1, \mathbf{u}_2$, and $\mathbf{u}_3$ form an orthonormal basis. Show that given any vector $\mathbf{A}$, we can write $\mathbf{A}$ as

$$\mathbf{A} = (\mathbf{A} \cdot \mathbf{u}_1)\mathbf{u}_1 + (\mathbf{A} \cdot \mathbf{u}_2)\mathbf{u}_2 + (\mathbf{A} \cdot \mathbf{u}_3)\mathbf{u}_3 \qquad (1.73)$$

Solution: Since the vectors $\mathbf{u}_1, \mathbf{u}_2$, and $\mathbf{u}_3$ form an orthonormal basis, any vector $\mathbf{A}$ can be expressed as a linear combination of $\mathbf{u}_1, \mathbf{u}_2$, and $\mathbf{u}_3$:

$$\mathbf{A} = \lambda_1\mathbf{u}_1 + \lambda_2\mathbf{u}_2 + \lambda_3\mathbf{u}_3 \qquad (1.74)$$

Taking the dot product of both sides of (1.74) with $\mathbf{u}_1$ and using (1.72), we obtain

$$\mathbf{A} \cdot \mathbf{u}_1 = (\lambda_1\mathbf{u}_1 + \lambda_2\mathbf{u}_2 + \lambda_3\mathbf{u}_3) \cdot \mathbf{u}_1 = \lambda_1\mathbf{u}_1 \cdot \mathbf{u}_1 + \lambda_2\mathbf{u}_2 \cdot \mathbf{u}_1 + \lambda_3\mathbf{u}_3 \cdot \mathbf{u}_1 = \lambda_1$$

Similarly, we find that $\mathbf{A} \cdot \mathbf{u}_2 = \lambda_2$ and $\mathbf{A} \cdot \mathbf{u}_3 = \lambda_3$. Hence (1.73) follows:

$$\mathbf{A} = (\mathbf{A} \cdot \mathbf{u}_1)\mathbf{u}_1 + (\mathbf{A} \cdot \mathbf{u}_2)\mathbf{u}_2 + (\mathbf{A} \cdot \mathbf{u}_3)\mathbf{u}_3$$

SUMMARY

1. A vector is a quantity that is characterized by both magnitude and direction.
2. The addition of two vectors $\mathbf{A}$ and $\mathbf{B}$ is performed by the triangle rule: Place the initial point of $\mathbf{B}$ at the terminal point of $\mathbf{A}$ and draw the vector $\mathbf{A} + \mathbf{B}$ from the initial point of $\mathbf{A}$ to the terminal point of $\mathbf{B}$.
3. The dot product of two vectors $\mathbf{A}$ and $\mathbf{B}$ is a scalar given by

$$\mathbf{A} \cdot \mathbf{B} = |\mathbf{A}|\,|\mathbf{B}| \cos \theta = AB \cos \theta$$

where θ is the angle between $\mathbf{A}$ and $\mathbf{B}$ and $0 \leqslant \theta \leqslant \pi$.
4. Properties of the dot product are

$$\mathbf{A} \cdot \mathbf{B} = \mathbf{B} \cdot \mathbf{A}$$
$$\mathbf{A} \cdot (\mathbf{B} + \mathbf{C}) = \mathbf{A} \cdot \mathbf{B} + \mathbf{A} \cdot \mathbf{C}$$
$$(m\mathbf{A}) \cdot \mathbf{B} = \mathbf{A} \cdot (m\mathbf{B}) = m(\mathbf{A} \cdot \mathbf{B})$$
$$\mathbf{A} \cdot \mathbf{A} = |\mathbf{A}|^2 = A^2 \quad \text{for every } \mathbf{A}$$
$$\mathbf{A} \cdot \mathbf{A} = 0 \quad \text{if and only if } \mathbf{A} = \mathbf{0}$$

5. Two vectors $\mathbf{A}$ and $\mathbf{B}$ are perpendicular to each other if and only if $\mathbf{A} \cdot \mathbf{B} = 0$.
6. The cross product of two vectors $\mathbf{A}$ and $\mathbf{B}$ is a vector given by

$$\mathbf{A} \times \mathbf{B} = |\mathbf{A}|\,|\mathbf{B}| \sin \theta \mathbf{u} = (AB \sin \theta)\mathbf{u}$$

where θ is the angle between $\mathbf{A}$ and $\mathbf{B}$, $0 \leqslant \theta \leqslant \pi$, and $\mathbf{u}$ is a unit vector whose direction is that of the advance of a right-hand screw as $\mathbf{A}$ rotates toward $\mathbf{B}$ through the angle θ. $\mathbf{A} \times \mathbf{B}$ is perpendicular to both $\mathbf{A}$ and $\mathbf{B}$.
7. Properties of the cross product are

$$\mathbf{A} \times \mathbf{B} = -\mathbf{B} \times \mathbf{A}$$
$$\mathbf{A} \times (\mathbf{B} + \mathbf{C}) = \mathbf{A} \times \mathbf{B} + \mathbf{A} \times \mathbf{C}$$
$$(m\mathbf{A}) \times \mathbf{B} = \mathbf{A} \times (m\mathbf{B}) = m(\mathbf{A} \times \mathbf{B})$$
$$\mathbf{A} \times \mathbf{A} = \mathbf{0}$$
$$\mathbf{A} \times \mathbf{0} = \mathbf{0} \quad \text{for any } \mathbf{A}$$

8. Two vectors $\mathbf{A}$ and $\mathbf{B}$ are parallel if and only if $\mathbf{A} \times \mathbf{B} = \mathbf{0}$ or if there exists a scalar m such that $\mathbf{B} = m\mathbf{A}$.
9. The fundamental identity for the scalar triple product is

$$\mathbf{A} \cdot \mathbf{B} \times \mathbf{C} = \mathbf{A} \times \mathbf{B} \cdot \mathbf{C}$$

That is, the positions of the dot and cross in the scalar triple product are immaterial.

10. The necessary and sufficient condition for the coplanarity of three vectors **A**, **B**, and **C** is $\mathbf{A} \cdot \mathbf{B} \times \mathbf{C} = 0$.

11. The fundamental identities for the vector triple product are

$$\mathbf{A} \times (\mathbf{B} \times \mathbf{C}) = (\mathbf{A} \cdot \mathbf{C})\mathbf{B} - (\mathbf{A} \cdot \mathbf{B})\mathbf{C}$$

$$(\mathbf{A} \times \mathbf{B}) \times \mathbf{C} = (\mathbf{A} \cdot \mathbf{C})\mathbf{B} - (\mathbf{B} \cdot \mathbf{C})\mathbf{A}$$

12. A set of three nonzero vectors $\mathbf{u}_1$, $\mathbf{u}_2$, $\mathbf{u}_3$ is called an orthonormal basis if and only if

$$\mathbf{u}_m \cdot \mathbf{u}_n = \delta_{mn} = \begin{cases} 1 & \text{if } m = n \\ 0 & \text{if } m \neq n \end{cases}$$

Using these, any vector **A** can be expressed as

$$\mathbf{A} = (\mathbf{A} \cdot \mathbf{u}_1)\mathbf{u}_1 + (\mathbf{A} \cdot \mathbf{u}_2)\mathbf{u}_2 + (\mathbf{A} \cdot \mathbf{u}_3)\mathbf{u}_3$$

RAISE YOUR GRADES

Can you explain...?

☑ how to add and subtract vectors
☑ how to define the dot product and the cross product of two vectors
☑ what the necessary and sufficient condition is for two vectors to be parallel, or for two vectors to be perpendicular to each other
☑ how to remember the fundamental identities for the vector triple product
☑ what the condition is for three given vectors to be noncoplanar
☑ what an orthonormal basis is

SOLVED PROBLEMS

Addition and Subtraction of Vectors

PROBLEM 1-1 Let $\mathbf{A} = \overrightarrow{OP}$ and $\mathbf{B} = \overrightarrow{OR}$, where $OPQR$ is a parallelogram (Figure 1-18). Express $\overrightarrow{OQ}$ and $\overrightarrow{PR}$ in terms of **A** and **B**.

Solution: Since $\overrightarrow{PQ}$ is parallel to $\overrightarrow{OR}$ and in the same direction,

$$\overrightarrow{PQ} = \overrightarrow{OR} = \mathbf{B}$$

Now

$$\overrightarrow{OQ} = \overrightarrow{OP} + \overrightarrow{PQ}$$

Hence

$$\overrightarrow{OQ} = \mathbf{A} + \mathbf{B}$$

(This is the **parallelogram law** for the sum of two vectors, which is equivalent to the *triangle law* of Figure 1-4.)

Next

$$\overrightarrow{OP} + \overrightarrow{PR} = \overrightarrow{OR}$$

Hence

$$\overrightarrow{PR} = \overrightarrow{OR} - \overrightarrow{OP} = \mathbf{B} - \mathbf{A}$$

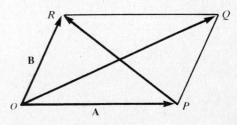

Figure 1-18

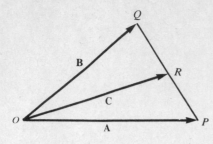

Figure 1-19

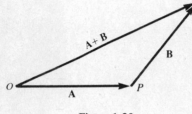

Figure 1-20

PROBLEM 1-2   Let O, Q, and P be three distinct points and let R be the midpoint of line segment PQ. If $\overrightarrow{OP} = \mathbf{A}$, $\overrightarrow{OQ} = \mathbf{B}$, and $\overrightarrow{OR} = \mathbf{C}$, show that $\mathbf{C} = \frac{1}{2}(\mathbf{A} + \mathbf{B})$. (See Figure 1-19.)

Solution: By the law of vector addition,

$$\mathbf{C} = \mathbf{B} + \overrightarrow{QR}$$

But

$$\overrightarrow{QP} = \mathbf{A} - \mathbf{B}$$

Since $\overrightarrow{QR} = \frac{1}{2}\overrightarrow{QP} = \frac{1}{2}(\mathbf{A} - \mathbf{B})$, we obtain

$$\mathbf{C} = \mathbf{B} + \overrightarrow{QR} = \mathbf{B} + \frac{1}{2}(\mathbf{A} - \mathbf{B}) = \frac{1}{2}(\mathbf{A} + \mathbf{B})$$

PROBLEM 1-3   If $m\mathbf{A} + n\mathbf{B} = \mathbf{0}$ but m and n are not both zero, show that $\mathbf{A}$ is parallel to $\mathbf{B}$.

Solution: If $m \neq 0$, then $\mathbf{A} = -(n/m)\mathbf{B}$. Thus, from the condition for parallelism (1.13), we conclude that $\mathbf{A}$ is parallel to $\mathbf{B}$.

PROBLEM 1-4   Show geometrically that

$$|\mathbf{A} + \mathbf{B}| \leqslant |\mathbf{A}| + |\mathbf{B}|$$

which is known as the **triangle inequality of vectors**.

Solution: Let $\overrightarrow{OP} = \mathbf{A}$ and $\overrightarrow{PQ} = \mathbf{B}$, where OPQ is a triangle (Figure 1-20). Then $\mathbf{A} + \mathbf{B} = \overrightarrow{OQ}$. From plane geometry we know that the length of one side of a triangle is less than or equal to the sum of the lengths of the other two sides; that is, $OQ \leqslant OP + PQ$. So we have

$$|\mathbf{A} + \mathbf{B}| \leqslant |\mathbf{A}| + |\mathbf{B}|$$

[*Note:* The equals sign holds when points O, P, and Q are collinear, that is, when $\mathbf{A}$ and $\mathbf{B}$ are parallel.]

Scalar Product

PROBLEM 1-5   If $\mathbf{A} \cdot \mathbf{B} = \mathbf{A} \cdot \mathbf{C}$, can we conclude that $\mathbf{B} = \mathbf{C}$?

Solution: Since $\mathbf{A} \cdot \mathbf{B} = \mathbf{A} \cdot \mathbf{C}$ can be written as

$$\mathbf{A} \cdot (\mathbf{B} - \mathbf{C}) = 0$$

we can say only that one of three things is true: that $\mathbf{A}$ is perpendicular to $\mathbf{B} - \mathbf{C}$, that $\mathbf{A} = \mathbf{0}$, or that $\mathbf{B} - \mathbf{C} = \mathbf{0}$, (i.e., $\mathbf{B} = \mathbf{C}$). So if $\mathbf{A} \cdot \mathbf{B} = \mathbf{A} \cdot \mathbf{C}$, it doesn't necessarily follow that $\mathbf{B} = \mathbf{C}$.

PROBLEM 1-6   Prove that

$$(\mathbf{A} + \mathbf{B}) \cdot (\mathbf{C} + \mathbf{D}) = \mathbf{A} \cdot \mathbf{C} + \mathbf{B} \cdot \mathbf{C} + \mathbf{A} \cdot \mathbf{D} + \mathbf{B} \cdot \mathbf{D}$$

Solution: By the distributive law (1.29) and the commutative law (1.28), we have

$$(\mathbf{A} + \mathbf{B}) \cdot (\mathbf{C} + \mathbf{D}) = (\mathbf{A} + \mathbf{B}) \cdot \mathbf{C} + (\mathbf{A} + \mathbf{B}) \cdot \mathbf{D}$$
$$= \mathbf{C} \cdot (\mathbf{A} + \mathbf{B}) + \mathbf{D} \cdot (\mathbf{A} + \mathbf{B})$$
$$= \mathbf{C} \cdot \mathbf{A} + \mathbf{C} \cdot \mathbf{B} + \mathbf{D} \cdot \mathbf{A} + \mathbf{D} \cdot \mathbf{B}$$
$$= \mathbf{A} \cdot \mathbf{C} + \mathbf{B} \cdot \mathbf{C} + \mathbf{A} \cdot \mathbf{D} + \mathbf{B} \cdot \mathbf{D}$$

PROBLEM 1-7   If $\mathbf{B} \neq \mathbf{0}$ and $\mathbf{A} = \mathbf{A}_1 + \mathbf{A}_2$, where $\mathbf{A}_1 \parallel \mathbf{B}$ and $\mathbf{A}_2 \perp \mathbf{B}$, show that

$$\mathbf{A}_1 = \frac{\mathbf{A} \cdot \mathbf{B}}{B^2} \mathbf{B} \tag{a}$$

$$A_2 = A - \frac{A \cdot B}{B^2} B \qquad \text{(b)}$$

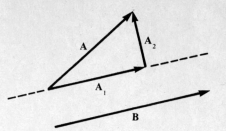

(See Figure 1-21.)

Solution: If $A_2 \perp B$, then $A_2 \cdot B = 0$. So

$$A \cdot B = A_1 \cdot B + A_2 \cdot B = A_1 \cdot B \qquad \text{(c)}$$

Figure 1-21

Since $A_1 \parallel B$, we can use the condition for parallelism (1.13): For some scalar m

$$A_1 = mB \qquad \text{(d)}$$

Dotting both sides of this equation by B, we obtain

$$A_1 \cdot B = mB \cdot B = mB^2$$

Solving this for m and using condition (c), we obtain

$$m = \frac{A_1 \cdot B}{B^2} = \frac{A \cdot B}{B^2} \qquad \text{(e)}$$

So, from eq. (d)

$$A_1 = \frac{A \cdot B}{B^2} B$$

And since $A = A_1 + A_2$,

$$A_2 = A - A_1 = A - \frac{A \cdot B}{B^2} B$$

Additionally, if we dot both sides of this equation with B, we obtain

$$A_2 \cdot B = A \cdot B - \frac{A \cdot B}{B^2} B \cdot B = A \cdot B - \frac{A \cdot B}{B^2} B^2 = A \cdot B - A \cdot B = 0$$

which implies that $A_2 \perp B$. So we see that eqs. (a) and (b) conversely imply that $A_1 \parallel B$ and $A_2 \perp B$.

PROBLEM 1-8 If A and B are two arbitrary vectors, show that

$$|A \cdot B| \leqslant |A| |B| \qquad \text{(a)}$$

This result is known as the **Cauchy–Schwarz inequality.**

Solution: From the definition of the scalar product (1.22),

$$|A \cdot B| = |AB \cos \theta| = AB|\cos \theta|$$

Since $|\cos \theta| \leqslant 1$,

$$|A \cdot B| = AB|\cos \theta| \leqslant AB = |A| |B|$$

Alternate Solution: Let m be any scalar. Then we can create the quantity $mA + B$ and observe that

$$
\begin{aligned}
|mA + B|^2 &= (mA + B) \cdot (mA + B) \\
&= m^2(A \cdot A) + 2m(A \cdot B) + (B \cdot B) \\
&= m^2 A^2 + 2m(A \cdot B) + B^2 \qquad \text{(b)}
\end{aligned}
$$

The right-hand side of eq. (b) is a quadratic in m except when $|A| = A = 0$. If $|A| = A \neq 0$, then adding $(A \cdot B)^2/A^2$ to eq. (b) and also subtracting the same quantity yields

$$|mA + B|^2 = m^2 A^2 + 2m(A \cdot B) + \frac{(A \cdot B)^2}{A^2} + B^2 - \frac{(A \cdot B)^2}{A^2}$$

$$= \left[mA + \frac{(A \cdot B)}{A} \right]^2 + \frac{1}{A^2} [A^2 B^2 - (A \cdot B)^2] \qquad \text{(c)}$$

The left-hand side of eq. (c) is nonnegative for any scalar m. Picking a particular value for m to simplify this result,

$$m = -(\mathbf{A} \cdot \mathbf{B})/A^2$$

we obtain

$$\frac{1}{A^2}[A^2B^2 - (\mathbf{A} \cdot \mathbf{B})^2] \geqslant 0$$

Since $A^2 = |\mathbf{A}|^2 > 0$, we have

$$(\mathbf{A} \cdot \mathbf{B})^2 \leqslant A^2 B^2 = |\mathbf{A}|^2 |\mathbf{B}|^2$$

or

$$|\mathbf{A} \cdot \mathbf{B}| \leqslant |\mathbf{A}||\mathbf{B}|$$

If $|\mathbf{A}| = A = 0$, then $\mathbf{A} = \mathbf{0}$. Then condition (a) is obviously true and the equality holds in this case.

PROBLEM 1-9 When is $|\mathbf{A} \cdot \mathbf{B}| = |\mathbf{A}||\mathbf{B}|$?

Solution: From the definition of the scalar product (1.22),

$$|\mathbf{A} \cdot \mathbf{B}| = |AB \cos \theta| = |\mathbf{A}||\mathbf{B}||\cos \theta| = |\mathbf{A}||\mathbf{B}|$$

Therefore $|\cos \theta| = 1$ or $\cos \theta = \pm 1$, so $\theta = 0°$ or $180°$. Hence $|\mathbf{A} \cdot \mathbf{B}| = |\mathbf{A}||\mathbf{B}|$ if $\mathbf{A}$ and $\mathbf{B}$ are parallel.

PROBLEM 1-10 Using vector methods, verify the **law of cosines**: If three sides of a triangle have lengths $|\mathbf{A}|$, $|\mathbf{B}|$, and $|\mathbf{C}|$, and if the angle opposite the side of length $|\mathbf{C}|$ is θ, then $C^2 = A^2 + B^2 - 2AB \cos \theta$.

Solution: As shown in Figure 1-22, construct a triangle with vectors, $\mathbf{A}$, $\mathbf{B}$, and $\mathbf{C}$ such that

$$\mathbf{C} = \mathbf{B} - \mathbf{A}$$

Since

$$\begin{aligned}
\mathbf{C} \cdot \mathbf{C} &= (\mathbf{B} - \mathbf{A}) \cdot (\mathbf{B} - \mathbf{A}) \\
&= \mathbf{B} \cdot \mathbf{B} - \mathbf{A} \cdot \mathbf{B} - \mathbf{B} \cdot \mathbf{A} + \mathbf{A} \cdot \mathbf{A} \\
&= \mathbf{A} \cdot \mathbf{A} + \mathbf{B} \cdot \mathbf{B} - 2\mathbf{A} \cdot \mathbf{B}
\end{aligned}$$

applying the definition of the dot product to each term, we obtain the law of cosines:

$$C^2 = A^2 + B^2 - 2AB \cos \theta$$

PROBLEM 1-11 Using the scalar product, verify the *triangle inequality*: If $\mathbf{A}$ and $\mathbf{B}$ are two arbitrary vectors, then

$$|\mathbf{A} + \mathbf{B}| \leqslant |\mathbf{A}| + |\mathbf{B}| \tag{a}$$

Solution: From the left-hand side of eq. (a),

$$\begin{aligned}
|\mathbf{A} + \mathbf{B}|^2 &= (\mathbf{A} + \mathbf{B}) \cdot (\mathbf{A} + \mathbf{B}) \\
&= (\mathbf{A} \cdot \mathbf{A}) + 2(\mathbf{A} \cdot \mathbf{B}) + (\mathbf{B} \cdot \mathbf{B}) \\
&= |\mathbf{A}|^2 + 2(\mathbf{A} \cdot \mathbf{B}) + |\mathbf{B}|^2
\end{aligned} \tag{b}$$

From the Cauchy–Schwarz inequality of Problem 1-8,

$$(\mathbf{A} \cdot \mathbf{B}) \leqslant |\mathbf{A}||\mathbf{B}|$$

So from eq. (b) we obtain

$$|\mathbf{A} + \mathbf{B}|^2 \leqslant |\mathbf{A}|^2 + 2|\mathbf{A}||\mathbf{B}| + |\mathbf{B}|^2 = (|\mathbf{A}| + |\mathbf{B}|)^2$$

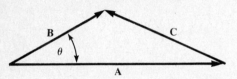

Figure 1-22

or

$$|\mathbf{A} + \mathbf{B}| \leqslant |\mathbf{A}| + |\mathbf{B}|$$

[*Note:* The geometric proof of eq. (a) is given in Problem 1-4.]

PROBLEM 1-12 Show that

$$|\mathbf{A} + \mathbf{B}|^2 = |\mathbf{A}|^2 + |\mathbf{B}|^2$$

if and only if **A** and **B** are orthogonal. This is known as the **Pythagorean theorem**.

Solution: From eq. (b) of Problem 1-11 we have

$$|\mathbf{A} + \mathbf{B}|^2 = |\mathbf{A}|^2 + 2(\mathbf{A} \cdot \mathbf{B}) + |\mathbf{B}|^2$$

So if

$$|\mathbf{A} + \mathbf{B}|^2 = |\mathbf{A}|^2 + |\mathbf{B}|^2$$

then

$$\mathbf{A} \cdot \mathbf{B} = 0$$

which implies that **A** and **B** are orthogonal (see Example 1-3). Now, if **A** and **B** are orthogonal, then $\mathbf{A} \cdot \mathbf{B} = 0$ and eq. (b) of Problem 1-11 reduces to

$$|\mathbf{A} + \mathbf{B}|^2 = |\mathbf{A}|^2 + |\mathbf{B}|^2$$

Vector Product

PROBLEM 1-13 Show that **A** is parallel to **B** if and only if $\mathbf{A} \times \mathbf{B} = \mathbf{0}$.

Solution: If $\mathbf{A} \parallel \mathbf{B}$, we know from the condition for parallelism (1.13) that there exists a scalar m such that $\mathbf{B} = m\mathbf{A}$. So, using the properties (1.41) and (1.42) of vector products,

$$\mathbf{A} \times \mathbf{B} = \mathbf{A} \times (m\mathbf{A}) = m\mathbf{A} \times \mathbf{A} = m\mathbf{0} = \mathbf{0}$$

If **A** is not parallel to **B**, then $\mathbf{A} \neq \mathbf{0}$, $\mathbf{B} \neq \mathbf{0}$, and $0 < \theta < \pi$. Thus $|\mathbf{A}| \neq 0$, $|\mathbf{B}| \neq 0$, and $\sin \theta \neq 0$. Hence, from the definition of a vector product (1.35),

$$\mathbf{A} \times \mathbf{B} = |\mathbf{A}||\mathbf{B}| \sin \theta \, \mathbf{u} \neq \mathbf{0}$$

So $\mathbf{A} \parallel \mathbf{B}$ if and only if $\mathbf{A} \times \mathbf{B} = \mathbf{0}$.

PROBLEM 1-14 Verify **Lagrange's identity**: If **A** and **B** are arbitrary vectors, then

$$|\mathbf{A} \times \mathbf{B}|^2 = |\mathbf{A}|^2 |\mathbf{B}|^2 - (\mathbf{A} \cdot \mathbf{B})^2$$

Solution: From the definition of the vector product (1.35)

$$\begin{aligned}
|\mathbf{A} \times \mathbf{B}|^2 &= |\mathbf{A}|^2 |\mathbf{B}|^2 \sin^2\theta \\
&= |\mathbf{A}|^2 |\mathbf{B}|^2 (1 - \cos^2\theta) \\
&= |\mathbf{A}|^2 |\mathbf{B}|^2 - |\mathbf{A}|^2 |\mathbf{B}|^2 \cos^2\theta
\end{aligned}$$

Then, using the definition of the scalar product (1.22),

$$|\mathbf{A} \times \mathbf{B}|^2 = |\mathbf{A}|^2 |\mathbf{B}|^2 - (\mathbf{A} \cdot \mathbf{B})^2$$

PROBLEM 1-15 Using vector methods, derive the **law of sines** for a triangle: If α, β, and γ are the interior angles opposite the three sides of a triangle whose lengths are $|\mathbf{A}|$, $|\mathbf{B}|$, and $|\mathbf{C}|$, (Figure 1-23), then

$$\frac{A}{\sin \alpha} = \frac{B}{\sin \beta} = \frac{C}{\sin \gamma}$$

Solution: From Figure 1-23, we can see that

$$\mathbf{C} = \mathbf{B} - \mathbf{A}$$

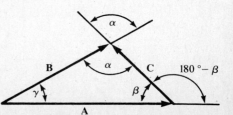

Figure 1-23

Then
$$C \times C = C \times (B - A)$$
or
$$0 = C \times B - C \times A$$
So
$$C \times A = C \times B$$

Using the definition of the vector product (1.35), $CA\sin(180° - \beta) = CB\sin\alpha$. Since $\sin(180° - \beta) = \sin\beta$, we have $CA\sin\beta = CB\sin\alpha$, or

$$\frac{A}{\sin\alpha} = \frac{B}{\sin\beta}$$

By similar reasoning we can show that

$$\frac{A}{\sin\alpha} = \frac{C}{\sin\gamma}$$

which gives the law of sines:

$$\frac{A}{\sin\alpha} = \frac{B}{\sin\beta} = \frac{C}{\sin\gamma}$$

Scalar Triple Product

PROBLEM 1-16   Show that

$$[ABC] = [BCA] = [CAB] = -[ACB] = -[BAC] = -[CBA] \qquad (a)$$

Solution: Using the commutative law of the scalar product (1.28) and the fundamental identity for the scalar triple product (1.53), we have

$$[ABC] = A \cdot B \times C = B \times C \cdot A$$
$$= B \cdot C \times A$$
$$= [BCA]$$

Using the anticommutative law of the vector product (1.39), we obtain

$$[ABC] = A \cdot B \times C = A \cdot [-(C \times B)]$$
$$= -A \cdot C \times B$$
$$= -[ACB]$$

We prove the other equations similarly.

We can see from eq. (a) that any cyclic permutation of the vectors A, B, and C in the triple scalar product will give the same result, whereas an acyclic permutation will give the negative result.

PROBLEM 1-17   Using the scalar triple product, prove the following (see Example 1-16): Let A and B be nonparallel vectors. Then vector C is coplanar with A and B if there exist scalars m and n such that

$$C = mA + nB$$

Solution: Suppose there exist scalars m and n such that $C = mA + nB$. Then by the fundamental identity (1.53), we have

$$A \cdot (B \times C) = (A \times B) \cdot C$$
$$= (A \times B) \cdot (mA + nB)$$
$$= m(A \times B) \cdot A + n(A \times B) \cdot B = 0$$

since $(A \times B) \cdot A = (A \times B) \cdot B = 0$ by identities (1.51). So from the coplanarity condition (1.50), we conclude that A, B, and C are coplanar.

PROBLEM 1-18   Using the scalar triple product, prove the distributive law of the vector product (1.40):

$$A \times (B + C) = A \times B + A \times C$$

Solution: Let vector U represent the difference of the two sides:

$$U = A \times (B + C) - A \times B - A \times C$$

We want to show that $U = 0$ and thus that the two sides are equal. So we take the dot product of U with an arbitrary vector V, which yields

$$V \cdot U = V \cdot [A \times (B + C)] - V \cdot A \times B - V \cdot A \times C$$

From the distributive law of the scalar product (1.29) and the fundamental identity (1.53), by interchanging the dot and cross, we have

$$V \cdot U = V \times A \cdot (B + C) - V \times A \cdot B - V \times A \cdot C$$
$$= V \times A \cdot B + V \times A \cdot C - V \times A \cdot B - V \times A \cdot C$$
$$= 0$$

which implies that $V \perp U$. Since V is arbitrary, we choose $V = U$. Thus $U \cdot U = 0$, hence $U = 0$, and the distributive law of the vector product is proved.

PROBLEM 1-19   Given vectors a_1, a_2, and a_3, let

$$b_1 = \frac{a_2 \times a_3}{[a_1 a_2 a_3]}, \qquad b_2 = \frac{a_3 \times a_1}{[a_1 a_2 a_3]}, \qquad b_3 = \frac{a_1 \times a_2}{[a_1 a_2 a_3]}$$

where $[a_1 a_2 a_3] \neq 0$. Then show that

$$a_m \cdot b_n = \delta_{mn}$$

where δ_{mn} is the Kronecker delta defined as

$$\delta_{mn} = \begin{cases} 1, & m = n \\ 0, & m \neq n \end{cases}$$

Solution: Using identities (1.51) and the result of Problem 1-16, we have

$$a_1 \cdot b_1 = \frac{a_1 \cdot (a_2 \times a_3)}{[a_1 a_2 a_3]} = \frac{[a_1 a_2 a_3]}{[a_1 a_2 a_3]} = 1$$

$$a_1 \cdot b_2 = \frac{a_1 \cdot (a_3 \times a_1)}{[a_1 a_2 a_3]} = \frac{0}{[a_1 a_2 a_3]} = 0$$

$$a_1 \cdot b_3 = \frac{a_1 \cdot (a_1 \times a_2)}{[a_1 a_2 a_3]} = \frac{0}{[a_1 a_2 a_3]} = 0$$

Similarly, we can prove that

$$a_2 \cdot b_2 = a_3 \cdot b_3 = 1$$
$$a_2 \cdot b_1 = a_2 \cdot b_3 = a_3 \cdot b_1 = a_3 \cdot b_2 = 0$$

Thus we conclude that

$$a_m \cdot b_n = \delta_{mn} = \begin{cases} 1, & m = n \\ 0, & m \neq n \end{cases}$$

PROBLEM 1-20   Show that $[b_1 b_2 b_3] \neq 0$, where b_1, b_2, and b_3 are the vectors defined in Problem 1-19.

Solution: We prove this by contradiction. If $[b_1 b_2 b_3] = 0$, then from condition (1.50) the vectors b_1, b_2, b_3 are coplanar. Then we can use the result of Problem 1-17 to write

$$b_3 = k_1 b_1 + k_2 b_2 \qquad\qquad (a)$$

for some scalars k_1 and k_2. Dotting both sides of eq. (a) with $\mathbf{a}_3$ and using the result of Problem 1-19, we have

$$1 = \mathbf{a}_3 \cdot \mathbf{b}_3 = k_1(\mathbf{a}_3 \cdot \mathbf{b}_1) + k_2(\mathbf{a}_3 \cdot \mathbf{b}_2) = 0$$

which is a contradiction. Hence we conclude that

$$[\mathbf{b}_1 \mathbf{b}_2 \mathbf{b}_3] \neq 0$$

PROBLEM 1-21   If

$$\mathbf{b}_1 = \frac{\mathbf{a}_2 \times \mathbf{a}_3}{[\mathbf{a}_1 \mathbf{a}_2 \mathbf{a}_3]}, \qquad \mathbf{b}_2 = \frac{\mathbf{a}_3 \times \mathbf{a}_1}{[\mathbf{a}_1 \mathbf{a}_2 \mathbf{a}_3]}, \qquad \mathbf{b}_3 = \frac{\mathbf{a}_1 \times \mathbf{a}_2}{[\mathbf{a}_1 \mathbf{a}_2 \mathbf{a}_3]}$$

show that

$$\mathbf{a}_1 = \frac{\mathbf{b}_2 \times \mathbf{b}_3}{[\mathbf{b}_1 \mathbf{b}_2 \mathbf{b}_3]}, \qquad \mathbf{a}_2 = \frac{\mathbf{b}_3 \times \mathbf{b}_1}{[\mathbf{b}_1 \mathbf{b}_2 \mathbf{b}_3]}, \qquad \mathbf{a}_3 = \frac{\mathbf{b}_1 \times \mathbf{b}_2}{[\mathbf{b}_1 \mathbf{b}_2 \mathbf{b}_3]}$$

(The sets $\mathbf{a}_1, \mathbf{a}_2, \mathbf{a}_3$ and $\mathbf{b}_1, \mathbf{b}_2, \mathbf{b}_3$ are called **reciprocal sets of vectors**.)

Solution: From the definition of the vector product, $\mathbf{a}_1$ is orthogonal to both $\mathbf{b}_2$ and $\mathbf{b}_3$. Hence we can write

$$\mathbf{a}_1 = \lambda(\mathbf{b}_2 \times \mathbf{b}_3)$$

Then from the result of Problem 1-19, we have

$$\mathbf{b}_1 \cdot \mathbf{a}_1 = \lambda \mathbf{b}_1 \cdot (\mathbf{b}_2 \times \mathbf{b}_3) = \lambda[\mathbf{b}_1 \mathbf{b}_2 \mathbf{b}_3] = 1$$

Thus

$$\lambda = \frac{1}{[\mathbf{b}_1 \mathbf{b}_2 \mathbf{b}_3]}$$

and we obtain

$$\mathbf{a}_1 = \frac{\mathbf{b}_2 \times \mathbf{b}_3}{[\mathbf{b}_1 \mathbf{b}_2 \mathbf{b}_3]}$$

Using similar reasoning, we can obtain the other results:

$$\mathbf{a}_2 = \frac{\mathbf{b}_3 \times \mathbf{b}_1}{[\mathbf{b}_1 \mathbf{b}_2 \mathbf{b}_3]}, \qquad \mathbf{a}_3 = \frac{\mathbf{b}_1 \times \mathbf{b}_2}{[\mathbf{b}_1 \mathbf{b}_2 \mathbf{b}_3]}$$

Vector Triple Product

PROBLEM 1-22   Verify the **Jacobi identity**: For arbitrary vectors $\mathbf{A}$, $\mathbf{B}$, and $\mathbf{C}$,

$$\mathbf{A} \times (\mathbf{B} \times \mathbf{C}) + \mathbf{B} \times (\mathbf{C} \times \mathbf{A}) + \mathbf{C} \times (\mathbf{A} \times \mathbf{B}) = \mathbf{0} \tag{a}$$

Solution: Using the fundamental identity (1.59) for the vector triple product, we have

$$\mathbf{A} \times (\mathbf{B} \times \mathbf{C}) = (\mathbf{A} \cdot \mathbf{C})\mathbf{B} - (\mathbf{A} \cdot \mathbf{B})\mathbf{C}$$

$$\mathbf{B} \times (\mathbf{C} \times \mathbf{A}) = (\mathbf{B} \cdot \mathbf{A})\mathbf{C} - (\mathbf{B} \cdot \mathbf{C})\mathbf{A}$$

$$\mathbf{C} \times (\mathbf{A} \times \mathbf{B}) = (\mathbf{C} \cdot \mathbf{B})\mathbf{A} - (\mathbf{C} \cdot \mathbf{A})\mathbf{B}$$

Adding these identities, the Jacobi identity (a) follows.

PROBLEM 1-23   Show that

$$(\mathbf{A} \times \mathbf{B}) \times (\mathbf{C} \times \mathbf{D}) = [\mathbf{ABD}]\mathbf{C} - [\mathbf{ABC}]\mathbf{D} \tag{a}$$

$$(\mathbf{A} \times \mathbf{B}) \times (\mathbf{C} \times \mathbf{D}) = [\mathbf{CDA}]\mathbf{B} - [\mathbf{CDB}]\mathbf{A} \tag{b}$$

Solution: Let $\mathbf{A} \times \mathbf{B} = \mathbf{E}$; then using the fundamental identity (1.59), we have

$$(\mathbf{A} \times \mathbf{B}) \times (\mathbf{C} \times \mathbf{D}) = \mathbf{E} \times (\mathbf{C} \times \mathbf{D})$$
$$= (\mathbf{E} \cdot \mathbf{D})\mathbf{C} - (\mathbf{E} \cdot \mathbf{C})\mathbf{D}$$
$$= (\mathbf{A} \times \mathbf{B} \cdot \mathbf{D})\mathbf{C} - (\mathbf{A} \times \mathbf{B} \cdot \mathbf{C})\mathbf{D}$$
$$= [\mathbf{ABD}]\mathbf{C} - [\mathbf{ABC}]\mathbf{D}$$

Similarly, if we let $\mathbf{C} \times \mathbf{D} = \mathbf{F}$, then using fundamental identity (1.60), we have

$$(\mathbf{A} \times \mathbf{B}) \times (\mathbf{C} \times \mathbf{D}) = (\mathbf{A} \times \mathbf{B}) \times \mathbf{F}$$
$$= (\mathbf{F} \cdot \mathbf{A})\mathbf{B} - (\mathbf{F} \cdot \mathbf{B})\mathbf{A}$$
$$= (\mathbf{C} \times \mathbf{D} \cdot \mathbf{A})\mathbf{B} - (\mathbf{C} \times \mathbf{D} \cdot \mathbf{B})\mathbf{A}$$
$$= (\mathbf{C} \cdot \mathbf{D} \times \mathbf{A})\mathbf{B} - (\mathbf{C} \cdot \mathbf{D} \times \mathbf{B})\mathbf{A}$$
$$= [\mathbf{CDA}]\mathbf{B} - [\mathbf{CDB}]\mathbf{A}$$

PROBLEM 1-24 Verify the **extended Lagrange identity**:

$$(\mathbf{A} \times \mathbf{B}) \cdot (\mathbf{C} \times \mathbf{D}) = (\mathbf{A} \cdot \mathbf{C})(\mathbf{B} \cdot \mathbf{D}) - (\mathbf{B} \cdot \mathbf{C})(\mathbf{A} \cdot \mathbf{D})$$

Solution: Considering the left-hand side as a scalar triple product and using identity (1.59), we have

$$(\mathbf{A} \times \mathbf{B}) \cdot (\mathbf{C} \times \mathbf{D}) = \mathbf{A} \cdot [\mathbf{B} \times (\mathbf{C} \times \mathbf{D})] = \mathbf{A} \cdot [(\mathbf{B} \cdot \mathbf{D})\mathbf{C} - (\mathbf{B} \cdot \mathbf{C})\mathbf{D}]$$
$$= (\mathbf{A} \cdot \mathbf{C})(\mathbf{B} \cdot \mathbf{D}) - (\mathbf{B} \cdot \mathbf{C})(\mathbf{A} \cdot \mathbf{D})$$

PROBLEM 1-25 Using the extended Lagrange identity of Problem 1-24, prove the Lagrange identity of Problem 1-14; that is,

$$|\mathbf{A} \times \mathbf{B}|^2 = |\mathbf{A}|^2|\mathbf{B}|^2 - (\mathbf{A} \cdot \mathbf{B})^2$$

Solution: By setting $\mathbf{C} = \mathbf{A}$ and $\mathbf{D} = \mathbf{B}$ in the extended Lagrange identity of Problem 1-24, we obtain

$$(\mathbf{A} \times \mathbf{B}) \cdot (\mathbf{A} \times \mathbf{B}) = (\mathbf{A} \cdot \mathbf{A})(\mathbf{B} \cdot \mathbf{B}) - (\mathbf{A} \cdot \mathbf{B})(\mathbf{A} \cdot \mathbf{B})$$

Applying scalar product property (1.31a), we obtain

$$|\mathbf{A} \times \mathbf{B}|^2 = |\mathbf{A}|^2|\mathbf{B}|^2 - (\mathbf{A} \cdot \mathbf{B})^2$$

PROBLEM 1-26 Show that

$$(\mathbf{A} \times \mathbf{B}) \cdot (\mathbf{B} \times \mathbf{C}) \times (\mathbf{C} \times \mathbf{A}) = [\mathbf{ABC}]^2$$

Solution: From the result eq. (a) of Problem 1-23, we have

$$(\mathbf{B} \times \mathbf{C}) \times (\mathbf{C} \times \mathbf{A}) = [\mathbf{BCA}]\mathbf{C} - [\mathbf{BCC}]\mathbf{A}$$
$$= [\mathbf{ABC}]\mathbf{C}$$

since $[\mathbf{BCA}] = [\mathbf{ABC}]$ from the result of Problem 1-16 and $[\mathbf{BCC}] = 0$ from identities (1.51). So

$$(\mathbf{A} \times \mathbf{B}) \cdot (\mathbf{B} \times \mathbf{C}) \times (\mathbf{C} \times \mathbf{A}) = (\mathbf{A} \times \mathbf{B}) \cdot [\mathbf{ABC}]\mathbf{C}$$
$$= [\mathbf{ABC}](\mathbf{A} \times \mathbf{B}) \cdot \mathbf{C}$$
$$= [\mathbf{ABC}][\mathbf{ABC}]$$
$$= [\mathbf{ABC}]^2$$

Basis

PROBLEM 1-27 Show that the vectors $\mathbf{A}$, $\mathbf{A} - \mathbf{B}$, $\mathbf{B}$, and $\mathbf{B} + \mathbf{C}$ are linearly dependent.

Solution: Using the given vectors, create a linear combination

$$1\mathbf{A} - 1(\mathbf{A} - \mathbf{B}) - 1\mathbf{B} + 0(\mathbf{B} + \mathbf{C}) = \mathbf{0}$$

where the coefficients are not all zero. Thus by the definition of linear dependence (1.68) this vector set is linearly dependent.

PROBLEM 1-28 Show that any vector $\mathbf{D}$ in three dimensions can be expressed as a linear combination of any three given noncoplanar vectors $\mathbf{A}$, $\mathbf{B}$, and $\mathbf{C}$.

Solution: Equating results (a) and (b) of Problem 1-23, we have

$$[\mathbf{ABD}]\mathbf{C} - [\mathbf{ABC}]\mathbf{D} = [\mathbf{CDA}]\mathbf{B} - [\mathbf{CDB}]\mathbf{A}$$

Since $\mathbf{A}$, $\mathbf{B}$, and $\mathbf{C}$ are noncoplanar, $[\mathbf{ABC}] \neq 0$. Solving algebraically for vector $\mathbf{D}$, we obtain

$$\mathbf{D} = \frac{1}{[\mathbf{ABC}]}([\mathbf{CDB}]\mathbf{A} - [\mathbf{CDA}]\mathbf{B} + [\mathbf{ABD}]\mathbf{C})$$

$$= \frac{1}{[\mathbf{ABC}]}([\mathbf{DBC}]\mathbf{A} + [\mathbf{DCA}]\mathbf{B} + [\mathbf{DAB}]\mathbf{C})$$

by using the result of Problem 1-16. This result shows that vector $\mathbf{D}$ can be expressed as a linear combination of any three given noncoplanar vectors $\mathbf{A}$, $\mathbf{B}$, and $\mathbf{C}$ and gives a formula for this expression.

PROBLEM 1-29 If $\mathbf{a}_1$, $\mathbf{a}_2$, and $\mathbf{a}_3$ are nonzero noncoplanar vectors, show that any vector $\mathbf{d}$ can be expressed as

$$\mathbf{d} = (\mathbf{d} \cdot \mathbf{b}_1)\mathbf{a}_1 + (\mathbf{d} \cdot \mathbf{b}_2)\mathbf{a}_2 + (\mathbf{d} \cdot \mathbf{b}_3)\mathbf{a}_3$$

where $\mathbf{a}_1$, $\mathbf{a}_2$, $\mathbf{a}_3$ and $\mathbf{b}_1$, $\mathbf{b}_2$, $\mathbf{b}_3$ are reciprocal sets of vectors, as defined in Problem 1-21.

Solution: From the result of Problem 1-28, we have

$$\mathbf{D} = \frac{1}{[\mathbf{ABC}]}\{[\mathbf{DBC}]\mathbf{A} + [\mathbf{DCA}]\mathbf{B} + [\mathbf{DAB}]\mathbf{C}\}$$

Let $\mathbf{A} = \mathbf{a}_1$, $\mathbf{B} = \mathbf{a}_2$, $\mathbf{C} = \mathbf{a}_3$, and $\mathbf{D} = \mathbf{d}$; then from the result of Problem 1-21, we obtain

$$\mathbf{d} = \frac{1}{[\mathbf{a}_1\mathbf{a}_2\mathbf{a}_3]}\{[\mathbf{da}_2\mathbf{a}_3]\mathbf{a}_1 + [\mathbf{da}_3\mathbf{a}_1]\mathbf{a}_2 + [\mathbf{da}_1\mathbf{a}_2]\mathbf{a}_3\}$$

$$= \frac{1}{[\mathbf{a}_1\mathbf{a}_2\mathbf{a}_3]}\{\mathbf{d} \cdot (\mathbf{a}_2 \times \mathbf{a}_3)\mathbf{a}_1 + \mathbf{d} \cdot (\mathbf{a}_3 \times \mathbf{a}_1)\mathbf{a}_2 + \mathbf{d} \cdot (\mathbf{a}_1 \times \mathbf{a}_2)\mathbf{a}_3\}$$

$$= \mathbf{d} \cdot \frac{\mathbf{a}_2 \times \mathbf{a}_3}{[\mathbf{a}_1\mathbf{a}_2\mathbf{a}_3]}\mathbf{a}_1 + \mathbf{d} \cdot \frac{(\mathbf{a}_3 \times \mathbf{a}_1)}{[\mathbf{a}_1\mathbf{a}_2\mathbf{a}_3]}\mathbf{a}_2 + \mathbf{d} \cdot \frac{(\mathbf{a}_1 \times \mathbf{a}_2)}{[\mathbf{a}_1\mathbf{a}_2\mathbf{a}_3]}\mathbf{a}_3$$

$$= (\mathbf{d} \cdot \mathbf{b}_1)\mathbf{a}_1 + (\mathbf{d} \cdot \mathbf{b}_2)\mathbf{a}_2 + (\mathbf{d} \cdot \mathbf{b}_3)\mathbf{a}_3$$

Supplementary Exercises

PROBLEM 1-30 If $\mathbf{A}$ and $\mathbf{B}$ are nonparallel vectors such that $\mathbf{C} = (m + n - 1)\mathbf{A} + (m + n)\mathbf{B}$ and $\mathbf{D} = (m - n)\mathbf{A} + (2m - n + 1)\mathbf{B}$, find m and n such that $\mathbf{C} = 3\mathbf{D}$.

Answer: $m = -\frac{2}{3}, n = -\frac{1}{12}$

PROBLEM 1-31   Show that $\mathbf{A} \cdot \mathbf{B} = \frac{1}{4}(|\mathbf{A} + \mathbf{B}|^2 - |\mathbf{A} - \mathbf{B}|^2)$.

PROBLEM 1-32   Prove that $(\mathbf{A} + \mathbf{B}) \cdot (\mathbf{A} - \mathbf{B}) = A^2 - B^2$ and give a geometric interpretation.

PROBLEM 1-33   If $\mathbf{a}$ and $\mathbf{b}$ are unit vectors and θ is the angle between them, show that

$$\tfrac{1}{2}|\mathbf{a} - \mathbf{b}| = |\sin \tfrac{1}{2}\theta|$$

[*Hint*: $|\mathbf{a} - \mathbf{b}|^2 = (\mathbf{a} - \mathbf{b}) \cdot (\mathbf{a} - \mathbf{b})$]

PROBLEM 1-34   Prove that $(\mathbf{A} - \mathbf{B}) \times (\mathbf{A} + \mathbf{B}) = 2\mathbf{A} \times \mathbf{B}$ and give a geometric interpretation.

PROBLEM 1-35   Let ABC be any triangle and P be any point, where $\mathbf{a} = \overrightarrow{PA}$, $\mathbf{b} = \overrightarrow{PB}$, and $\mathbf{c} = \overrightarrow{PC}$. Show that the area of triangle ABC is equal to $\frac{1}{2}|\mathbf{a} \times \mathbf{b} + \mathbf{b} \times \mathbf{c} + \mathbf{c} \times \mathbf{a}|$.

PROBLEM 1-36   Prove that

$$(\mathbf{a} - \mathbf{d}) \times (\mathbf{b} - \mathbf{c}) + (\mathbf{b} - \mathbf{d}) \times (\mathbf{c} - \mathbf{a}) + (\mathbf{c} - \mathbf{d}) \times (\mathbf{a} - \mathbf{b}) = 2(\mathbf{a} \times \mathbf{b} + \mathbf{b} \times \mathbf{c} + \mathbf{c} \times \mathbf{a})$$

and give a geometric interpretation.

PROBLEM 1-37   Prove that if $\mathbf{A}$, $\mathbf{B}$, and $\mathbf{C}$ are nonparallel vectors and $\mathbf{A} \times \mathbf{B} = \mathbf{B} \times \mathbf{C} = \mathbf{C} \times \mathbf{A}$, then $\mathbf{A} + \mathbf{B} + \mathbf{C} = 0$. Give a geometric interpretation.

PROBLEM 1-38   If $\mathbf{A} \times \mathbf{B} = \mathbf{A} \times \mathbf{C}$, can we conclude that $\mathbf{B} = \mathbf{C}$?

PROBLEM 1-39   Show that $(\mathbf{A} + \mathbf{B}) \cdot (\mathbf{B} + \mathbf{C}) \times (\mathbf{C} + \mathbf{A}) = 2[\mathbf{ABC}]$.

PROBLEM 1-40   Show that
$$(\mathbf{A} \times \mathbf{B}) \cdot (\mathbf{C} \times \mathbf{D}) + (\mathbf{B} \times \mathbf{C}) \cdot (\mathbf{A} \times \mathbf{D}) + (\mathbf{C} \times \mathbf{A}) \cdot (\mathbf{B} \times \mathbf{D}) = 0.$$

PROBLEM 1-41   If $\mathbf{A}$, $\mathbf{B}$, $\mathbf{C}$, and $\mathbf{D}$ are coplanar, show that $(\mathbf{A} \times \mathbf{B}) \times (\mathbf{C} \times \mathbf{D}) = 0$. Is the converse true?

PROBLEM 1-42   Prove that a necessary and sufficient condition for $\mathbf{A} \times (\mathbf{B} \times \mathbf{C}) = (\mathbf{A} \times \mathbf{B}) \times \mathbf{C}$ is $(\mathbf{A} \times \mathbf{C}) \times \mathbf{B} = 0$.

PROBLEM 1-43   Prove that if $\mathbf{A}$, $\mathbf{B}$, and $\mathbf{C}$ are noncoplanar and $\mathbf{A} \times (\mathbf{B} \times \mathbf{C}) = (\mathbf{A} \times \mathbf{B}) \times \mathbf{C} = 0$, then $\mathbf{A}$, $\mathbf{B}$, and $\mathbf{C}$ are mutually perpendicular.

PROBLEM 1-44   If $\mathbf{a}_1, \mathbf{a}_2, \mathbf{a}_3$ and $\mathbf{b}_1, \mathbf{b}_2, \mathbf{b}_3$ are reciprocal sets of vectors, show that $\mathbf{a}_2 \times \mathbf{a}_3, \mathbf{a}_3 \times \mathbf{a}_1, \mathbf{a}_1 \times \mathbf{a}_2$ and $\mathbf{b}_2 \times \mathbf{b}_3, \mathbf{b}_3 \times \mathbf{b}_1, \mathbf{b}_1 \times \mathbf{b}_2$ are also reciprocal sets.

PROBLEM 1-45   If $\mathbf{a}_1, \mathbf{a}_2, \mathbf{a}_3$ and $\mathbf{b}_1, \mathbf{b}_2, \mathbf{b}_3$ are reciprocal sets of vectors, show that $\mathbf{a}_1 \times \mathbf{b}_1 + \mathbf{a}_2 \times \mathbf{b}_2 + \mathbf{a}_3 \times \mathbf{b}_3 = 0$.

PROBLEM 1-46   Show that for two reciprocal sets of vectors $\mathbf{a}_1, \mathbf{a}_2, \mathbf{a}_3$ and $\mathbf{b}_1, \mathbf{b}_2, \mathbf{b}_3$,

$$[\mathbf{a}_1 \mathbf{a}_2 \mathbf{a}_3] = \frac{1}{[\mathbf{b}_1 \mathbf{b}_2 \mathbf{b}_3]}$$

2 VECTORS IN THE RECTANGULAR COORDINATE SYSTEM

THIS CHAPTER IS ABOUT

- ☑ **Analytic Representation of Vectors**
- ☑ **Vector Algebra Using Analytic Representation**
- ☑ **Base Vectors and Position Vectors**
- ☑ **Scalar Product**
- ☑ **Vector Product**
- ☑ **Analytic Characterization of Vectors**
- ☑ **Triple Products**
- ☑ **Basis**

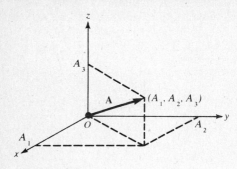

Figure 2-1
Vector in a rectangular coordinate system.

2-1. Analytic Representation of Vectors

In Chapter 1, a vector was represented as a directed line segment. Now if we place the initial point of vector **A** at the origin of a rectangular coordinate system, we can specify vector **A** by the rectangular coordinates (A_1, A_2, A_3) of the terminal point, as shown in Figure 2-1. Thus there is a one-to-one correspondence between the *set of triples of numbers* (coordinates of points in three-dimensional space) and the *set of vectors* whose initial points are at the origin. Consequently, a vector **A** in three dimensions is often defined as an *ordered triple of real numbers*; that is,

ANALYTIC REPRESENTATION OF A VECTOR
$$\mathbf{A} = [A_1, A_2, A_3] \tag{2.1}$$

where $A_1, A_2,$ and A_3 are called the *components* of **A**, relative to the given coordinate system, and are real numbers. Expression (2.1) is called the *analytic representation* of **A**.

Note: There are infinitely many possible coordinate systems, so the same vector will have different components in different systems. But in the following we'll consider only the standard rectangular coordinate system.

The zero vector **0** is a vector whose components are all zero; that is,

ZERO VECTOR
$$\mathbf{0} = [0, 0, 0] \tag{2.2}$$

2-2. Vector Algebra Using Analytic Representation

A. Analytic definitions

Two vectors are equal if and only if their corresponding components are equal. Analytically, if $\mathbf{A} = [A_1, A_2, A_3]$ and $\mathbf{B} = [B_1, B_2, B_3]$, we can write

EQUALITY OF TWO VECTORS (ANALYTIC)
$$\mathbf{A} = \mathbf{B} \Leftrightarrow A_1 = B_1, A_2 = B_2, A_3 = B_3 \tag{2.3}$$

The product of a vector **A** and a scalar m is obtained by multiplication of the components of **A** by m; that is,

SCALAR MULTIPLICATION (ANALYTIC)
$$m\mathbf{A} = m[A_1, A_2, A_3] = [mA_1, mA_2, mA_3] \tag{2.4}$$

By setting $m = -1$, the negative of $\mathbf{A}$ is

NEGATIVE OF
A VECTOR
(ANALYTIC)
$$-\mathbf{A} = (-1)\mathbf{A} = [-A_1, -A_2, -A_3] \qquad (2.5)$$

and by setting $m = 0$, the zero vector is

ZERO
VECTOR
(ANALYTIC)
$$\mathbf{0} = 0\mathbf{A} = [0, 0, 0] \qquad (2.6)$$

The sum or resultant of two vectors $\mathbf{A}$ and $\mathbf{B}$ is a vector $\mathbf{C}$ obtained by adding the corresponding components of $\mathbf{A}$ and $\mathbf{B}$; that is, $\mathbf{C} = \mathbf{A} + \mathbf{B}$, where the resultant is defined analytically as

VECTOR
ADDITION
(ANALYTIC)
$$\mathbf{C} = [C_1, C_2, C_3] = [A_1 + B_1, A_2 + B_2, A_3 + B_3] \qquad (2.7)$$

The difference $\mathbf{A} - \mathbf{B}$ of two vectors $\mathbf{A}$ and $\mathbf{B}$ is a vector $\mathbf{C}$ obtained by addition of $\mathbf{A}$ and $-\mathbf{B}$; that is, $\mathbf{C} = \mathbf{A} - \mathbf{B} = \mathbf{A} + (-1)\mathbf{B}$, expressed analytically as

VECTOR
SUBTRACTION
(ANALYTIC)
$$\mathbf{C} = [C_1, C_2, C_3] = [A_1 - B_1, A_2 - B_2, A_3 - B_3] \qquad (2.8)$$

B. Properties of vector addition

$$\mathbf{A} + \mathbf{B} = \mathbf{B} + \mathbf{A} \qquad \text{[Commutative law]} \qquad (2.9)$$
$$\mathbf{A} + (\mathbf{B} + \mathbf{C}) = (\mathbf{A} + \mathbf{B}) + \mathbf{C} \qquad \text{[Associative law]} \qquad (2.10)$$
$$m(\mathbf{A} + \mathbf{B}) = m\mathbf{A} + m\mathbf{B} \qquad \text{[Distributive law]} \qquad (2.11)$$
$$(m + n)\mathbf{A} = m\mathbf{A} + n\mathbf{A} \qquad \text{[Scalar distributive law]} \qquad (2.12)$$
$$\mathbf{A} + \mathbf{0} = \mathbf{A} \qquad (2.13)$$
$$\mathbf{A} + (-\mathbf{A}) = \mathbf{0} \qquad \text{or} \qquad \mathbf{A} - \mathbf{A} = \mathbf{0} \qquad (2.14)$$

EXAMPLE 2-1: Verify the properties of vector addition (2.9), (2.12), and (2.14) analytically.

Solution: From the analytic definition of vector addition (2.7) we have

$$\mathbf{A} + \mathbf{B} = [A_1 + B_1, A_2 + B_2, A_3 + B_3]$$
$$= [B_1 + A_1, B_2 + A_2, B_3 + A_3]$$
$$= \mathbf{B} + \mathbf{A} \qquad (2.9)$$

Then

$$(m + n)\mathbf{A} = [(m + n)A_1, (m + n)A_2, (m + n)A_3]$$
$$= [mA_1 + nA_1, mA_2 + nA_2, mA_3 + nA_3]$$
$$= m\mathbf{A} + n\mathbf{A} \qquad (2.12)$$

Finally,

$$\mathbf{A} - \mathbf{A} = [A_1 - A_1, A_2 - A_2, A_3 - A_3]$$
$$= [0, 0, 0] = \mathbf{0} \qquad (2.14)$$

2-3. Base Vectors and Position Vectors

A. Base Vectors

The three unit vectors

BASE VECTORS 　$\mathbf{i} = [1, 0, 0], \qquad \mathbf{j} = [0, 1, 0], \qquad \mathbf{k} = [0, 0, 1] \qquad (2.15)$

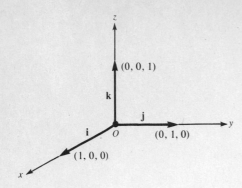
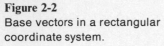

Figure 2-2
Base vectors in a rectangular coordinate system.

are called the **base vectors** of the rectangular coordinate system. Geometrically, these base vectors are unit vectors in the positive direction along the *x*-, *y*-, and *z*-axes, respectively as shown in Figure 2-2. It is clear that **i**, **j**, and **k** are mutually perpendicular.

EXAMPLE 2-2: Show that any vector in three dimensions can be expressed as a linear combination of the base vectors **i**, **j**, and **k**.

Solution: Given vector $\mathbf{A} = [A_1, A_2, A_3]$, we can write

$$[A_1, A_2, A_3] = [A_1 + 0 + 0, 0 + A_2 + 0, 0 + 0 + A_3]$$
$$= [A_1, 0, 0] + [0, A_2, 0] + [0, 0, A_3]$$
$$= A_1[1, 0, 0] + A_2[0, 1, 0] + A_3[0, 0, 1] \qquad (2.16)$$

or

$$\mathbf{A} = A_1\mathbf{i} + A_2\mathbf{j} + A_3\mathbf{k} \qquad (2.17)$$

From (2.16) or (2.17) (and because the coefficients of **i**, **j**, and **k** are precisely the components of **A**), we conclude that any vector in three dimensions can be expressed as a linear combination of the base vectors. This is why we call **i**, **j**, and **k** the "base" vectors of the rectangular coordinate system.

B. Position vectors

For each point (x, y, z) in three-dimensional space there exists a unique **position vector r** defined as

POSITION VECTOR $\qquad \mathbf{r} = [x, y, z] = x\mathbf{i} + y\mathbf{j} + z\mathbf{k} \qquad (2.18)$

EXAMPLE 2-3: Show that a vector **R** with its initial point at (x_1, y_1, z_1) and its terminal point at (x_2, y_2, z_2) is given by

$$\mathbf{R} = \mathbf{r}_2 - \mathbf{r}_1$$
$$= [x_2 - x_1, y_2 - y_1, z_2 - z_1]$$
$$= (x_2 - x_1)\mathbf{i} + (y_2 - y_1)\mathbf{j} + (z_2 - z_1)\mathbf{k} \qquad (2.19)$$

Solution: Points (x_1, y_1, z_1) and (x_2, y_2, z_2) in three-dimensional space determine the position vectors

$$\mathbf{r}_1 = x_1\mathbf{i} + y_1\mathbf{j} + z_1\mathbf{k}$$
$$\mathbf{r}_2 = x_2\mathbf{i} + y_2\mathbf{j} + z_2\mathbf{k}$$

If we construct $\mathbf{R} = \mathbf{r}_2 - \mathbf{r}_1$ (see Figure 2-3), we obtain

$$\mathbf{R} = \mathbf{r}_2 - \mathbf{r}_1$$
$$= (x_2\mathbf{i} + y_2\mathbf{j} + z_2\mathbf{k}) - (x_1\mathbf{i} + y_1\mathbf{j} + z_1\mathbf{k})$$
$$= (x_2 - x_1)\mathbf{i} + (y_2 - y_1)\mathbf{j} + (z_2 - z_1)\mathbf{k}$$

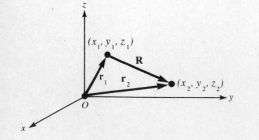

Figure 2-3

2-4. Scalar Product

A. Analytic definition of the scalar product

Let $\mathbf{A} = [A_1, A_2, A_3]$ and $\mathbf{B} = [B_1, B_2, B_3]$. Then the **scalar** (or **dot** or **inner**) **product** of vectors **A** and **B** is a scalar given by

SCALAR PRODUCT (ANALYTIC) $\qquad \mathbf{A} \cdot \mathbf{B} = A_1 B_1 + A_2 B_2 + A_3 B_3 \qquad (2.20)$

EXAMPLE 2-4: If **i**, **j**, and **k** are the base vectors defined in (2.15), verify the following relations:

$$\mathbf{i} \cdot \mathbf{i} = \mathbf{j} \cdot \mathbf{j} = \mathbf{k} \cdot \mathbf{k} = 1 \qquad (2.21\text{a})$$

$$\mathbf{i} \cdot \mathbf{j} = \mathbf{j} \cdot \mathbf{k} = \mathbf{k} \cdot \mathbf{i} = 0 \qquad (2.21\text{b})$$

$$\mathbf{j} \cdot \mathbf{i} = \mathbf{k} \cdot \mathbf{j} = \mathbf{i} \cdot \mathbf{k} = 0 \qquad (2.21\text{c})$$

Solution: Since $\mathbf{i} = [1, 0, 0]$, $\mathbf{j} = [0, 1, 0]$, and $\mathbf{k} = [0, 0, 1]$, by the analytic definition of the scalar product (2.20), we have

$$\mathbf{i} \cdot \mathbf{i} = (1)(1) + (0)(0) + (0)(0) = 1$$

$$\mathbf{j} \cdot \mathbf{j} = (0)(0) + (1)(1) + (0)(0) = 1$$

$$\mathbf{k} \cdot \mathbf{k} = (0)(0) + (0)(0) + (1)(1) = 1$$

$$\mathbf{i} \cdot \mathbf{j} = (1)(0) + (0)(1) + (0)(0) = 0$$

$$\mathbf{j} \cdot \mathbf{k} = (0)(0) + (1)(0) + (0)(1) = 0$$

$$\mathbf{k} \cdot \mathbf{i} = (0)(1) + (0)(0) + (1)(0) = 0$$

$$\mathbf{j} \cdot \mathbf{i} = (0)(1) + (1)(0) + (0)(0) = 0$$

$$\mathbf{k} \cdot \mathbf{j} = (0)(0) + (0)(1) + (1)(0) = 0$$

$$\mathbf{i} \cdot \mathbf{k} = (1)(0) + (0)(0) + (0)(1) = 0$$

B. Properties of the scalar product

$$\mathbf{A} \cdot \mathbf{B} = \mathbf{B} \cdot \mathbf{A} \qquad \text{[Commutative law]} \qquad (2.22)$$

$$\mathbf{A} \cdot (\mathbf{B} + \mathbf{C}) = \mathbf{A} \cdot \mathbf{B} + \mathbf{A} \cdot \mathbf{C} \qquad \text{[Distributive law]} \qquad (2.23)$$

$$(m\mathbf{A}) \cdot \mathbf{B} = \mathbf{A} \cdot (m\mathbf{B}) = m(\mathbf{A} \cdot \mathbf{B}) \qquad (2.24)$$

$$\mathbf{A} \cdot \mathbf{A} = |\mathbf{A}|^2 = A^2 \geqslant 0 \quad \text{for every } \mathbf{A}$$

$$\mathbf{A} \cdot \mathbf{A} = 0 \quad \text{if and only if} \quad \mathbf{A} = \mathbf{0} \qquad (2.25)$$

where m is an arbitrary scalar.

EXAMPLE 2-5: Verify the commutative law (2.22).

Proof: By the analytic definition of the scalar product (2.20),

$$\mathbf{A} \cdot \mathbf{B} = A_1 B_1 + A_2 B_2 + A_3 B_3$$

$$= B_1 A_1 + B_2 A_2 + B_3 A_3$$

$$= \mathbf{B} \cdot \mathbf{A}$$

EXAMPLE 2-6: Verify the distributive law (2.23).

Proof: From the analytic definition of vector addition (2.7) we have

$$\mathbf{B} + \mathbf{C} = [B_1 + C_1, B_2 + C_2, B_3 + C_3]$$

Hence by the analytic definition of the scalar product (2.20) we obtain

$$\mathbf{A} \cdot (\mathbf{B} + \mathbf{C}) = A_1(B_1 + C_1) + A_2(B_2 + C_2) + A_3(B_3 + C_3)$$

$$= (A_1 B_1 + A_2 B_2 + A_3 B_3) + (A_1 C_1 + A_2 C_2 + A_3 C_3)$$

$$= \mathbf{A} \cdot \mathbf{B} + \mathbf{A} \cdot \mathbf{C}$$

2-5. Vector Product

A. Analytic definition of the vector product

The **vector** (or **cross** or **outer**) **product** of vectors $\mathbf{A}$ and $\mathbf{B}$ is a vector

VECTOR PRODUCT (ANALYTIC) $\quad \mathbf{A} \times \mathbf{B} = [A_2 B_3 - A_3 B_2, A_3 B_1 - A_1 B_3, A_1 B_2 - A_2 B_1] \quad (2.26)$

or using the base vectors **i**, **j**, **k**,

$$\mathbf{A} \times \mathbf{B} = (A_2 B_3 - A_3 B_2)\mathbf{i} + (A_3 B_1 - A_1 B_3)\mathbf{j} + (A_1 B_2 - A_2 B_1)\mathbf{k} \quad (2.27)$$

In determinant form (2.27) can be written as

$$\mathbf{A} \times \mathbf{B} = \begin{vmatrix} \mathbf{i} & \mathbf{j} & \mathbf{k} \\ A_1 & A_2 & A_3 \\ B_1 & B_2 & B_3 \end{vmatrix} \quad (2.28)$$

EXAMPLE 2-7: If **i**, **j**, **k** are the base vectors of three-dimensional space (2.15), verify the following relations:

$$\mathbf{i} \times \mathbf{i} = \mathbf{j} \times \mathbf{j} = \mathbf{k} \times \mathbf{k} = \mathbf{0} \quad (2.29a)$$
$$\mathbf{i} \times \mathbf{j} = \mathbf{k}, \qquad \mathbf{j} \times \mathbf{k} = \mathbf{i}, \qquad \mathbf{k} \times \mathbf{i} = \mathbf{j}, \quad (2.29b)$$
$$\mathbf{j} \times \mathbf{i} = -\mathbf{k}, \qquad \mathbf{k} \times \mathbf{j} = -\mathbf{i}, \qquad \mathbf{i} \times \mathbf{k} = -\mathbf{j} \quad (2.29c)$$

Proof: Since $\mathbf{i} = [1, 0, 0]$, $\mathbf{j} = [0, 1, 0]$, and $\mathbf{k} = [0, 0, 1]$, by the analytic definition of the vector product (2.26), we have

$$\mathbf{i} \times \mathbf{i} = [(0)(0) - (0)(0), (0)(1) - (1)(0), (1)(0) - (0)(1)] = [0, 0, 0] = \mathbf{0}$$

Similarly

$$\mathbf{j} \times \mathbf{j} = \mathbf{0} \quad \text{and} \quad \mathbf{k} \times \mathbf{k} = \mathbf{0}$$

and (2.29a) is verified.

Next we verify (2.29b):

$$\mathbf{i} \times \mathbf{j} = [(0)(0) - (0)(0), (0)(0) - (1)(0), (1)(1) - (0)(0)] = [0, 0, 1] = \mathbf{k}$$
$$\mathbf{j} \times \mathbf{k} = [(1)(1) - (0)(0), (0)(0) - (0)(1), (0)(0) - (1)(0)] = [1, 0, 0] = \mathbf{i}$$
$$\mathbf{k} \times \mathbf{i} = [(0)(0) - (1)(0), (1)(1) - (0)(0), (0)(0) - (0)(1)] = [0, 1, 0] = \mathbf{j}$$

Then we can verify (2.29c) similarly.

$$\mathbf{j} \times \mathbf{i} = -\mathbf{k}, \qquad \mathbf{k} \times \mathbf{j} = -\mathbf{i}, \qquad \mathbf{i} \times \mathbf{k} = -\mathbf{j}$$

EXAMPLE 2-8: Find $[-1, 2, 3] \times [5, -1, 1]$.

Solution: Using the determinant form of the vector product definition (2.28), we have

$$\begin{vmatrix} \mathbf{i} & \mathbf{j} & \mathbf{k} \\ -1 & 2 & 3 \\ 5 & -1 & 1 \end{vmatrix} = \mathbf{i} \begin{vmatrix} 2 & 3 \\ -1 & 1 \end{vmatrix} - \mathbf{j} \begin{vmatrix} -1 & 3 \\ 5 & 1 \end{vmatrix} + \mathbf{k} \begin{vmatrix} -1 & 2 \\ 5 & -1 \end{vmatrix}$$

$$= 5\mathbf{i} + 16\mathbf{j} - 9\mathbf{k} = [5, 16, -9]$$

B. Properties of the vector product

$$\mathbf{A} \times \mathbf{B} = -\mathbf{B} \times \mathbf{A} \qquad \text{[Anticommutative law]} \quad (2.30)$$
$$\mathbf{A} \times (\mathbf{B} + \mathbf{C}) = \mathbf{A} \times \mathbf{B} + \mathbf{A} \times \mathbf{C} \qquad \text{[Distributive law]} \quad (2.31)$$
$$(m\mathbf{A}) \times \mathbf{B} = \mathbf{A} \times (m\mathbf{B}) = m\mathbf{A} \times \mathbf{B} \quad (2.32)$$
$$\mathbf{A} \times \mathbf{A} = \mathbf{0} \quad (2.33)$$

EXAMPLE 2-9: Verify the anticommutative law (2.30).

Proof: From the determinant form (2.28) of the vector product of **A** and **B** we write

$$\mathbf{A} \times \mathbf{B} = \begin{vmatrix} \mathbf{i} & \mathbf{j} & \mathbf{k} \\ A_1 & A_2 & A_3 \\ B_1 & B_2 & B_3 \end{vmatrix} = - \begin{vmatrix} \mathbf{i} & \mathbf{j} & \mathbf{k} \\ B_1 & B_2 & B_3 \\ A_1 & A_2 & A_3 \end{vmatrix} = -\mathbf{B} \times \mathbf{A}$$

since the interchange of the second and third rows changes the sign of the determinant.

EXAMPLE 2-10: Verify the distributive law (2.31).

Proof: Again using the determinant form (2.28),

$$\mathbf{A} \times (\mathbf{B} + \mathbf{C}) = \begin{vmatrix} \mathbf{i} & \mathbf{j} & \mathbf{k} \\ A_1 & A_2 & A_3 \\ B_1 + C_1 & B_2 + C_2 & B_3 + C_3 \end{vmatrix}$$

$$= \begin{vmatrix} \mathbf{i} & \mathbf{j} & \mathbf{k} \\ A_1 & A_2 & A_3 \\ B_1 & B_2 & B_3 \end{vmatrix} + \begin{vmatrix} \mathbf{i} & \mathbf{j} & \mathbf{k} \\ A_1 & A_2 & A_3 \\ C_1 & C_2 & C_3 \end{vmatrix}$$

$$= \mathbf{A} \times \mathbf{B} + \mathbf{A} \times \mathbf{C}$$

EXAMPLE 2-11: Verify property (2.33).

Proof: Using the definition of the vector product (2.26) and setting $\mathbf{A} = \mathbf{B}$,

$$\mathbf{A} \times \mathbf{A} = [A_2 A_3 - A_3 A_2, A_3 A_1 - A_1 A_3, A_1 A_2 - A_2 A_1] = [0,0,0] = \mathbf{0}$$

2-6. Analytic Characterization of Vectors

In Chapter 1 we defined a vector to be a quantity that has both magnitude and direction. We've seen that a vector $\mathbf{A}$ can be represented analytically by its components $[A_1, A_2, A_3]$. Now we can give analytic representations of these properties and express the magnitude and direction of a vector in terms of its components.

A. Magnitude of a vector

Since by (2.25), $\mathbf{A} \cdot \mathbf{A} = |\mathbf{A}|^2 = A^2$, the **magnitude** or length of the vector $\mathbf{A} = [A_1, A_2, A_3]$, denoted by $|\mathbf{A}|$ or A, is given by

MAGNITUDE OF A VECTOR (ANALYTIC)
$$|\mathbf{A}| = A = (A_1^2 + A_2^2 + A_3^2)^{1/2} \qquad (2.34)$$

Then the unit vector $\mathbf{e}_A$ along $\mathbf{A}$ is

UNIT VECTOR (ANALYTIC)
$$\mathbf{e}_A = \frac{1}{|\mathbf{A}|} \mathbf{A} = \frac{1}{A} [A_1, A_2, A_3] = \left[\frac{A_1}{A}, \frac{A_2}{A}, \frac{A_3}{A} \right] \qquad (2.35)$$

EXAMPLE 2-12: Show that for a vector $\mathbf{A}$ and a scalar m,
$$|m\mathbf{A}| = |m| \, |\mathbf{A}| \qquad (2.36)$$

Solution: Since $m\mathbf{A} = [mA_1, mA_2, mA_3]$, from the definition of magnitude (2.34) we have

$$|m\mathbf{A}| = [(mA_1)^2 + (mA_2)^2 + (mA_3)^2]^{1/2}$$
$$= [m^2(A_1^2 + A_2^2 + A_3^2)]^{1/2}$$
$$= |m|(A_1^2 + A_2^2 + A_3^2)^{1/2}$$
$$= |m| \, |\mathbf{A}|$$

B. The angle between two vectors

Let θ denote the angle between two nonzero vectors $\mathbf{A} = [A_1, A_2, A_3]$ and $\mathbf{B} = [B_1, B_2, B_3]$. Then the angle θ is determined from

ANGLE BETWEEN TWO VECTORS
$$\cos\theta = \frac{A_1B_1 + A_2B_2 + A_3B_3}{(A_1^2 + A_2^2 + A_3^2)^{1/2}(B_1^2 + B_2^2 + B_3^2)^{1/2}} \qquad (2.37)$$

EXAMPLE 2-13: Verify (2.37).

Proof: Let $\mathbf{e}_A$ and $\mathbf{e}_B$ be the two unit vectors along $\mathbf{A}$ and $\mathbf{B}$, respectively. Then from the definition of a unit vector (2.35)

$$\mathbf{e}_A = \left[\frac{A_1}{A}, \frac{A_2}{A}, \frac{A_3}{A}\right] = [a_1, a_2, a_3]$$

$$\mathbf{e}_B = \left[\frac{B_1}{B}, \frac{B_2}{B}, \frac{B_3}{B}\right] = [b_1, b_2, b_3]$$

where

$$A = |\mathbf{A}| = (A_1^2 + A_2^2 + A_3^2)^{1/2}, \qquad B = |\mathbf{B}| = (B_1^2 + B_2^2 + B_3^2)^{1/2}$$

Now represent $\mathbf{e}_A$ and $\mathbf{e}_B$ by the line segments $\overrightarrow{OP}$ and $\overrightarrow{OQ}$, as shown in Figure 2-4a. Then the magnitude of $\overrightarrow{PQ}$ is given by the positive square root of

$$\begin{aligned}|\overrightarrow{PQ}|^2 &= (a_1 - b_1)^2 + (a_2 - b_2)^2 + (a_3 - b_3)^2 \\ &= a_1^2 + a_2^2 + a_3^2 + b_1^2 + b_2^2 + b_3^2 - 2(a_1b_1 + a_2b_2 + a_3b_3) \\ &= 2 - 2(a_1b_1 + a_2b_2 + a_3b_3)\end{aligned} \qquad (2.38)$$

since $\mathbf{e}_A$ and $\mathbf{e}_B$ are unit vectors, for which $a_1^2 + a_2^2 + a_3^2 = b_1^2 + b_2^2 + b_3^2 = 1$.

Next, consider a triangle ORS in the xy-plane with vertices at $(0, 0, 0)$, $(1, 0, 0)$, and $(\cos\theta, \sin\theta, 0)$, as shown in Figure 2-4b. Then the triangles ORS and OPQ are congruent. Thus since $\cos^2\theta + \sin^2\theta = 1$,

$$\begin{aligned}|\overrightarrow{PQ}|^2 = |\overrightarrow{RS}|^2 &= (1 - \cos\theta)^2 + (0 - \sin\theta)^2 + (0 - 0)^2 \\ &= 1 - 2\cos\theta + \cos^2\theta + \sin^2\theta \\ &= 2 - 2\cos\theta\end{aligned} \qquad (2.39)$$

Equating (2.38) and (2.39), we obtain

$$\cos\theta = a_1b_1 + a_2b_2 + a_3b_3 \qquad (2.40)$$

$$= \mathbf{e}_A \cdot \mathbf{e}_B \qquad (2.41)$$

$$= \frac{\mathbf{A} \cdot \mathbf{B}}{AB} \qquad (2.42)$$

$$= \frac{A_1B_1 + A_2B_2 + A_3B_3}{(A_1^2 + A_2^2 + A_3^2)^{1/2}(B_1^2 + B_2^2 + B_3^2)^{1/2}}$$

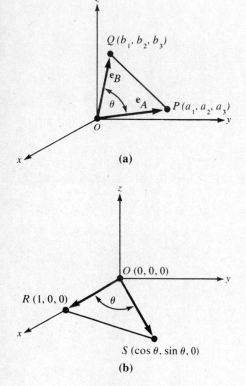

(a)

(b)

Figure 2-4

C. Direction cosines of a vector

The **direction cosines** of a vector $\mathbf{A}$ are defined as the three numbers $\cos\alpha$, $\cos\beta$, and $\cos\gamma$, where α, β, and γ are the angles between $\mathbf{A}$ and the positive x-, y-, and z-axes, respectively, of the rectangular coordinate system (see Figure 2-5). From (2.41) and this definition of the direction cosines we have

DIRECTION COSINES
$$\cos\alpha = \mathbf{i} \cdot \mathbf{e}_A, \qquad \cos\beta = \mathbf{j} \cdot \mathbf{e}_A, \qquad \cos\gamma = \mathbf{k} \cdot \mathbf{e}_A \qquad (2.43)$$

where $\mathbf{i}$, $\mathbf{j}$, and $\mathbf{k}$ are the base vectors defined in (2.15), and $\mathbf{e}_A$ is the unit vector along $\mathbf{A}$.

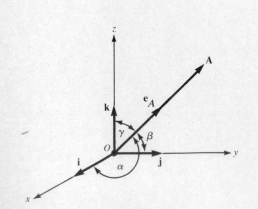

Figure 2-5
Direction cosines.

EXAMPLE 2-14: Find the direction cosines of $\mathbf{A}$ in terms of its components and its magnitude.

Solution: By using the formula for determining the angle between two vectors (2.37), we obtain

$$\cos \alpha = \frac{A_1}{A}, \quad \cos \beta = \frac{A_2}{A}, \quad \cos \gamma = \frac{A_3}{A} \tag{2.44}$$

Equations (2.44) are an analytic characterization of the direction of **A**. The direction cosines of the zero vector are undefined.

2-7. Triple Products

A. Scalar triple product

The **scalar triple product** of three vectors **A**, **B**, and **C** is the scalar $\mathbf{A} \cdot \mathbf{B} \times \mathbf{C}$, which is denoted by $[\mathbf{ABC}]$. In determinant form it is expressed as

SCALAR TRIPLE PRODUCT (DETERMINANT FORM)
$$[\mathbf{ABC}] = \mathbf{A} \cdot \mathbf{B} \times \mathbf{C} = \begin{vmatrix} A_1 & A_2 & A_3 \\ B_1 & B_2 & B_3 \\ C_1 & C_2 & C_3 \end{vmatrix} \tag{2.45}$$

EXAMPLE 2-15: Verify (2.45).

Proof: From the determinant form of the definition of the vector product (2.28), we get

$$\mathbf{B} \times \mathbf{C} = \begin{vmatrix} \mathbf{i} & \mathbf{j} & \mathbf{k} \\ B_1 & B_2 & B_3 \\ C_1 & C_2 & C_3 \end{vmatrix}$$

$$= (B_2 C_3 - B_3 C_2)\mathbf{i} + (B_3 C_1 - B_1 C_3)\mathbf{j} + (B_1 C_2 - B_2 C_1)\mathbf{k}$$

$$= [B_2 C_3 - B_3 C_2, B_3 C_1 - B_1 C_3, B_1 C_2 - B_2 C_1]$$

Hence from the analytic definition of the scalar product (2.20), we obtain

$$[\mathbf{ABC}] = \mathbf{A} \cdot \mathbf{B} \times \mathbf{C}$$
$$= A_1(B_2 C_3 - B_3 C_2) + A_2(B_3 C_1 - B_1 C_3) + A_3(B_1 C_2 - B_2 C_1)$$

$$= A_1 \begin{vmatrix} B_2 & B_3 \\ C_2 & C_3 \end{vmatrix} - A_2 \begin{vmatrix} B_1 & B_3 \\ C_1 & C_3 \end{vmatrix} + A_3 \begin{vmatrix} B_1 & B_2 \\ C_1 & C_2 \end{vmatrix}$$

$$= \begin{vmatrix} A_1 & A_2 & A_3 \\ B_1 & B_2 & B_3 \\ C_1 & C_2 & C_3 \end{vmatrix}$$

EXAMPLE 2-16: Show that

$$[\mathbf{ABC}] = [\mathbf{BCA}] = [\mathbf{CAB}] = -[\mathbf{ACB}] = -[\mathbf{BAC}] = -[\mathbf{CBA}] \tag{2.46}$$

Solution: Because the interchange of two rows of a determinant always changes its sign, it follows from definition (2.45) that

$$\begin{vmatrix} A_1 & A_2 & A_3 \\ B_1 & B_2 & B_3 \\ C_1 & C_2 & C_3 \end{vmatrix} = - \begin{vmatrix} A_1 & A_2 & A_3 \\ C_1 & C_2 & C_3 \\ B_1 & B_2 & B_3 \end{vmatrix} = \begin{vmatrix} C_1 & C_2 & C_3 \\ A_1 & A_2 & A_3 \\ B_1 & B_2 & B_3 \end{vmatrix}$$

$$= - \begin{vmatrix} C_1 & C_2 & C_3 \\ B_1 & B_2 & B_3 \\ A_1 & A_2 & A_3 \end{vmatrix} = \begin{vmatrix} B_1 & B_2 & B_3 \\ C_1 & C_2 & C_3 \\ A_1 & A_2 & A_3 \end{vmatrix} = - \begin{vmatrix} B_1 & B_2 & B_3 \\ A_1 & A_2 & A_3 \\ C_1 & C_2 & C_3 \end{vmatrix}$$

In other words,

$$[\mathbf{ABC}] = -[\mathbf{ACB}] = [\mathbf{CAB}] = -[\mathbf{CBA}] = [\mathbf{BCA}] = -[\mathbf{BAC}]$$

(see Problem 1-16).

B. Vector triple product

The **vector triple product** of three vectors $\mathbf{A}$, $\mathbf{B}$, and $\mathbf{C}$ is defined as the vector $\mathbf{A} \times (\mathbf{B} \times \mathbf{C})$.

EXAMPLE 2-17: Using the component representation of vectors, show that

$$\mathbf{A} \times (\mathbf{B} \times \mathbf{C}) = (\mathbf{A} \cdot \mathbf{C})\mathbf{B} - (\mathbf{A} \cdot \mathbf{B})\mathbf{C} \tag{2.47}$$

Solution: Let

$$\mathbf{A} = A_1\mathbf{i} + A_2\mathbf{j} + A_3\mathbf{k} \qquad \mathbf{B} = B_1\mathbf{i} + B_2\mathbf{j} + B_3\mathbf{k}$$

and

$$\mathbf{C} = C_1\mathbf{i} + C_2\mathbf{j} + C_3\mathbf{k}.$$

Then

$$\mathbf{B} \times \mathbf{C} = (B_2C_3 - B_3C_2)\mathbf{i} + (B_3C_1 - B_1C_3)\mathbf{j} + (B_1C_2 - B_2C_1)\mathbf{k}$$

and

$$\mathbf{A} \times (\mathbf{B} \times \mathbf{C}) = A_1\mathbf{i} \times (\mathbf{B} \times \mathbf{C}) + A_2\mathbf{j} \times (\mathbf{B} \times \mathbf{C}) + A_3\mathbf{k} \times (\mathbf{B} \times \mathbf{C})$$

Using identities (2.29), $\mathbf{i} \times \mathbf{i} = 0$, $\mathbf{i} \times \mathbf{j} = \mathbf{k}$, $\mathbf{i} \times \mathbf{k} = -\mathbf{j}$, we have

$$A_1\mathbf{i} \times (\mathbf{B} \times \mathbf{C}) = A_1(B_3C_1 - B_1C_3)\mathbf{k} - A_1(B_1C_2 - B_2C_1)\mathbf{j}$$
$$= A_1C_1B_2\mathbf{j} + A_1C_1B_3\mathbf{k} - A_1B_1C_2\mathbf{j} - A_1B_1C_3\mathbf{k}$$

On the right-hand side we now add and subtract the term $A_1B_1C_1\mathbf{i}$:

$$A_1\mathbf{i} \times (\mathbf{B} \times \mathbf{C}) =$$
$$= A_1C_1B_1\mathbf{i} + A_1C_1B_2\mathbf{j} + A_1C_1B_3\mathbf{k} - A_1B_1C_1\mathbf{i} - A_1B_1C_2\mathbf{j} - A_1B_1C_3\mathbf{k}$$
$$= A_1C_1(B_1\mathbf{i} + B_2\mathbf{j} + B_3\mathbf{k}) - A_1B_1(C_1\mathbf{i} + C_2\mathbf{j} + C_3\mathbf{k})$$
$$= A_1C_1\mathbf{B} - A_1B_1\mathbf{C} \tag{2.48}$$

In a similar manner we obtain

$$A_2\mathbf{j} \times (\mathbf{B} \times \mathbf{C}) = A_2C_2\mathbf{B} - A_2B_2\mathbf{C} \tag{2.49}$$
$$A_3\mathbf{k} \times (\mathbf{B} \times \mathbf{C}) = A_3C_3\mathbf{B} - A_3B_3\mathbf{C} \tag{2.50}$$

Adding (2.48), (2.49), and (2.50), we obtain

$$(A_1\mathbf{i} + A_2\mathbf{j} + A_3\mathbf{k}) \times (\mathbf{B} \times \mathbf{C})$$
$$= (A_1C_1 + A_2C_2 + A_3C_3)\mathbf{B} - (A_1B_1 + A_2B_2 + A_3B_3)\mathbf{C}$$

or using definition (2.20),

$$\mathbf{A} \times (\mathbf{B} \times \mathbf{C}) = (\mathbf{A} \cdot \mathbf{C})\mathbf{B} - (\mathbf{A} \cdot \mathbf{B})\mathbf{C}$$

Alternate Solution: Using the determinant form of the definition of the vector product (2.28),

$$\mathbf{B} \times \mathbf{C} = \begin{vmatrix} \mathbf{i} & \mathbf{j} & \mathbf{k} \\ B_1 & B_2 & B_3 \\ C_1 & C_2 & C_3 \end{vmatrix}$$

$$= (B_2C_3 - B_3C_2)\mathbf{i} + (B_3C_1 - B_1C_3)\mathbf{j} + (B_1C_2 - B_2C_1)\mathbf{k}$$

Then

$$\mathbf{A} \times (\mathbf{B} \times \mathbf{C}) = \begin{vmatrix} \mathbf{i} & \mathbf{j} & \mathbf{k} \\ A_1 & A_2 & A_3 \\ (B_2C_3 - B_3C_2) & (B_3C_1 - B_1C_3) & (B_1C_2 - B_2C_1) \end{vmatrix}$$

$$= [A_2(B_1 C_2 - B_2 C_1) - A_3(B_3 C_1 - B_1 C_3)]\mathbf{i}$$
$$+ [A_3(B_2 C_3 - B_3 C_2) - A_1(B_1 C_2 - B_2 C_1)]\mathbf{j}$$
$$+ [A_1(B_3 C_1 - B_1 C_3) - A_2(B_2 C_3 - B_3 C_2)]\mathbf{k}$$
$$= (A_2 B_1 C_2 - A_2 B_2 C_1 - A_3 B_3 C_1 + A_3 B_1 C_3)\mathbf{i}$$
$$+ (A_3 B_2 C_3 - A_3 B_3 C_2 - A_1 B_1 C_2 + A_1 B_2 C_1)\mathbf{j}$$
$$+ (A_1 B_3 C_1 - A_1 B_1 C_3 - A_2 B_2 C_3 + A_2 B_3 C_2)\mathbf{k} \qquad (2.51)$$

Since

$$(\mathbf{A} \cdot \mathbf{C})\mathbf{B} = (A_1 C_1 + A_2 C_2 + A_3 C_3)(B_1 \mathbf{i} + B_2 \mathbf{j} + B_3 \mathbf{k})$$
$$= (A_1 B_1 C_1 + A_2 B_1 C_2 + A_3 B_1 C_3)\mathbf{i}$$
$$+ (A_1 B_2 C_1 + A_2 B_2 C_2 + A_3 B_2 C_3)\mathbf{j}$$
$$+ (A_1 B_3 C_1 + A_2 B_3 C_2 + A_3 B_3 C_3)\mathbf{k},$$
$$(\mathbf{A} \cdot \mathbf{B})\mathbf{C} = (A_1 B_1 + A_2 B_2 + A_3 B_3)(C_1 \mathbf{i} + C_2 \mathbf{j} + C_3 \mathbf{k})$$
$$= (A_1 B_1 C_1 + A_2 B_2 C_1 + A_3 B_3 C_1)\mathbf{i}$$
$$+ (A_1 B_1 C_2 + A_2 B_2 C_2 + A_3 B_3 C_2)\mathbf{j}$$
$$+ (A_1 B_1 C_3 + A_2 B_2 C_3 + A_3 B_3 C_3)\mathbf{k}$$

we have

$$(\mathbf{A} \cdot \mathbf{C})\mathbf{B} - (\mathbf{A} \cdot \mathbf{B})\mathbf{C} = (A_2 B_1 C_2 + A_3 B_1 C_3 - A_2 B_2 C_1 - A_3 B_3 C_1)\mathbf{i}$$
$$+ (A_1 B_2 C_1 + A_3 B_2 C_3 - A_1 B_1 C_2 - A_3 B_3 C_2)\mathbf{j}$$
$$+ (A_1 B_3 C_1 + A_2 B_3 C_2 - A_1 B_1 C_3 - A_2 B_2 C_3)\mathbf{k} \qquad (2.52)$$

Comparing results (2.51) and (2.52), we conclude that

$$\mathbf{A} \times (\mathbf{B} \times \mathbf{C}) = (\mathbf{A} \cdot \mathbf{C})\mathbf{B} - (\mathbf{A} \cdot \mathbf{B})\mathbf{C}$$

2-8. Basis

A. Reciprocal basis

As we have shown in Problem 1-29, if $\mathbf{a}_1$, $\mathbf{a}_2$, and $\mathbf{a}_3$ are nonzero noncoplanar vectors, any vector $\mathbf{d}$ can be represented as

$$\mathbf{d} = (\mathbf{d} \cdot \mathbf{b}_1)\mathbf{a}_1 + (\mathbf{d} \cdot \mathbf{b}_2)\mathbf{a}_2 + (\mathbf{d} \cdot \mathbf{b}_3)\mathbf{a}_3 \qquad (2.53)$$

where the set $\mathbf{b}_1, \mathbf{b}_2, \mathbf{b}_3$ is the reciprocal set of vectors to $\mathbf{a}_1, \mathbf{a}_2, \mathbf{a}_3$, given by

RECIPROCAL BASIS
$$\mathbf{b}_1 = \frac{\mathbf{a}_2 \times \mathbf{a}_3}{[\mathbf{a}_1 \mathbf{a}_2 \mathbf{a}_3]}, \quad \mathbf{b}_2 = \frac{\mathbf{a}_3 \times \mathbf{a}_1}{[\mathbf{a}_1 \mathbf{a}_2 \mathbf{a}_3]}, \quad \mathbf{b}_3 = \frac{\mathbf{a}_1 \times \mathbf{a}_2}{[\mathbf{a}_1 \mathbf{a}_2 \mathbf{a}_3]} \qquad (2.54)$$

Thus the set $\mathbf{a}_1, \mathbf{a}_2, \mathbf{a}_3$ constitutes a basis. The second basis $\mathbf{b}_1, \mathbf{b}_2, \mathbf{b}_3$ is said to be a **reciprocal basis** to $\mathbf{a}_1, \mathbf{a}_2, \mathbf{a}_3$.

EXAMPLE 2-18: Show that $\mathbf{i}, \mathbf{j}$, and $\mathbf{k}$ form a self-reciprocal basis.

Solution: By the definition of the scalar triple product (2.45) we have

$$[\mathbf{i}\,\mathbf{j}\,\mathbf{k}] = \begin{vmatrix} 1 & 0 & 0 \\ 0 & 1 & 0 \\ 0 & 0 & 1 \end{vmatrix} = 1$$

and from (2.29b) we have $\mathbf{i} \times \mathbf{j} = \mathbf{k}, \mathbf{j} \times \mathbf{k} = \mathbf{i}, \mathbf{k} \times \mathbf{i} = \mathbf{j}$. Now suppose $\mathbf{i}', \mathbf{j}'$, and $\mathbf{k}'$ form a reciprocal set to $\mathbf{i}, \mathbf{j}$, and $\mathbf{k}$. Then from the definition formulas for reciprocal basis (2.54), we have

$$\mathbf{i}' = \frac{\mathbf{j} \times \mathbf{k}}{[\mathbf{i}\,\mathbf{j}\,\mathbf{k}]} = \mathbf{i}, \quad \mathbf{j}' = \frac{\mathbf{k} \times \mathbf{i}}{[\mathbf{i}\,\mathbf{j}\,\mathbf{k}]} = \mathbf{k}, \quad \mathbf{k}' = \frac{\mathbf{i} \times \mathbf{j}}{[\mathbf{i}\,\mathbf{j}\,\mathbf{k}]} = \mathbf{k} \qquad (2.55)$$

Hence we see that $\mathbf{i}$, $\mathbf{j}$, and $\mathbf{k}$ form a self-reciprocal basis.

Using (2.55), we can reduce the representation of $\mathbf{d}$ (2.53), to

$$\mathbf{d} = (\mathbf{d} \cdot \mathbf{i})\mathbf{i} + (\mathbf{d} \cdot \mathbf{j})\mathbf{j} + (\mathbf{d} \cdot \mathbf{k})\mathbf{k} \tag{2.56}$$
$$= d_1\mathbf{i} + d_2\mathbf{j} + d_3\mathbf{k}$$

B. Orthonormal basis

If $\mathbf{u}_1$, $\mathbf{u}_2$, and $\mathbf{u}_3$ form an orthonormal basis (as we saw in Example 1-21), then any given vector $\mathbf{A}$ can be expressed as

$$\mathbf{A} = (\mathbf{A} \cdot \mathbf{u}_1)\mathbf{u}_1 + (\mathbf{A} \cdot \mathbf{u}_2)\mathbf{u}_2 + (\mathbf{A} \cdot \mathbf{u}_3)\mathbf{u}_3 \tag{2.57}$$

where the three nonzero vectors $\mathbf{u}_1, \mathbf{u}_2, \mathbf{u}_3$ satisfy the condition

ORTHONORMAL BASIS
$$\mathbf{u}_m \cdot \mathbf{u}_n = \delta_{mn} = \begin{cases} 1 & \text{if } m = n \\ 0 & \text{if } m \neq n \end{cases} \tag{2.58}$$

We note that the base vectors $\mathbf{i}, \mathbf{j}, \mathbf{k}$ form an orthonormal basis. So again, with this basis, expression (2.57) reduces to representation (2.56).

C. Gram–Schmidt orthogonalization process

The process by which an orthonormal set of three vectors is constructed from a given set of three nonzero noncoplanar vectors is called the **Gram–Schmidt orthogonalization process**, as shown in Example 2-19.

EXAMPLE 2-19: Show that for a given set of linearly independent vectors $\mathbf{a}_1, \mathbf{a}_2, \mathbf{a}_3$, an orthonormal basis $\mathbf{u}_1, \mathbf{u}_2, \mathbf{u}_3$ can be constructed such that $\mathbf{u}_1$ is a scalar multiple of $\mathbf{a}_1$, $\mathbf{u}_2$ is a linear combination of $\mathbf{a}_1$ and $\mathbf{a}_2$, and $\mathbf{u}_3$ is a linear combination of $\mathbf{a}_1$, $\mathbf{a}_2$, and $\mathbf{a}_3$.

Solution: Since $\mathbf{a}_1, \mathbf{a}_2$, and $\mathbf{a}_3$ are linearly independent, none of them can be zero vectors. Now since $\mathbf{u}_1$ is a unit vector, we construct $\mathbf{u}_1$ so that

$$\mathbf{u}_1 = m\mathbf{a}_1 = \frac{1}{|\mathbf{a}_1|}\mathbf{a}_1 \tag{2.59}$$

Then since $\mathbf{u}_2$ is to be a linear combination of $\mathbf{a}_1$ and $\mathbf{a}_2$, it must also be a linear combination of $\mathbf{u}_1$ and $\mathbf{a}_2$ and be orthogonal to $\mathbf{u}_1$. We first consider a vector $\mathbf{v}_2$ such that

$$\mathbf{v}_2 = \mathbf{a}_2 + m_1\mathbf{u}_1 \tag{2.60}$$

where m_1 is determined so that $\mathbf{v}_2$ is orthogonal to $\mathbf{u}_1$. Setting $\mathbf{v}_2 \cdot \mathbf{u}_1 = 0$,

$$0 = (\mathbf{a}_2 \cdot \mathbf{u}_1) + m_1(\mathbf{u}_1 \cdot \mathbf{u}_1) = (\mathbf{a}_2 \cdot \mathbf{u}_1) + m_1$$

since $\mathbf{u}_1 \cdot \mathbf{u}_1 = 1$. Hence $m_1 = -(\mathbf{a}_2 \cdot \mathbf{u}_1)$, and (2.60) becomes

$$\mathbf{v}_2 = \mathbf{a}_2 - (\mathbf{a}_2 \cdot \mathbf{u}_1)\mathbf{u}_1 \tag{2.61}$$

The vector $\mathbf{v}_2$ is now orthogonal to $\mathbf{u}_1$, but it is not a unit vector in general. So $\mathbf{u}_2$ can be found by setting

$$\mathbf{u}_2 = \frac{1}{|\mathbf{v}_2|}\mathbf{v}_2 \tag{2.62}$$

[Note that $\mathbf{v}_2 \neq \mathbf{0}$ in (2.61); if $\mathbf{v}_2 = \mathbf{0}$, then $\mathbf{a}_2$ would be a multiple of $\mathbf{u}_1$ and hence of $\mathbf{a}_1$, which is impossible because they are linearly independent.]

To determine $\mathbf{u}_3$, we proceed in a similar manner. We construct a vector $\mathbf{v}_3$ such that

$$\mathbf{v}_3 = \mathbf{a}_3 + m_1\mathbf{u}_1 + m_2\mathbf{u}_2 \tag{2.63}$$

where m_1 and m_2 are so determined that $\mathbf{v}_3$ is orthogonal to $\mathbf{u}_1$ and $\mathbf{u}_2$. Then setting $\mathbf{v}_3 \cdot \mathbf{u}_1 = 0$ and $\mathbf{v}_3 \cdot \mathbf{u}_2 = 0$, we obtain, respectively,

$$\mathbf{a}_3 \cdot \mathbf{u}_1 + m_1 = 0 \quad \text{and} \quad \mathbf{a}_3 \cdot \mathbf{u}_2 + m_2 = 0$$

since $\mathbf{u}_1 \cdot \mathbf{u}_1 = \mathbf{u}_2 \cdot \mathbf{u}_2 = 1$ and $\mathbf{u}_1 \cdot \mathbf{u}_2 = \mathbf{u}_2 \cdot \mathbf{u}_1 = 0$. Hence $m_1 = -(\mathbf{a}_3 \cdot \mathbf{u}_1)$ and $m_2 = -(\mathbf{a}_3 \; \mathbf{u}_2)$, so we may write $\mathbf{v}_3$ as

$$\mathbf{v}_3 = \mathbf{a}_3 - (\mathbf{a}_3 \cdot \mathbf{u}_1)\mathbf{u}_1 - (\mathbf{a}_3 \cdot \mathbf{u}_2)\mathbf{u}_2 \qquad (2.64)$$

[Note that, as before, $\mathbf{v}_3 \neq \mathbf{0}$ in (2.64); if $\mathbf{v}_3 = \mathbf{0}$, then $\mathbf{a}_3$ would be a linear combination of $\mathbf{u}_1$ and $\mathbf{u}_2$ and hence of $\mathbf{a}_1$ and $\mathbf{a}_2$.] The unit vector $\mathbf{u}_3$ is given by

$$\mathbf{u}_3 = \frac{1}{|\mathbf{v}_3|} \mathbf{v}_3 \qquad (2.65)$$

SUMMARY

1. A vector $\mathbf{A}$ in a rectangular coordinate system can be represented by an ordered triple of real numbers

$$\mathbf{A} = [A_1, A_2, A_3]$$

where A_1, A_2, and A_3 are the components of $\mathbf{A}$ relative to the given coordinate system.

2. The base vectors of the rectangular coordinate system are

$$\mathbf{i} = [1, 0, 0], \qquad \mathbf{j} = [0, 1, 0], \qquad \mathbf{k} = [0, 0, 1]$$

3. The addition of two vectors $\mathbf{A} = [A_1, A_2, A_3]$ and $\mathbf{B} = [B_1, B_2, B_3]$ is given by

$$\mathbf{A} + \mathbf{B} = [A_1 + B_1, A_2 + B_2, A_3 + B_3]$$

4. The dot product of vectors $\mathbf{A}$ and $\mathbf{B}$ is given by

$$\mathbf{A} \cdot \mathbf{B} = A_1 B_1 + A_2 B_2 + A_3 B_3$$

5. The cross product of vectors $\mathbf{A}$ and $\mathbf{B}$ is given by

$$\mathbf{A} \times \mathbf{B} = [A_2 B_3 - A_3 B_2, A_3 B_1 - A_1 B_3, A_1 B_2 - A_2 B_1]$$

or

$$\mathbf{A} \times \mathbf{B} = \begin{vmatrix} \mathbf{i} & \mathbf{j} & \mathbf{k} \\ A_1 & A_2 & A_3 \\ B_1 & B_2 & B_3 \end{vmatrix}$$

$$= (A_2 B_3 - A_3 B_2)\mathbf{i} + (A_3 B_1 - A_1 B_3)\mathbf{j} + (A_1 B_2 - A_2 B_1)\mathbf{k}$$

6. The angle θ between two nonzero vectors $\mathbf{A}$ and $\mathbf{B}$ can be found from

$$\cos \theta = \frac{\mathbf{A} \cdot \mathbf{B}}{AB} = \frac{A_1 B_1 + A_2 B_2 + A_3 B_3}{[A_1^2 + A_2^2 + A_3^2]^{1/2}[B_1^2 + B_2^2 + B_3^2]^{1/2}}$$

7. The direction cosines of a vector $\mathbf{A}$ are defined by

$$\cos \alpha = \mathbf{i} \cdot \mathbf{e}_A, \qquad \cos \beta = \mathbf{j} \cdot \mathbf{e}_A, \qquad \cos \gamma = \mathbf{k} \cdot \mathbf{e}_A$$

where $\mathbf{e}_A$ is the unit vector along $\mathbf{A}$.

8. If $\mathbf{a}_1$, $\mathbf{a}_2$, and $\mathbf{a}_3$ are nonzero noncoplanar vectors, an orthonormal basis $\mathbf{u}_1, \mathbf{u}_2, \mathbf{u}_3$ can be constructed from $\mathbf{a}_1, \mathbf{a}_2, \mathbf{a}_3$ by the Gram–Schmidt orthogonalization process.

RAISE YOUR GRADES

Can you explain...?

☑ what the analytic representation of a vector is
☑ how to add and subtract vectors in the analytic form

☑ how to take the dot product and the cross product of two vectors in the analytic form

☑ how to find the angle between two vectors

☑ how to construct an orthonormal set of three vectors from a given set of three noncoplanar vectors

SOLVED PROBLEMS

PROBLEM 2-1   If $A \neq 0$ and $B \neq 0$, determine the condition under which the vector $A = [A_1, A_2, A_3]$ will be parallel to the vector $B = [B_1, B_2, B_3]$.

Solution: If the vectors A and B are to be parallel, there must be a scalar m such that

$$B = mA$$

Equating components, we obtain $B_1 = mA_1$, $B_2 = mA_2$, and $B_3 = mA_3$ for the same scalar m. So the required condition is

$$\frac{B_1}{A_1} = \frac{B_2}{A_2} = \frac{B_3}{A_3}$$

or

$$A_1 : A_2 : A_3 = B_1 : B_2 : B_3$$

In other words, the corresponding components are proportional.

PROBLEM 2-2   Let $A = [2, -3, 4]$ and $B = [1, 2, 1]$. Find (a) $A + B$ and (b) $A - B$.

Solution: From the analytic definitions of the vector sum (2.7) and the vector difference (2.8), we have

(a) $A + B = [2 + 1, -3 + 2, 4 + 1] = [3, -1, 5]$
(b) $A - B = [2 - 1, -3 - 2, 4 - 1] = [1, -5, 3]$

PROBLEM 2-3   Find a unit vector parallel to the vector that is the sum of the vectors $A = [2, -3, 4]$ and $B = [1, 2, -2]$.

Solution:

$$A + B = [2 + 1, -3 + 2, 4 - 2] = [3, -1, 2]$$

From the result of Problem 2-1, all vectors parallel to $A + B$ are of the form

$$m(A + B) = m[3, -1, 2]$$

From definition (2.34) the magnitude of these vectors is $\sqrt{3^2 + (-1)^2 + 2^2}\, m = \sqrt{14}\, m$ if $m > 0$ and $-\sqrt{14}\, m$ if $m < 0$. Hence the required unit vector is obtained by setting $\sqrt{14}\, m = 1$ or $-\sqrt{14}\, m = 1$:

$$\frac{1}{\sqrt{14}}[3, -1, 2] = \left[\frac{3}{\sqrt{14}}, \frac{-1}{\sqrt{14}}, \frac{2}{\sqrt{14}}\right]$$

or

$$-\frac{1}{\sqrt{14}}[3, -1, 2] = \left[\frac{-3}{\sqrt{14}}, \frac{1}{\sqrt{14}}, \frac{-2}{\sqrt{14}}\right]$$

PROBLEM 2-4 Given two vectors

$$\mathbf{A} = [2, 4, 6] = 2\mathbf{i} + 4\mathbf{j} + 6\mathbf{k}, \qquad \mathbf{B} = [1, -3, 2] = \mathbf{i} - 3\mathbf{j} + 2\mathbf{k}$$

compute the scalar and vector product $\mathbf{A} \cdot \mathbf{B}$ and $\mathbf{A} \times \mathbf{B}$.

Solution: The scalar product is, by definition (2.20),

$$\mathbf{A} \cdot \mathbf{B} = (2)(1) + (4)(-3) + (6)(2) = 2$$

From the determinant form of the definition of the vector product (2.28),

$$\mathbf{A} \times \mathbf{B} = \begin{vmatrix} \mathbf{i} & \mathbf{j} & \mathbf{k} \\ 2 & 4 & 6 \\ 1 & -3 & 2 \end{vmatrix}$$

$$= [(4)(2) - (6)(-3)]\mathbf{i} - [(2)(2) - (6)(1)]\mathbf{j} + [(2)(-3) - (4)(1)]\mathbf{k}$$

$$= 26\mathbf{i} + 2\mathbf{j} - 10\mathbf{k}$$

$$= [26, 2, -10]$$

PROBLEM 2-5 Show that

$$(A_1 B_1 + A_2 B_2 + A_3 B_3)^2 \leqslant (A_1^2 + A_2^2 + A_3^2)(B_1^2 + B_2^2 + B_3^2)$$

Solution: Let $\mathbf{A} = [A_1, A_2, A_3]$ and $\mathbf{B} = [B_1, B_2, B_3]$. Then from (2.37) and squaring, we obtain

$$(A_1 B_1 + A_2 B_2 + A_3 B_3)^2$$
$$= (A_1^2 + A_2^2 + A_3^2)(B_1^2 + B_2^2 + B_3^2)\cos^2\theta$$
$$\leqslant (A_1^2 + A_2^2 + A_3^2)(B_1^2 + B_2^2 + B_3^2)$$

because $\cos^2\theta \leqslant 1$.

PROBLEM 2-6 Find the angle θ between the vectors $\mathbf{A} = [2, 4, 6]$ and $\mathbf{B} = [1, -3, 2]$.

Solution: From the definition formula (2.37) for the angle between two vectors, we have

$$\cos \theta = \frac{(2)(1) + (4)(-3) + (6)(2)}{(4 + 16 + 36)^{1/2}(1 + 9 + 4)^{1/2}} = \frac{2}{\sqrt{56}\sqrt{14}} = \frac{1}{14}$$

Hence $\theta = \cos^{-1}(1/14) = 85° 54'$.

PROBLEM 2-7 Show that the vectors $\mathbf{A} = [2, 3, -1]$ and $\mathbf{B} = [1, 0, 2]$ are perpendicular.

Solution: From the fact that $\cos \theta = \mathbf{A} \cdot \mathbf{B}/AB$ (2.42), we conclude that $\mathbf{A}$ and $\mathbf{B}$ must be perpendicular if $\mathbf{A} \cdot \mathbf{B} = 0$. Now since

$$\mathbf{A} \cdot \mathbf{B} = (2)(1) + (3)(0) + (-1)(2) = 2 + 0 - 2 = 0$$

the two vectors are perpendicular.

PROBLEM 2-8 Show that the vector $\mathbf{A} \times \mathbf{B}$ is perpendicular to both $\mathbf{A}$ and $\mathbf{B}$.

Solution: From the analytic definition of the vector product (2.26)

$$\mathbf{A} \times \mathbf{B} = [A_2 B_3 - A_3 B_2, A_3 B_1 - A_1 B_3, A_1 B_2 - A_2 B_1]$$

Calculating the dot product of $\mathbf{A} \times \mathbf{B}$ with $\mathbf{A}$ and $\mathbf{B}$, we obtain

$$\mathbf{A} \cdot (\mathbf{A} \times \mathbf{B}) = A_1(A_2 B_3 - A_3 B_2) + A_2(A_3 B_1 - A_1 B_3) + A_3(A_1 B_2 - A_2 B_1)$$
$$= A_1 A_2 B_3 - A_1 A_3 B_2 + A_2 A_3 B_1 - A_2 A_1 B_3 + A_3 A_1 B_2 - A_3 A_2 B_1$$
$$= 0$$

$$\mathbf{B} \cdot (\mathbf{A} \times \mathbf{B}) = B_1(A_2 B_3 - A_3 B_2) + B_2(A_3 B_1 - A_1 B_3) + B_3(A_1 B_2 - A_2 B_1)$$
$$= B_1 A_2 B_3 - B_1 A_3 B_2 + B_2 A_3 B_1 - B_2 A_1 B_3 + B_3 A_1 B_2 - B_3 A_2 B_1$$
$$= 0$$

So $\mathbf{A} \times \mathbf{B}$ is perpendicular to both $\mathbf{A}$ and $\mathbf{B}$.

PROBLEM 2-9   Find a unit vector $\mathbf{u}$ that is perpendicular to both $\mathbf{A} = [2, 1, 1]$ and $\mathbf{B} = [1, -1, 2]$.

Solution: We know that $\mathbf{A} \times \mathbf{B}$ is perpendicular to $\mathbf{A}$ and $\mathbf{B}$ from the result of Problem 2-8. Hence the vector perpendicular to both $\mathbf{A}$ and $\mathbf{B}$ is

$$\mathbf{A} \times \mathbf{B} = \begin{vmatrix} \mathbf{i} & \mathbf{j} & \mathbf{k} \\ 2 & 1 & 1 \\ 1 & -1 & 2 \end{vmatrix}$$

$$= 3\mathbf{i} - 3\mathbf{j} - 3\mathbf{k}$$

$$= [3, -3, -3]$$

and the magnitude of $\mathbf{A} \times \mathbf{B}$ is

$$|\mathbf{A} \times \mathbf{B}| = [3^2 + (-3)^2 + (-3)^2]^{1/2} = \sqrt{27} = 3\sqrt{3}$$

So a unit vector $\mathbf{u}$ that is perpendicular to both $\mathbf{A}$ and $\mathbf{B}$ is

$$\mathbf{u} = \frac{\mathbf{A} \times \mathbf{B}}{|\mathbf{A} \times \mathbf{B}|} = \frac{1}{3\sqrt{3}} [3, -3, -3] = \left[\frac{1}{\sqrt{3}}, -\frac{1}{\sqrt{3}}, -\frac{1}{\sqrt{3}} \right]$$

Note: $-\mathbf{u} = [-1/\sqrt{3}, 1/\sqrt{3}, 1/\sqrt{3}]$ is also a unit vector perpendicular to $\mathbf{A}$ and $\mathbf{B}$.

PROBLEM 2-10   Given the points $A(2, -1, 1)$, $B(1, 1, 0)$, $C(1, -1, 2)$, and $D(2, 2, 2)$, determine if $\overrightarrow{AC}$ is perpendicular to $\overrightarrow{BD}$.

Solution: Points A, B, C, and D determine the position vectors

$$\mathbf{r}_A = [2, -1, 1], \qquad \mathbf{r}_B = [1, 1, 0], \qquad \mathbf{r}_C = [1, -1, 2], \qquad \mathbf{r}_D = [2, 2, 2]$$

Referring to Example 2-3 and Figure 2-3, we have

$$\overrightarrow{AC} = \mathbf{r}_C - \mathbf{r}_A = [1 - 2, -1 - (-1), 2 - 1] = [-1, 0, 1]$$
$$\overrightarrow{BD} = \mathbf{r}_D - \mathbf{r}_B = [2 - 1, 2 - 1, 2 - 0] = [1, 1, 0]$$

Thus by the analytic definition of the scalar product (2.20),

$$\overrightarrow{AC} \cdot \overrightarrow{BD} = (-1) \cdot 1 + 0 \cdot 1 + 1 \cdot 0 = -1 \neq 0$$

So $\overrightarrow{AC}$ is not perpendicular to $\overrightarrow{BD}$.

PROBLEM 2-11   Show that

$$\cos^2 \alpha + \cos^2 \beta + \cos^2 \gamma = 1$$

where $\cos \alpha$, $\cos \beta$, and $\cos \gamma$ are the direction cosines of a vector.

Solution: Using the formulas (2.44) for finding the direction cosines, we have

$$\cos^2 \alpha + \cos^2 \beta + \cos^2 \gamma = \frac{A_1^2}{A^2} + \frac{A_2^2}{A^2} + \frac{A_3^2}{A^2} = \frac{A_1^2 + A_2^2 + A_3^2}{A^2} = \frac{A^2}{A^2} = 1$$

PROBLEM 2-12   Using the base vector relations (2.21), verify the analytic definition of the dot product (2.20).

Solution: Let $\mathbf{A} = A_1 \mathbf{i} + A_2 \mathbf{j} + A_3 \mathbf{k}$ and $\mathbf{B} = B_1 \mathbf{i} + B_2 \mathbf{j} + \mathbf{B}_3 \mathbf{k}$. Then

$$\begin{aligned}
\mathbf{A} \cdot \mathbf{B} &= (A_1\mathbf{i} + A_2\mathbf{j} + A_3\mathbf{k}) \cdot (B_1\mathbf{i} + B_2\mathbf{j} + B_3\mathbf{k}) \\
&= A_1 B_1 \mathbf{i} \cdot \mathbf{i} + A_1 B_2 \mathbf{i} \cdot \mathbf{j} + A_1 B_3 \mathbf{i} \cdot \mathbf{k} \\
&\quad + A_2 B_1 \mathbf{j} \cdot \mathbf{i} + A_2 B_2 \mathbf{j} \cdot \mathbf{j} + A_2 B_3 \mathbf{j} \cdot \mathbf{k} \\
&\quad + A_3 B_1 \mathbf{k} \cdot \mathbf{i} + A_3 B_2 \mathbf{k} \cdot \mathbf{j} + A_3 B_3 \mathbf{k} \cdot \mathbf{k} \\
&= A_1 B_1 + A_2 B_2 + A_3 B_3
\end{aligned}$$

PROBLEM 2-13 Using the base vector relations (2.29), verify the analytic definition of the cross product (2.26).

Solution: Let $\mathbf{A} = A_1\mathbf{i} + A_2\mathbf{j} + A_3\mathbf{k}$ and $\mathbf{B} = B_1\mathbf{i} + B_2\mathbf{j} + B_3\mathbf{k}$. Then

$$\begin{aligned}
\mathbf{A} \times \mathbf{B} &= (A_1\mathbf{i} + A_2\mathbf{j} + A_3\mathbf{k}) \times (B_1\mathbf{i} + B_2\mathbf{j} + B_3\mathbf{k}) \\
&= A_1 B_1 \mathbf{i} \times \mathbf{i} + A_1 B_2 \mathbf{i} \times \mathbf{j} + A_1 B_3 \mathbf{i} \times \mathbf{k} \\
&\quad + A_2 B_1 \mathbf{j} \times \mathbf{i} + A_2 B_2 \mathbf{j} \times \mathbf{j} + A_2 B_3 \mathbf{j} \times \mathbf{k} \\
&\quad + A_3 B_1 \mathbf{k} \times \mathbf{i} + A_3 B_2 \mathbf{k} \times \mathbf{j} + A_3 B_3 \mathbf{k} \times \mathbf{k} \\
&= A_1 B_2 \mathbf{k} - A_1 B_3 \mathbf{j} - A_2 B_1 \mathbf{k} + A_2 B_3 \mathbf{i} + A_3 B_1 \mathbf{j} - A_3 B_2 \mathbf{i} \\
&= (A_2 B_3 - A_3 B_2)\mathbf{i} + (A_3 B_1 - A_1 B_3)\mathbf{j} + (A_1 B_2 - A_2 B_1)\mathbf{k} \\
&= [(A_2 B_3 - A_3 B_2), (A_3 B_1 - A_1 B_3), (A_1 B_2 - A_2 B_1)]
\end{aligned}$$

PROBLEM 2-14 If any two vectors in a scalar triple product are equal, show that the product is zero.

Solution: We can show that this is true by expressing the scalar triple product in determinant form as in (2.45) and then using one of the properties of determinants—if any two rows of a determinant are the same, then the value of the determinant is always zero.

PROBLEM 2-15 Show that

$$\mathbf{A} \cdot \mathbf{B} \times \mathbf{C} = \mathbf{A} \times \mathbf{B} \cdot \mathbf{C}$$

Solution: From the result of Example 2-16, we have

$$[\mathbf{ABC}] = [\mathbf{CAB}]$$

or

$$\mathbf{A} \cdot \mathbf{B} \times \mathbf{C} = \mathbf{C} \cdot \mathbf{A} \times \mathbf{B}$$

From the commutative law of the scalar product (2.22), we know that

$$\mathbf{C} \cdot \mathbf{A} \times \mathbf{B} = \mathbf{A} \times \mathbf{B} \cdot \mathbf{C}$$

So

$$\mathbf{A} \cdot \mathbf{B} \times \mathbf{C} = \mathbf{A} \times \mathbf{B} \cdot \mathbf{C}$$

PROBLEM 2-16 Show that

$$[\mathbf{i}\,\mathbf{j}\,\mathbf{k}] = [\mathbf{j}\,\mathbf{k}\,\mathbf{i}] = [\mathbf{k}\,\mathbf{i}\,\mathbf{j}] = 1$$
$$[\mathbf{i}\,\mathbf{k}\,\mathbf{j}] = [\mathbf{k}\,\mathbf{j}\,\mathbf{i}] = [\mathbf{j}\,\mathbf{i}\,\mathbf{k}] = -1$$

Solution: From the determinant form of the definition of the scalar triple product (2.45) we obtain

$$[\mathbf{i}\,\mathbf{j}\,\mathbf{k}] = \begin{vmatrix} 1 & 0 & 0 \\ 0 & 1 & 0 \\ 0 & 0 & 1 \end{vmatrix} = 1$$

The other scalar triple products follow from relation (2.46).

PROBLEM 2-17 Show that the three vectors $\mathbf{A} = [2,0,1]$, $\mathbf{B} = [0,3,4]$, and $\mathbf{C} = [8,-3,0]$ are coplanar. Then express $\mathbf{C}$ as a linear combination of $\mathbf{A}$ and $\mathbf{B}$.

Solution: In Example 1-13, it was shown that the absolute value of $[\mathbf{ABC}]$ is the volume of the parallelepiped defined by $\mathbf{A}$, $\mathbf{B}$, and $\mathbf{C}$. So if $\mathbf{A}$, $\mathbf{B}$, and $\mathbf{C}$ are coplanar, the volume is 0 and $[\mathbf{ABC}] = 0$. Conversely, if $[\mathbf{ABC}] = 0$, the vectors are coplanar. Now from (2.45)

$$[\mathbf{ABC}] = \begin{vmatrix} 2 & 0 & 1 \\ 0 & 3 & 4 \\ 8 & -3 & 0 \end{vmatrix} = 0$$

Hence $\mathbf{A}$, $\mathbf{B}$, and $\mathbf{C}$ are coplanar.

Next let $\mathbf{C}$ be expressed as $\mathbf{C} = \lambda_1 \mathbf{A} + \lambda_2 \mathbf{B}$, for some scalars λ_1 and λ_2. Then

$$[8,-3,0] = \lambda_1 [2,0,1] + \lambda_2 [0,3,4]$$
$$= [2\lambda_1, 0, \lambda_1] + [0, 3\lambda_2, 4\lambda_2]$$
$$= [2\lambda_1, 3\lambda_2, \lambda_1 + 4\lambda_2]$$

Thus by the analytic definition of the equality of vectors (2.3),

$$8 = 2\lambda_1, \qquad -3 = 3\lambda_2, \qquad 0 = \lambda_1 + 4\lambda_2$$

Hence $\lambda_1 = 4$, $\lambda_2 = -1$, and $\mathbf{C} = 4\mathbf{A} - \mathbf{B}$.

PROBLEM 2-18 Let $\mathbf{A} = [1,2,-3]$, $\mathbf{B} = [2,-1,1]$, and $\mathbf{C} = [-1,1,-1]$.

(a) Calculate $\mathbf{A} \times (\mathbf{B} \times \mathbf{C})$ and verify (2.47):

$$\mathbf{A} \times (\mathbf{B} \times \mathbf{C}) = (\mathbf{A} \cdot \mathbf{C})\mathbf{B} - (\mathbf{A} \cdot \mathbf{B})\mathbf{C}$$

(b) Calculate $(\mathbf{A} \times \mathbf{B}) \times \mathbf{C}$ and verify

$$(\mathbf{A} \times \mathbf{B}) \times \mathbf{C} = (\mathbf{A} \cdot \mathbf{C})\mathbf{B} - (\mathbf{B} \cdot \mathbf{C})\mathbf{A}$$

Solution:
(a) To calculate $\mathbf{A} \times (\mathbf{B} \times \mathbf{C})$, first find

$$\mathbf{B} \times \mathbf{C} = \begin{vmatrix} \mathbf{i} & \mathbf{j} & \mathbf{k} \\ 2 & -1 & 1 \\ -1 & 1 & -1 \end{vmatrix} = 0\mathbf{i} + 1\mathbf{j} + 1\mathbf{k} = [0,1,1]$$

Then substitute $[0,1,1]$ for $\mathbf{B} \times \mathbf{C}$ and find

$$\mathbf{A} \times (\mathbf{B} \times \mathbf{C}) = \begin{vmatrix} \mathbf{i} & \mathbf{j} & \mathbf{k} \\ 1 & 2 & -3 \\ 0 & 1 & 1 \end{vmatrix} = 5\mathbf{i} - 1\mathbf{j} + 1\mathbf{k} = [5,-1,1]$$

Having found the value of the left-hand side, you then calculate the right-hand side:

$$(\mathbf{A} \cdot \mathbf{C})\mathbf{B} = [(1)(-1) + (2)(1) + (-3)(-1)][2,-1,1]$$
$$= 4[2,-1,1] = [8,-4,4]$$
$$(\mathbf{A} \cdot \mathbf{B})\mathbf{C} = [(1)(2) + (2)(-1) + (-3)(1)][-1,1,-1]$$
$$= -3[-1,1,-1] = [3,-3,3]$$

and find that

$$(\mathbf{A} \cdot \mathbf{C})\mathbf{B} - (\mathbf{A} \cdot \mathbf{B})\mathbf{C} = [8,-4,4] - [3,-3,3]$$
$$= [5,-1,1]$$
$$= \mathbf{A} \times (\mathbf{B} \times \mathbf{C})$$

(b) To calculate $(\mathbf{A} \times \mathbf{B}) \times \mathbf{C}$, first calculate

$$\mathbf{A} \times \mathbf{B} = \begin{vmatrix} \mathbf{i} & \mathbf{j} & \mathbf{k} \\ 1 & 2 & -3 \\ 2 & -1 & 1 \end{vmatrix} = -1\mathbf{i} - 7\mathbf{j} - 5\mathbf{k} = [-1, -7, -5]$$

and hence

$$(\mathbf{A} \times \mathbf{B}) \times \mathbf{C} = \begin{vmatrix} \mathbf{i} & \mathbf{j} & \mathbf{k} \\ -1 & -7 & -5 \\ -1 & 1 & -1 \end{vmatrix} = 12\mathbf{i} + 4\mathbf{j} - 8\mathbf{k} = [12, 4, -8]$$

From part (a) we have $(\mathbf{A} \cdot \mathbf{C})\mathbf{B} = [8, -4, 4]$. Then

$$\begin{aligned} (\mathbf{B} \cdot \mathbf{C})\mathbf{A} &= [(2)(-1) + (-1)(1) + (1)(-1)], [1, 2, -3] \\ &= -4[1, 2, -3] \\ &= [-4, -8, 12] \end{aligned}$$

Consequently

$$\begin{aligned} (\mathbf{A} \cdot \mathbf{C})\mathbf{B} - (\mathbf{B} \cdot \mathbf{C})\mathbf{A} &= [8, -4, 4] - [-4, -8, 12] \\ &= [12, 4, -8] \\ &= (\mathbf{A} \times \mathbf{B}) \times \mathbf{C} \end{aligned}$$

PROBLEM 2-19 Let $\mathbf{a}_1 = [-1, 1, 1]$, $\mathbf{a}_2 = [1, -1, 1]$ and $\mathbf{a}_3 = [1, 1, -1]$.
(a) Show that $\mathbf{a}_1, \mathbf{a}_2,$ and $\mathbf{a}_3$ constitute a basis. (b) Obtain a reciprocal basis to $\mathbf{a}_1,$ $\mathbf{a}_2, \mathbf{a}_3$.

Solution:
(a) Since

$$[\mathbf{a}_1 \mathbf{a}_2 \mathbf{a}_3] = \begin{vmatrix} -1 & 1 & 1 \\ 1 & -1 & 1 \\ 1 & 1 & -1 \end{vmatrix} = 4 \neq 0$$

the vectors $\mathbf{a}_1, \mathbf{a}_2,$ and $\mathbf{a}_3$ are noncoplanar and hence constitute a basis.
(b) From definition (2.54), the reciprocal basis $\mathbf{b}_1, \mathbf{b}_2, \mathbf{b}_3$ is given by

$$\mathbf{b}_1 = \frac{\mathbf{a}_2 \times \mathbf{a}_3}{[\mathbf{a}_1 \mathbf{a}_2 \mathbf{a}_3]} = \frac{1}{4} \begin{vmatrix} \mathbf{i} & \mathbf{j} & \mathbf{k} \\ 1 & -1 & 1 \\ 1 & 1 & -1 \end{vmatrix} = \tfrac{1}{4}(0\mathbf{i} + 2\mathbf{j} + 2\mathbf{k})$$

$$= \tfrac{1}{4}[0, 2, 2]$$

$$= [0, \tfrac{1}{2}, \tfrac{1}{2}]$$

$$\mathbf{b}_2 = \frac{\mathbf{a}_3 \times \mathbf{a}_1}{[\mathbf{a}_1 \mathbf{a}_2 \mathbf{a}_3]} = \frac{1}{4} \begin{vmatrix} \mathbf{i} & \mathbf{j} & \mathbf{k} \\ 1 & 1 & -1 \\ -1 & 1 & 1 \end{vmatrix} = \tfrac{1}{4}(2\mathbf{i} + 0\mathbf{j} + 2\mathbf{k})$$

$$= \tfrac{1}{4}[2, 0, 2]$$

$$= [\tfrac{1}{2}, 0, \tfrac{1}{2}]$$

$$\mathbf{b}_3 = \frac{\mathbf{a}_1 \times \mathbf{a}_2}{[\mathbf{a}_1 \mathbf{a}_2 \mathbf{a}_3]} = \frac{1}{4} \begin{vmatrix} \mathbf{i} & \mathbf{j} & \mathbf{k} \\ -1 & 1 & 1 \\ 1 & -1 & 1 \end{vmatrix} = \tfrac{1}{4}(2\mathbf{i} + 2\mathbf{j} + 0\mathbf{k})$$

$$= \tfrac{1}{4}[2, 2, 0]$$

$$= [\tfrac{1}{2}, \tfrac{1}{2}, 0]$$

PROBLEM 2-20 Express $A = [1, 2, 3]$ as a linear combination of the set of vectors $\mathbf{a}_1 = [-1, 1, 1]$, $\mathbf{a}_2 = [1, -1, 1]$, and $\mathbf{a}_3 = [1, 1, -1]$, given in Problem 2-19.

Solution: From (2.53), we have

$$\mathbf{A} = (\mathbf{A} \cdot \mathbf{b}_1)\mathbf{a}_1 + (\mathbf{A} \cdot \mathbf{b}_2)\mathbf{a}_2 + (\mathbf{A} \cdot \mathbf{b}_3)\mathbf{a}_3$$

where $\mathbf{b}_1$, $\mathbf{b}_2$, and $\mathbf{b}_3$ form the reciprocal basis to $\mathbf{a}_1, \mathbf{a}_2, \mathbf{a}_3$. Using the results of Problem 2-19(b), we have

$$\mathbf{A} \cdot \mathbf{b}_1 = (1)(0) + (2)(\tfrac{1}{2}) + (3)(\tfrac{1}{2}) = \tfrac{5}{2}$$
$$\mathbf{A} \cdot \mathbf{b}_2 = (1)(\tfrac{1}{2}) + (2)(0) + (3)(\tfrac{1}{2}) = 2$$
$$\mathbf{A} \cdot \mathbf{b}_3 = (1)(\tfrac{1}{2}) + (2)(\tfrac{1}{2}) + (e)(0) = \tfrac{3}{2}$$

So

$$\mathbf{A} = \tfrac{5}{2}\mathbf{a}_1 + 2\mathbf{a}_2 + \tfrac{3}{2}\mathbf{a}_3$$

To check the solution, substitute and find

$$\begin{aligned}
\tfrac{5}{2}\mathbf{a}_1 + 2\mathbf{a}_2 + \tfrac{3}{2}\mathbf{a}_3 &= \tfrac{5}{2}[-1, 1, 1] + 2[1, -1, 1] + \tfrac{3}{2}[1, 1, -1] \\
&= [-\tfrac{5}{2}, \tfrac{5}{2}, \tfrac{5}{2}] + [2, -2, 2] + [\tfrac{3}{2}, \tfrac{3}{2}, -\tfrac{3}{2}] \\
&= [-\tfrac{5}{2} + 2 + \tfrac{3}{2}, \tfrac{5}{2} - 2 + \tfrac{3}{2}, \tfrac{5}{2} + 2 - \tfrac{3}{2}] \\
&= [1, 2, 3] \\
&= \mathbf{A}
\end{aligned}$$

PROBLEM 2-21 From the vectors

$$\mathbf{a}_1 = [-1, 1, 1], \mathbf{a}_2 = [1, -1, 1], \text{ and } \mathbf{a}_3 = [1, 1, -1]$$

construct an orthonormal basis $\mathbf{u}_1, \mathbf{u}_2, \mathbf{u}_3$.

Solution: Following the same basic steps as in Example 2-19, we use the Gram–Schmidt process starting with (2.59):

$$\mathbf{u}_1 = \frac{1}{|\mathbf{a}_1|}\mathbf{a}_1 = \frac{1}{\sqrt{3}}[-1, 1, 1] = \left[-\frac{1}{\sqrt{3}}, \frac{1}{\sqrt{3}}, \frac{1}{\sqrt{3}}\right]$$

From (2.61) we obtain

$$\mathbf{v}_2 = \mathbf{a}_2 - (\mathbf{a}_2 \cdot \mathbf{u}_1)\mathbf{u}_1$$

$$= [1, -1, 1] - \left(-\frac{1}{\sqrt{3}}\right)\left[-\frac{1}{\sqrt{3}}, \frac{1}{\sqrt{3}}, \frac{1}{\sqrt{3}}\right]$$

$$= [1, -1, 1] - [\tfrac{1}{3}, -\tfrac{1}{3}, -\tfrac{1}{3}]$$

$$= [\tfrac{2}{3}, -\tfrac{2}{3}, \tfrac{4}{3}] = \tfrac{1}{3}[2, -2, 4]$$

and from the definition of magnitude (2.34) we have

$$|\mathbf{v}_2| = \tfrac{1}{3}\sqrt{4 + 4 + 16} = \tfrac{2}{3}\sqrt{6}$$

Hence

$$\mathbf{u}_2 = \frac{1}{|\mathbf{v}_2|}\mathbf{v}_2 = \frac{3}{2\sqrt{6}}\left[\frac{2}{3}, -\frac{2}{3}, \frac{4}{3}\right] = \left[\frac{1}{\sqrt{6}}, -\frac{1}{\sqrt{6}}, \frac{2}{\sqrt{6}}\right]$$

Next, from (2.64),

$$\mathbf{v}_3 = \mathbf{a}_3 - (\mathbf{a}_3 \cdot \mathbf{u}_1)\mathbf{u}_1 - (\mathbf{a}_3 \cdot \mathbf{u}_2)\mathbf{u}_2$$

$$= [1, 1, -1] - \left(-\frac{1}{\sqrt{3}}\right)\left[-\frac{1}{\sqrt{3}}, \frac{1}{\sqrt{3}}, \frac{1}{\sqrt{3}}\right] - \left(-\frac{2}{\sqrt{6}}\right)\left[\frac{1}{\sqrt{6}}, -\frac{1}{\sqrt{6}}, \frac{2}{\sqrt{6}}\right]$$

$$= [1, 1, -1] - [\tfrac{1}{3}, -\tfrac{1}{3}, -\tfrac{1}{3}] - [-\tfrac{2}{6}, \tfrac{2}{6}, -\tfrac{4}{6}]$$

$$= [1, 1, 0]$$

and its magnitude (2.34) is

$$|\mathbf{v}_3| = \sqrt{1 + 1 + 0} = \sqrt{2}$$

Hence

$$\mathbf{u}_3 = \frac{1}{|\mathbf{v}_3|}\mathbf{v}_3 = \frac{1}{\sqrt{2}}[1,1,0]$$

$$= \left[\frac{1}{\sqrt{2}}, \frac{1}{\sqrt{2}}, 0\right]$$

Thus we obtain the orthonormal basis

$$\mathbf{u}_1 = \left[-\frac{1}{\sqrt{3}}, \frac{1}{\sqrt{3}}, \frac{1}{\sqrt{3}}\right], \quad \mathbf{u}_2 = \left[\frac{1}{\sqrt{6}}, -\frac{1}{\sqrt{6}}, \frac{2}{\sqrt{6}}\right], \quad \mathbf{u}_3 = \left[\frac{1}{\sqrt{2}}, \frac{1}{\sqrt{2}}, 0\right]$$

PROBLEM 2-22 Express $\mathbf{A} = [1,2,3]$ in terms of the orthonormal basis $\mathbf{u}_1, \mathbf{u}_2, \mathbf{u}_3$ found in Problem 2-21.

Solution: From (2.57), we know that if $\mathbf{u}_1$, $\mathbf{u}_2$, and $\mathbf{u}_3$ form an orthonormal basis, any vector $\mathbf{A}$ can be expressed as

$$\mathbf{A} = (\mathbf{A} \cdot \mathbf{u}_1)\mathbf{u}_1 + (\mathbf{A} \cdot \mathbf{u}_2)\mathbf{u}_2 + (\mathbf{A} \cdot \mathbf{u}_3)\mathbf{u}_3$$

Now substitute $\mathbf{A} = [1,2,3]$ and the constructed orthonormal basis $\mathbf{u}_1, \mathbf{u}_2, \mathbf{u}_3$ obtained in Problem 2-21 to get

$$\mathbf{A} \cdot \mathbf{u}_1 = \frac{4}{\sqrt{3}}, \quad \mathbf{A} \cdot \mathbf{u}_2 = \frac{5}{\sqrt{6}}, \quad \mathbf{A} \cdot \mathbf{u}_3 = \frac{3}{\sqrt{2}}$$

So

$$\mathbf{A} = [1,2,3] = \frac{4}{\sqrt{3}}\mathbf{u}_1 + \frac{5}{\sqrt{6}}\mathbf{u}_2 + \frac{3}{\sqrt{2}}\mathbf{u}_3$$

Now check our solution:

$$\frac{4}{\sqrt{3}}\mathbf{u}_1 + \frac{5}{\sqrt{6}}\mathbf{u}_2 + \frac{3}{\sqrt{2}}\mathbf{u}_3$$

$$= \frac{4}{\sqrt{3}}\left[-\frac{1}{\sqrt{3}}, \frac{1}{\sqrt{3}}, \frac{1}{\sqrt{3}}\right] + \frac{5}{\sqrt{6}}\left[\frac{1}{\sqrt{6}}, -\frac{1}{\sqrt{6}}, \frac{2}{\sqrt{6}}\right] + \frac{3}{\sqrt{2}}\left[\frac{1}{\sqrt{2}}, \frac{1}{\sqrt{2}}, 0\right]$$

$$= [-\tfrac{4}{3}, \tfrac{4}{3}, \tfrac{4}{3}] + [\tfrac{5}{6}, -\tfrac{5}{6}, \tfrac{10}{6}] + [\tfrac{3}{2}, \tfrac{3}{2}, 0]$$
$$= [-\tfrac{4}{3} + \tfrac{5}{6} + \tfrac{3}{2}, \tfrac{4}{3} - \tfrac{5}{6} + \tfrac{3}{2}, \tfrac{4}{3} + \tfrac{10}{6} + 0]$$
$$= [1,2,3]$$
$$= \mathbf{A}$$

PROBLEM 2-23 Construct the reciprocal basis of $\mathbf{u}_1, \mathbf{u}_2, \mathbf{u}_3$ obtained in Problem 2-21.

Solution: Let $\mathbf{u}_1', \mathbf{u}_2', \mathbf{u}_3'$ form the reciprocal basis of $\mathbf{u}_1, \mathbf{u}_2, \mathbf{u}_3$. Then using the formulas defining a reciprocal basis (2.54), we have

$$\mathbf{u}_1' = \frac{\mathbf{u}_2 \times \mathbf{u}_3}{[\mathbf{u}_1\mathbf{u}_2\mathbf{u}_3]}, \quad \mathbf{u}_2' = \frac{\mathbf{u}_3 \times \mathbf{u}_1}{[\mathbf{u}_1\mathbf{u}_2\mathbf{u}_3]}, \quad \mathbf{u}_3' = \frac{\mathbf{u}_1 \times \mathbf{u}_2}{[\mathbf{u}_1\mathbf{u}_2\mathbf{u}_3]}$$

From Problem 2-21, we have

$$\mathbf{u}_1 = \left[-\frac{1}{\sqrt{3}}, \frac{1}{\sqrt{3}}, \frac{1}{\sqrt{3}}\right], \quad \mathbf{u}_2 = \left[\frac{1}{\sqrt{6}}, -\frac{1}{\sqrt{6}}, \frac{2}{\sqrt{6}}\right], \quad \mathbf{u}_3 = \left[\frac{1}{\sqrt{2}}, \frac{1}{\sqrt{2}}, 0\right]$$

Because

$$[\mathbf{u}_1\mathbf{u}_2\mathbf{u}_3] = \begin{vmatrix} -\dfrac{1}{\sqrt{3}} & \dfrac{1}{\sqrt{3}} & \dfrac{1}{\sqrt{3}} \\[2mm] \dfrac{1}{\sqrt{6}} & -\dfrac{1}{\sqrt{6}} & \dfrac{2}{\sqrt{6}} \\[2mm] \dfrac{1}{\sqrt{2}} & \dfrac{1}{\sqrt{2}} & 0 \end{vmatrix}$$

$$= \left(\frac{1}{\sqrt{6}}\right)\left(\frac{1}{\sqrt{2}}\right)\left(\frac{1}{\sqrt{3}}\right) + \left(\frac{1}{\sqrt{3}}\right)\left(\frac{2}{\sqrt{6}}\right)\left(\frac{1}{\sqrt{2}}\right)$$

$$- \left(\frac{1}{\sqrt{3}}\right)\left(-\frac{1}{\sqrt{6}}\right)\left(\frac{1}{\sqrt{2}}\right) - \left(-\frac{1}{\sqrt{3}}\right)\left(\frac{1}{\sqrt{2}}\right)\left(\frac{2}{\sqrt{6}}\right)$$

$$= 6\left(\frac{1}{\sqrt{36}}\right)$$

$$= 6\left(\frac{1}{6}\right)$$

$$= 1$$

$$\mathbf{u}_2 \times \mathbf{u}_3 = \begin{vmatrix} \mathbf{i} & \mathbf{j} & \mathbf{k} \\[2mm] \dfrac{1}{\sqrt{6}} & -\dfrac{1}{\sqrt{6}} & \dfrac{2}{\sqrt{6}} \\[2mm] \dfrac{1}{\sqrt{2}} & \dfrac{1}{\sqrt{2}} & 0 \end{vmatrix}$$

$$= -\left(\frac{2}{\sqrt{6}}\right)\left(\frac{1}{\sqrt{2}}\right)\mathbf{i} + \left(\frac{2}{\sqrt{6}}\right)\left(\frac{1}{\sqrt{2}}\right)\mathbf{j}$$

$$+ \left[\left(\frac{1}{\sqrt{6}}\right)\left(\frac{1}{\sqrt{2}}\right) - \left(-\frac{1}{\sqrt{6}}\right)\left(\frac{1}{\sqrt{2}}\right)\right]\mathbf{k}$$

$$= -\frac{1}{\sqrt{3}}\mathbf{i} + \frac{1}{\sqrt{3}}\mathbf{j} + \frac{1}{\sqrt{3}}\mathbf{k}$$

$$= \left[-\frac{1}{\sqrt{3}}, \frac{1}{\sqrt{3}}, \frac{1}{\sqrt{3}}\right]$$

$$\mathbf{u}_3 \times \mathbf{u}_1 = \begin{vmatrix} \mathbf{i} & \mathbf{j} & \mathbf{k} \\[2mm] \dfrac{1}{\sqrt{2}} & \dfrac{1}{\sqrt{2}} & 0 \\[2mm] -\dfrac{1}{\sqrt{3}} & \dfrac{1}{\sqrt{3}} & \dfrac{1}{\sqrt{3}} \end{vmatrix}$$

$$= \frac{1}{\sqrt{6}}\mathbf{i} - \frac{1}{\sqrt{6}}\mathbf{j} + \frac{2}{\sqrt{6}}\mathbf{k}$$

$$= \left[\frac{1}{\sqrt{6}}, -\frac{1}{\sqrt{6}}, \frac{2}{\sqrt{6}}\right]$$

$$\mathbf{u}_1 \times \mathbf{u}_2 = \begin{vmatrix} \mathbf{i} & \mathbf{j} & \mathbf{k} \\ -\dfrac{1}{\sqrt{3}} & \dfrac{1}{\sqrt{3}} & \dfrac{1}{\sqrt{3}} \\ \dfrac{1}{\sqrt{6}} & -\dfrac{1}{\sqrt{6}} & \dfrac{2}{\sqrt{6}} \end{vmatrix}$$

$$= \frac{3}{\sqrt{18}}\mathbf{i} - \frac{3}{\sqrt{18}}\mathbf{j} + 0\mathbf{k}$$

$$= \frac{1}{\sqrt{2}}\mathbf{i} - \frac{1}{\sqrt{2}}\mathbf{j} + 0\mathbf{k}$$

$$= \left[\frac{1}{\sqrt{2}}, -\frac{1}{\sqrt{2}}, 0\right]$$

the reciprocal basis is

$$\mathbf{u}_1' = \frac{\mathbf{u}_2 \times \mathbf{u}_3}{[\mathbf{u}_1\mathbf{u}_2\mathbf{u}_3]} = \left[-\frac{1}{\sqrt{3}}, \frac{1}{\sqrt{3}}, \frac{1}{\sqrt{3}}\right] = \mathbf{u}_1$$

$$\mathbf{u}_2' = \frac{\mathbf{u}_3 \times \mathbf{u}_1}{[\mathbf{u}_1\mathbf{u}_2\mathbf{u}_3]} = \left[\frac{1}{\sqrt{6}}, -\frac{1}{\sqrt{6}}, \frac{2}{\sqrt{6}}\right] = \mathbf{u}_2$$

$$\mathbf{u}_3' = \frac{\mathbf{u}_1 \times \mathbf{u}_2}{[\mathbf{u}_1\mathbf{u}_2\mathbf{u}_3]} = \left[\frac{1}{\sqrt{2}}, -\frac{1}{\sqrt{2}}, 0\right] = \mathbf{u}_3$$

Note: This example illustrates the fact that an orthonormal basis is automatically self-reciprocal.

PROBLEM 2-24 Using the Gram–Schmidt orthogonalization process, determine whether or not vectors $[1, 2, -1]$, $[2, 7, 1]$, and $[1, 8, 5]$ are linearly dependent. If so, express one of them as a linear combination of the other two.

Solution: Let $\mathbf{a}_1 = [1, 2, -1]$, $\mathbf{a}_2 = [2, 7, 1]$, and $\mathbf{a}_3 = [1, 8, 5]$. Then define

$$\mathbf{v}_1 = \mathbf{a}_1 = [1, 2, -1] \tag{a}$$

and let

$$\mathbf{v}_2 = \mathbf{a}_2 + m_1\mathbf{v}_1 \tag{b}$$

where m_1 is a scalar determined so that $\mathbf{v}_2$ is orthogonal to $\mathbf{v}_1$. In other words, $\mathbf{v}_2 \cdot \mathbf{v}_1 = 0$. Dotting both sides of eq. (b) with $\mathbf{v}_1$,

$$0 = (\mathbf{a}_2 \cdot \mathbf{v}_1) + m_1(\mathbf{v}_1 \cdot \mathbf{v}_1)$$

Hence

$$m_1 = -\frac{(\mathbf{a}_2 \cdot \mathbf{v}_1)}{(\mathbf{v}_1 \cdot \mathbf{v}_1)}$$

and

$$\mathbf{v}_2 = \mathbf{a}_2 - \frac{(\mathbf{a}_2 \cdot \mathbf{v}_1)}{(\mathbf{v}_1 \cdot \mathbf{v}_1)}\mathbf{v}_1$$

Since

$$\mathbf{a}_2 \cdot \mathbf{v}_1 = \mathbf{a}_2 \cdot \mathbf{a}_1 = (2)(1) + (7)(2) + (1)(-1) = 15$$

and

$$\mathbf{v}_1 \cdot \mathbf{v}_1 = \mathbf{a}_1 \cdot \mathbf{a}_1 = 1^2 + 2^2 + (-1)^2 = 6$$

the vector $\mathbf{v}_2$ is

$$\mathbf{v}_2 = \mathbf{a}_2 - \tfrac{15}{6}\mathbf{v}_1 = \mathbf{a}_2 - \tfrac{5}{2}\mathbf{v}_1$$

$$= [2, 7, 1] - \tfrac{5}{2}[1, 2, -1] = [-\tfrac{1}{2}, 2, \tfrac{7}{2}] = \tfrac{1}{2}[-1, 4, 7] \qquad \text{(c)}$$

Next let $\mathbf{v}_3 = \mathbf{a}_3 + n_1\mathbf{v}_1 + n_2\mathbf{v}_2$, where n_1 and n_2 are determined so that $\mathbf{v}_3$ is orthogonal to $\mathbf{v}_1$ and $\mathbf{v}_2$. Setting $\mathbf{v}_3 \cdot \mathbf{v}_1 = 0$ and $\mathbf{v}_3 \cdot \mathbf{v}_2 = 0$, we have

$$0 = (\mathbf{a}_3 \cdot \mathbf{v}_1) + n_1(\mathbf{v}_1 \cdot \mathbf{v}_1) + n_2(\mathbf{v}_2 \cdot \mathbf{v}_1), \quad 0 = (\mathbf{a}_3 \cdot \mathbf{v}_2) + n_1(\mathbf{v}_1 \cdot \mathbf{v}_2) + n_2(\mathbf{v}_2 \cdot \mathbf{v}_2)$$

Since $(\mathbf{v}_1 \cdot \mathbf{v}_2) = (\mathbf{v}_2 \cdot \mathbf{v}_1) = 0$

$$n_1 = -\frac{(\mathbf{a}_3 \cdot \mathbf{v}_1)}{(\mathbf{v}_1 \cdot \mathbf{v}_1)}, \qquad n_2 = -\frac{(\mathbf{a}_3 \cdot \mathbf{v}_2)}{(\mathbf{v}_2 \cdot \mathbf{v}_2)}$$

Hence

$$\mathbf{v}_3 = \mathbf{a}_3 - \frac{(\mathbf{a}_3 \cdot \mathbf{v}_1)}{(\mathbf{v}_1 \cdot \mathbf{v}_1)}\mathbf{v}_1 - \frac{(\mathbf{a}_3 \cdot \mathbf{v}_2)}{(\mathbf{v}_2 \cdot \mathbf{v}_2)}\mathbf{v}_2$$

Because

$$\mathbf{a}_3 \cdot \mathbf{v}_1 = \mathbf{a}_3 \cdot \mathbf{a}_1 = (1)(1) + (8)(2) + (5)(-1) = 12$$
$$\mathbf{a}_3 \cdot \mathbf{v}_2 = (1)(-\tfrac{1}{2}) + (8)(2) + (5)(\tfrac{7}{2}) = 33$$
$$\mathbf{v}_2 \cdot \mathbf{v}_2 = (-\tfrac{1}{2})^2 + (2)^2 + (\tfrac{7}{2})^2 = \tfrac{33}{2}$$

the vector $\mathbf{v}_3$ is

$$\mathbf{v}_3 = \mathbf{a}_3 - \tfrac{12}{6}\mathbf{v}_1 - 2\mathbf{v}_2 = \mathbf{a}_3 - 2\mathbf{v}_1 - 2\mathbf{v}_2 \qquad \text{(d)}$$

That is,

$$\mathbf{v}_3 = [1, 8, 5] - 2[1, 2, -1] - 2[-\tfrac{1}{2}, 2, \tfrac{7}{2}]$$
$$= [0, 0, 0]$$
$$= \mathbf{0} \qquad \text{(e)}$$

Thus, the vectors $\mathbf{a}_1$, $\mathbf{a}_2$, and $\mathbf{a}_3$ are linearly dependent.

By (combining eqs.) (a), (c), (d), and (e), we can express $\mathbf{a}_3$ as a linear combination of $\mathbf{a}_1$ and $\mathbf{a}_2$:

$$\mathbf{a}_3 = 2\mathbf{v}_1 + 2\mathbf{v}_2 = 2\mathbf{a}_1 + 2(\mathbf{a}_2 - \tfrac{5}{2}\mathbf{a}_1) = -3\mathbf{a}_1 + 2\mathbf{a}_2$$

Supplementary Exercises

PROBLEM 2-25 Let $\mathbf{A} = [2, -3, 6]$ and $\mathbf{B} = [1, 8, -4]$. Find (a) $|\mathbf{A}|$ and $|\mathbf{B}|$, (b) $|\mathbf{A} - \mathbf{B}|$, (c) $\mathbf{A} \cdot \mathbf{B}$, (d) the direction cosine of $\mathbf{A}$, (e) the angle between $\mathbf{A}$ and $\mathbf{B}$, and (f) $\mathbf{A} \times \mathbf{B}$.

Answer: (a) 7, 9   (b) $\sqrt{222}$ (c) -46 (d) 2/7, $-3/7$, 6/7   (e) 136° 54′
(f) $[-36, 14, 19]$

PROBLEM 2-26 Let $\mathbf{A} = [1, 2, 1]$, $\mathbf{B} = [2, 0, -1]$, and $\mathbf{C} = [0, -1, 2]$. Evaluate (a) $[\mathbf{ABC}]$, (b) $\mathbf{A} \times (\mathbf{B} \times \mathbf{C})$, (c) $(\mathbf{A} \times \mathbf{B}) \times \mathbf{C}$, (d) $(\mathbf{A} \times \mathbf{B}) \times (\mathbf{B} \times \mathbf{C})$, and (e) $(\mathbf{B} \cdot \mathbf{C})(\mathbf{A} \times \mathbf{B})$.

Answer: (a) -11 (b) $[0, 1, -2]$ (c) $[2, 4, 2]$ (d) $[-22, 0, 11]$ (e) $[4, -6, 8]$

PROBLEM 2-27 Find a unit vector parallel to the sum of the vectors $\mathbf{A} = [2, 4, -5]$ and $\mathbf{B} = [1, 2, 3]$.

Answer: $\left[\dfrac{3}{7}, \dfrac{6}{7}, -\dfrac{2}{7}\right]$ or $\left[-\dfrac{3}{7}, -\dfrac{6}{7}, \dfrac{2}{7}\right]$

PROBLEM 2-28 Find the value of m such that $\mathbf{A} = [m, -2, 1]$ and $\mathbf{B} = [2m, m, -4]$ are perpendicular.

Answer: 2 or -1

PROBLEM 2-29 Find a unit vector that makes an angle of 45° with the vector $\mathbf{A} = [2, 2, -1]$ and an angle of 60° with $\mathbf{B} = [0, 1, -1]$.

Answer: $\left[\dfrac{1}{\sqrt{2}}, 0, -\dfrac{1}{\sqrt{2}}\right]$ or $\left[\dfrac{1}{3\sqrt{2}}, \dfrac{4}{3\sqrt{2}}, \dfrac{1}{3\sqrt{2}}\right]$

PROBLEM 2-30 Find a unit vector parallel to the xy-plane and perpendicular to the vector $[4, -3, -1]$.

Answer: $\left[\dfrac{3}{5}, \dfrac{4}{5}, 0\right]$ or $\left[-\dfrac{3}{5}, -\dfrac{4}{5}, 0\right]$

PROBLEM 2-31 Find the area of the triangle with vertices $(3, 5, 2)$, $(1, -1, 6)$, and $(-2, 1, 4)$ by evaluating the magnitude of a vector product.

Answer: $3\sqrt{21}$

PROBLEM 2-32 If $\mathbf{A} = [1, 2, 1]$, $\mathbf{B} = [2, 0, -1]$, and $\mathbf{C} = [0, -1, 2]$, verify

$$\mathbf{A} \times (\mathbf{B} \times \mathbf{C}) = (\mathbf{A} \cdot \mathbf{C})\mathbf{B} - (\mathbf{A} \cdot \mathbf{B})\mathbf{C}, \text{ and } (\mathbf{A} \times \mathbf{B}) \times \mathbf{C} = (\mathbf{A} \cdot \mathbf{C})\mathbf{B} - (\mathbf{B} \cdot \mathbf{C})\mathbf{A}$$

PROBLEM 2-33 Find a unit vector perpendicular to both vectors $[2, 1, -1]$ and $[3, 4, -1]$.

Answer: $\left[\dfrac{3}{\sqrt{35}}, -\dfrac{1}{\sqrt{35}}, \dfrac{5}{\sqrt{35}}\right]$ or $\left[-\dfrac{3}{\sqrt{35}}, \dfrac{1}{\sqrt{35}}, -\dfrac{5}{\sqrt{35}}\right]$

PROBLEM 2-34 Find the value of m that makes vectors $\mathbf{A} = [1, 1, -1]$, $\mathbf{B} = [2, -1, 1]$, and $\mathbf{C} = [m, -1, m]$ coplanar.

Answer: $m = 1$

PROBLEM 2-35 Determine whether the following vectors are linearly dependent or independent: (a) $[1, 1, 0]$, $[0, 1, 1]$, $[1, 0, 1]$; (b) $[1, -6, 2]$, $[0, 2, 7]$, $[-2, 12, -4]$.

Answer: (a) independent   (b) dependent

PROBLEM 2-36 Using components, verify the following identities:

(a) $(\mathbf{A} \cdot \mathbf{B})^2 + (\mathbf{A} \times \mathbf{B})^2 = (AB)^2$

(b) $(\mathbf{A} \times \mathbf{B}) \cdot (\mathbf{C} \times \mathbf{D}) = (\mathbf{A} \cdot \mathbf{C})(\mathbf{B} \cdot \mathbf{D}) - (\mathbf{A} \cdot \mathbf{D})(\mathbf{B} \cdot \mathbf{C})$

(c) $(\mathbf{A} \times \mathbf{B}) \times (\mathbf{C} \times \mathbf{D}) = [\mathbf{ABD}]\mathbf{C} - [\mathbf{ABC}]\mathbf{D}$

(d) $\mathbf{A} \times (\mathbf{B} \times \mathbf{C}) + \mathbf{B} \times (\mathbf{C} \times \mathbf{A}) + \mathbf{C} \times (\mathbf{A} \times \mathbf{B}) = \mathbf{0}$

PROBLEM 2-37 Prove the identity

$$(a^2 + b^2 + c^2)(a'^2 + b'^2 + c'^2) = (aa' + bb' + cc')^2 + (ab' - a'b)^2 + (bc' - b'c)^2 + (ca' - c'a)^2$$

[*Hint*: Let $\mathbf{A} = [a, b, c]$, and $\mathbf{B} = [a', b', c']$ and use the result of Problem 2-36(a).]

PROBLEM 2-38 Let $\mathbf{u}_1$ and $\mathbf{u}_2$ be unit vectors in the xy-plane making angles α and β, respectively, with the positive x-axis. Prove the following:

(a) $\mathbf{u}_1 = \cos \alpha \mathbf{i} + \sin \alpha \mathbf{j}$, $\mathbf{u}_2 = \cos \beta \mathbf{i} + \sin \beta \mathbf{j}$

(b) $$\cos(\alpha - \beta) = \cos\alpha\cos\beta + \sin\alpha\sin\beta$$

(c) $$\sin(\alpha - \beta) = \sin\alpha\cos\beta - \cos\alpha\sin\beta$$

[*Hint*: Consider $\mathbf{u}_1 \cdot \mathbf{u}_2$ and $\mathbf{u}_1 \times \mathbf{u}_2$.]

PROBLEM 2-39 Let $\mathbf{u}_1$ and $\mathbf{u}_2$ be unit vectors in the xy-plane making angles α and $-\beta$, respectively, with the positive x-axis. Verify the following identities:

(a) $$\mathbf{u}_1 = \cos\alpha\mathbf{i} + \sin\alpha\mathbf{j}, \qquad \mathbf{u}_2 = \cos\beta\mathbf{i} - \sin\beta\mathbf{j}$$

(b) $$\cos(\alpha + \beta) = \cos\alpha\cos\beta - \sin\alpha\sin\beta$$

(c) $$\sin(\alpha + \beta) = \sin\alpha\cos\beta + \cos\alpha\sin\beta$$

PROBLEM 2-40 If $\mathbf{a}_1 = [1, 1, 0]$, $\mathbf{a}_2 = [0, 1, 0]$, and $\mathbf{a}_3 = [1, 1, 1]$, find their reciprocal set of vectors.

Answer: $\mathbf{b}_1 = [1, 0, -1]$, $\mathbf{b}_2 = [-1, 1, 0]$, and $\mathbf{b}_3 = [0, 0, 1]$.

PROBLEM 2-41 Using the vectors from Problem 2-40, verify that

$$[\mathbf{a}_1\mathbf{a}_2\mathbf{a}_3] = \frac{1}{[\mathbf{b}_1\mathbf{b}_2\mathbf{b}_3]}$$

PROBLEM 2-42 Express $\mathbf{d} = [5, 3, -1]$ as a linear combination of the set of vectors $\mathbf{a}_1$, $\mathbf{a}_2$, and $\mathbf{a}_3$ from Problem 2-40.

Answer: $6\mathbf{a}_1 - 2\mathbf{a}_2 - \mathbf{a}_3$

PROBLEM 2-43 Express $\mathbf{d} = [5, 3, -1]$ as a linear combination of the set of vectors $\mathbf{b}_1$, $\mathbf{b}_2$, and $\mathbf{b}_3$ obtained in Problem 2-40.

Answer: $8\mathbf{b}_1 + 3\mathbf{b}_2 + 7\mathbf{b}_3$

PROBLEM 2-44 Construct an orthonormal set from $\mathbf{A} = [1, 3, 0]$ and $\mathbf{B} = [-1, 1, 0]$.

Answer: $\mathbf{u}_1 = \left[\dfrac{1}{\sqrt{10}}, \dfrac{3}{\sqrt{10}}, 0\right]$, $\quad \mathbf{u}_2 = \left[\dfrac{-3}{\sqrt{10}}, \dfrac{1}{\sqrt{10}}, 0\right]$, $\quad \mathbf{u}_3 = [0, 0, 1]$

PROBLEM 2-45 Express the vector $[2, 2, 2]$ as a linear combination of the orthonormal set obtained in Problem 2-44.

Answer: $\dfrac{4\sqrt{10}}{5}\mathbf{u}_1 - \dfrac{2\sqrt{10}}{5}\mathbf{u}_2 + 2\mathbf{u}_3$

PROBLEM 2-46 Let $\mathbf{u}_1$, $\mathbf{u}_2$, and $\mathbf{u}_3$ form an orthonormal basis. If

$$\mathbf{A} = a_1\mathbf{u}_1 + a_2\mathbf{u}_2 + a_3\mathbf{u}_3 \qquad \text{and} \qquad \mathbf{B} = b_1\mathbf{u}_1 + b_2\mathbf{u}_2 + b_3\mathbf{u}_3$$

show that

$$\mathbf{A} \cdot \mathbf{B} = a_1 b_1 + a_2 b_2 + a_3 b_3$$

PROBLEM 2-47 Let $\mathbf{u}_1$, $\mathbf{u}_2$, and $\mathbf{u}_3$ form an orthonormal basis that constitutes a right-hand orthonormal system. If $\mathbf{A} = a_1\mathbf{u}_1 + a_2\mathbf{u}_2 + a_3\mathbf{u}_3$ and $\mathbf{B} = b_1\mathbf{u}_1 + b_2\mathbf{u}_2 + b_3\mathbf{u}_3$, show that

$$\mathbf{A} \times \mathbf{B} = (a_2 b_3 - a_3 b_2)\mathbf{u}_1 + (a_3 b_1 - a_1 b_3)\mathbf{u}_2 + (a_1 b_2 - a_2 b_1)\mathbf{u}_3$$

3 APPLICATIONS TO GEOMETRY

THIS CHAPTER IS ABOUT

☑ **Application of Vector Methods to Elementary Geometry**
☑ **Vector Equation of a Straight Line**
☑ **Vector Equation of a Plane**
☑ **Distance Formulas**

3-1. Application of Vector Methods to Elementary Geometry

Many elementary geometry theorems can be proved easily by using vector methods.

- **Method 1:** Representing the sides of geometric figures by vectors

EXAMPLE 3-1: Prove that the line segment that connects the midpoints of two sides of a triangle is parallel to and one-half the length of the third side.

Solution: In the triangle ABC shown in Figure 3-1, points D and E are the respective midpoints of the sides BA and CA. If $\overrightarrow{BC} = \mathbf{a}$ and $\overrightarrow{BA} = \mathbf{b}$, we can use the definitions of vector addition and the negative of a vector to write

$$\overrightarrow{CA} - \overrightarrow{CB} + \overrightarrow{BA} - -\mathbf{a} + \mathbf{b} - \mathbf{b} - \mathbf{a}$$

Now $\overrightarrow{BD} = \frac{1}{2}\mathbf{b}$ and $\overrightarrow{CE} = \frac{1}{2}\overrightarrow{CA} = \frac{1}{2}(\mathbf{b} - \mathbf{a})$. So the line segment that joins D and E, denoted by $\overrightarrow{DE}$, can also be represented by a path from D to B to C to E, giving

$$\begin{aligned}
\overrightarrow{DE} &= -\overrightarrow{BD} + \overrightarrow{BC} + \overrightarrow{CE} \\
&= \overrightarrow{BC} + \overrightarrow{CE} - \overrightarrow{BD} \\
&= \mathbf{a} + \tfrac{1}{2}(\mathbf{b} - \mathbf{a}) - \tfrac{1}{2}\mathbf{b} \\
&= \tfrac{1}{2}\mathbf{a} \\
&= \tfrac{1}{2}\overrightarrow{BC}
\end{aligned}$$

which shows that $\overrightarrow{DE}$ is parallel to $\overrightarrow{BC}$. and has one-half its magnitude.

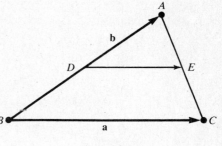

Figure 3-1

- **Method 2:** Forming vectors by joining reference points to the vertices of geometric figures

EXAMPLE 3-2: Prove that the altitudes of a triangle are concurrent, that is, pass through a common point.

Solution: Let the altitudes AD and BE of the triangle ABC meet at O, as shown in Figure 3-2. If $\overrightarrow{OA} = \mathbf{a}$, $\overrightarrow{OB} = \mathbf{b}$, and $\overrightarrow{OC} = \mathbf{c}$, then

$$\overrightarrow{BC} = \mathbf{c} - \mathbf{b}, \qquad \overrightarrow{AC} = \mathbf{c} - \mathbf{a}, \qquad \overrightarrow{BA} = \mathbf{a} - \mathbf{b}$$

Since $\overrightarrow{OA} \perp \overrightarrow{BC}$, we can say that $\overrightarrow{OA} \cdot \overrightarrow{BC} = 0$; that is,

$$\mathbf{a} \cdot (\mathbf{c} - \mathbf{b}) = 0 \quad \text{or} \quad \mathbf{a} \cdot \mathbf{c} = \mathbf{a} \cdot \mathbf{b}$$

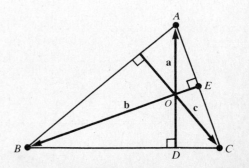

Figure 3-2

Similarly, because $\overrightarrow{OB} \perp \overrightarrow{AC}$, we know that $\overrightarrow{OB} \cdot \overrightarrow{AC} = 0$; that is,

$$\mathbf{b} \cdot (\mathbf{c} - \mathbf{a}) = 0 \quad \text{or} \quad \mathbf{b} \cdot \mathbf{c} = \mathbf{b} \cdot \mathbf{a}$$

Since the dot product is commutative, $\mathbf{b} \cdot \mathbf{a} = \mathbf{a} \cdot \mathbf{b}$. Hence

$$\mathbf{a} \cdot \mathbf{c} = \mathbf{b} \cdot \mathbf{c} \quad \text{or} \quad (\mathbf{a} - \mathbf{b}) \cdot \mathbf{c} = 0$$

which implies that $\overrightarrow{OC} \perp \overrightarrow{BA}$. So the altitude from C to AB passes through O. Thus all the altitudes are concurrent.

3-2. Vector Equation of a Straight Line

The vector equation of the straight line that passes through a point whose position vector is $\mathbf{A}$ and that is parallel to the vector $\mathbf{B}$ (see Figure 3-3) is given by

VECTOR EQUATION OF A LINE
$$\mathbf{r} = \mathbf{A} + t\mathbf{B} \tag{3.1}$$

where $\mathbf{r}$ is the position vector of any point on the line and t is a scalar parameter.

In terms of the components of the vectors, the vector equation of a line (3.1) can be written as

$$x = A_1 + tB_1, \qquad y = A_2 + tB_2, \qquad z = A_3 + tB_3 \tag{3.2}$$

which are the **parametric equations** for the straight line that passes through (A_1, A_2, A_3) and is parallel to $\mathbf{B} = [B_1, B_2, B_3]$.

Eliminating the common parameter t from the parametric equations (3.2), we obtain

$$\frac{x - A_1}{B_1} = \frac{y - A_2}{B_2} = \frac{z - A_3}{B_3} \tag{3.3}$$

which is the **nonparametric equation** of the line.

EXAMPLE 3-3: Use vector properties to verify the vector equation of a line (3.1).

Solution: From Figure 3-3, we see that the point (x, y, z) represented by the position vector $\mathbf{r}$ will lie on the desired straight line if and only if $\mathbf{r} - \mathbf{A}$ is parallel to $\mathbf{B}$. Now from the condition for parallelism (1.13), two vectors are parallel if and only if they are scalar multiples of each other. This implies that

$$\mathbf{r} - \mathbf{A} = t\mathbf{B} \tag{3.4}$$

where t is some scalar. Rewriting (3.4), we obtain the vector equation of a line (3.1):

$$\mathbf{r} = \mathbf{A} + t\mathbf{B}$$

EXAMPLE 3-4: Find the vector equation that represents the straight line L that passes through the points $Q = (A_1, A_2, A_3)$ and $R = (B_1, B_2, B_3)$.

Solution: Represent the points Q and R by the position vectors $\mathbf{A} = [A_1, A_2, A_3]$ and $\mathbf{B} = [B_1, B_2, B_3]$, respectively. Let a point $P = (x, y, z)$ on L be represented by the position vector $\mathbf{r} = [x, y, z]$. Then we see from Figure 3-4 that there must be a scalar t such that

$$\overrightarrow{QP} = t\overrightarrow{QR}$$

Now $\overrightarrow{QP} = \mathbf{r} - \mathbf{A}$ and $\overrightarrow{QR} = \mathbf{B} - \mathbf{A}$. So

$$\mathbf{r} - \mathbf{A} = t(\mathbf{B} - \mathbf{A}) \tag{3.5}$$

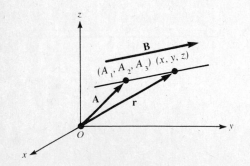

Figure 3-3

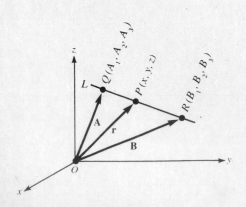

Figure 3-4

or, upon simplification,

$$\mathbf{r} = \mathbf{A} + t(\mathbf{B} - \mathbf{A}) = (1 - t)\mathbf{A} + t\mathbf{B} \qquad (3.6)$$

which gives us the vector equation of the line through Q and R.

3-3. Vector Equation of a Plane

The vector equation of a plane that is perpendicular to a nonzero vector $\mathbf{B}$ and that passes through a point whose position vector is $\mathbf{A}$ (see Figure 3-5) is given by

VECTOR EQUATION OF A PLANE

$$(\mathbf{r} - \mathbf{A}) \cdot \mathbf{B} = 0 \qquad (3.7)$$

EXAMPLE 3-5: Use geometric properties to verify the vector equation of a plane (3.7).

Proof: Let $\mathbf{r}$ be the position vector of any point in space, as in Figure 3-5. Points will lie in the plane if and only if $\mathbf{r} - \mathbf{A}$ is perpendicular to $\mathbf{B}$; that is,

$$(\mathbf{r} - \mathbf{A}) \cdot \mathbf{B} = 0$$

(See Example 1-3.)

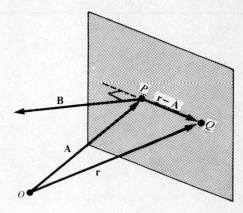

Figure 3-5

3-4. Distance Formulas

A. Distance from a point to a line

Let the arbitrary points P and Q be specified by the position vectors $\mathbf{r}_0$ and $\mathbf{A}$, relative to the point O (see Figure 3-6). The distance d from P to a line that is parallel to $\mathbf{B}$ and passes through Q is given by

DISTANCE FROM A POINT TO A LINE

$$d = \frac{|(\mathbf{r}_0 - \mathbf{A}) \times \mathbf{B}|}{B} \qquad (3.8)$$

where $B = |\mathbf{B}|$.

EXAMPLE 3-6: Verify the distance formula (3.8) using geometric properties.

Solution: In Figure 3-6, if $\overrightarrow{OQ} = \mathbf{A}$, then $\overrightarrow{QP} = \mathbf{r}_0 - \mathbf{A}$. Now if θ is the angle between $\overrightarrow{QP}$ and $\overrightarrow{QR}$,

$$d = |\overrightarrow{PR}| = |\overrightarrow{QP}| \sin \theta$$

But from the definition of the magnitude of the cross product of two vectors (1.44),

$$|(\mathbf{r}_0 - \mathbf{A}) \times \mathbf{B}| = |\mathbf{B}||\overrightarrow{QP}| \sin \theta = B|\overrightarrow{QP}| \sin \theta = Bd$$

Hence

$$d = \frac{|(\mathbf{r}_0 - \mathbf{A}) \times \mathbf{B}|}{B}$$

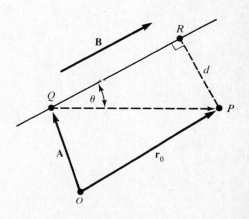

Figure 3-6

B. Distance between two lines

EXAMPLE 3-7: Find the shortest distance d between the two straight lines that pass through the two distinct points P and Q, whose position vectors are $\mathbf{A}$ and $\mathbf{C}$ with the directions $\mathbf{B}$ and $\mathbf{D}$, respectively.

Solution: Assume that $\mathbf{B}$ and $\mathbf{D}$ are not collinear, that is, $\mathbf{B} \times \mathbf{D} \neq \mathbf{0}$, and that there exist points M and N on the two lines such that $\overrightarrow{MN}$ is perpendicular to both lines (see Figure 3-7). Then the shortest distance d between these two

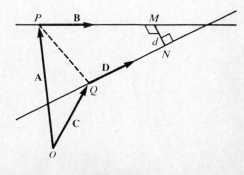

Figure 3-7

straight lines is $|\overrightarrow{MN}|$. Since $\overrightarrow{MN}$ is perpendicular to both vectors **B** and **D**, the vector

$$\frac{\mathbf{B} \times \mathbf{D}}{|\mathbf{B} \times \mathbf{D}|}$$

is a unit vector in the direction of $\overrightarrow{MN}$.

But $\overrightarrow{MN}$ is the component of $\overrightarrow{PQ}$ in the direction of $\overrightarrow{MN}$. So $\overrightarrow{PQ} = \mathbf{C} - \mathbf{A}$, and hence

DISTANCE BETWEEN TWO LINES
$$d = |\overrightarrow{MN}| = \left| \overrightarrow{PQ} \cdot \frac{\mathbf{B} \times \mathbf{D}}{|\mathbf{B} \times \mathbf{D}|} \right| = \frac{|(\mathbf{C} - \mathbf{A}) \cdot \mathbf{B} \times \mathbf{D}|}{|\mathbf{B} \times \mathbf{D}|} \qquad (3.9)$$

Note: If $d = 0$, that is, if $|(\mathbf{C} - \mathbf{A}) \cdot \mathbf{B} \times \mathbf{D}| = |(\mathbf{A} - \mathbf{C}) \cdot \mathbf{B} \times \mathbf{D}| = 0$, the two straight lines intersect. Hence $(\mathbf{A} - \mathbf{C}) \cdot \mathbf{B} \times \mathbf{D} = 0$ is the condition that the two lines intersect. (See Problem 3-9.)

C. Distance from a point to a plane

In Figure 3-8 the arbitrary points P and Q are specified by the position vector $\mathbf{r}_0$ and **A** relative to the point O. The distance d from P to the plane that is perpendicular to the nonzero vector **B** and that passes through Q is given by

DISTANCE FROM A POINT TO A PLANE
$$d = \frac{|(\mathbf{r}_0 - \mathbf{A}) \cdot \mathbf{B}|}{B} \qquad (3.10)$$

EXAMPLE 3-8: Geometrically verify the distance formula (3.10).

Solution: Let **r** be the position vector of the base of the perpendicular from P to the plane, as shown in Figure 3-8. Then for some scalar λ,

$$\mathbf{r} = \mathbf{r}_0 + \lambda \mathbf{B} \qquad (3.11)$$

The desired distance d is then

$$d = |\mathbf{r} - \mathbf{r}_0| = |\lambda \mathbf{B}| = |\lambda| B \qquad (3.12)$$

where $B = |\mathbf{B}|$. Now from (3.11),

$$\mathbf{r} - \mathbf{A} = \mathbf{r}_0 - \mathbf{A} + \lambda \mathbf{B}$$

Taking the dot product of both sides with **B** and then using the definition of the vector equation of a plane (3.7), we get

$$(\mathbf{r} - \mathbf{A}) \cdot \mathbf{B} = (\mathbf{r}_0 - \mathbf{A}) \cdot \mathbf{B} + \lambda \mathbf{B} \cdot \mathbf{B} = (\mathbf{r}_0 - \mathbf{A}) \cdot \mathbf{B} + \lambda B^2 = 0$$

Hence

$$\lambda = -\frac{(\mathbf{r}_0 - \mathbf{A}) \cdot \mathbf{B}}{B^2}$$

Substituting into (3.12), the distance is

$$d = |\lambda| B = \frac{|(\mathbf{r}_0 - \mathbf{A}) \cdot \mathbf{B}|}{B}$$

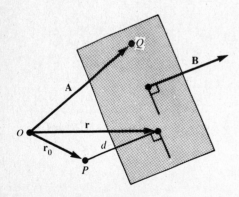

Figure 3-8

SUMMARY

1. Many elementary geometry theorems can be proved using two basic vector methods: by representing the sides of geometric figures as vectors, or by joining a reference point to the vertices of geometric figures with vectors.

2. The vector equation of the straight line that passes through a point (A_1, A_2, A_3) and is parallel to **B** is given by

$$\mathbf{r} = \mathbf{A} + t\mathbf{B}$$

where **r** is the position vector of any point on the line and t is a scalar parameter and $\mathbf{A} = [A_1, A_2, A_3]$.

3. The vector equation of a plane that is perpendicular to a vector **B** and passes through (A_1, A_2, A_3) is given by

$$(\mathbf{r} - \mathbf{A}) \cdot \mathbf{B} = 0$$

RAISE YOUR GRADES

Can you explain...?

☑ how to use vector methods to prove geometric theorems
☑ how to find the shortest distance from a point to a line by vector methods
☑ how to find the shortest distance from a point to a plane by vector methods

SOLVED PROBLEMS

Application of Vector Methods to Elementary Geometry

PROBLEM 3-1 Prove that the medians of a triangle meet at a point that divides each median into a 2:1 ratio.

Solution: In the triangle ABC shown in Figure 3-9, we see that $\overrightarrow{BC} = \mathbf{a}$ and $\overrightarrow{BA} = \mathbf{b}$. Then from the definitions of vector addition and of the negative of a vector, you know that $\overrightarrow{CA} = \mathbf{b} - \mathbf{a}$. Now let the medians AF and BE intersect at G. Then

$$\overrightarrow{BE} = \overrightarrow{BC} + \overrightarrow{CE} = \mathbf{a} + \tfrac{1}{2}(\mathbf{b} - \mathbf{a}) = \tfrac{1}{2}(\mathbf{a} + \mathbf{b})$$

Because G lies on the median BE, there exists a scalar m such that

$$\overrightarrow{BG} = m\overrightarrow{BE} = \frac{m}{2}(\mathbf{a} + \mathbf{b})$$

Because

$$\overrightarrow{FA} = \overrightarrow{FB} + \overrightarrow{BA} = -\tfrac{1}{2}\mathbf{a} + \mathbf{b} = \mathbf{b} - \tfrac{1}{2}\mathbf{a}$$

and G lies on the median FA, there exists a scalar n such that

$$\overrightarrow{FG} = n\overrightarrow{FA} = n\left(\mathbf{b} - \frac{1}{2}\mathbf{a}\right)$$

But $\overrightarrow{BG} = \overrightarrow{BF} + \overrightarrow{FG}$; that is,

$$\frac{m}{2}(\mathbf{a} + \mathbf{b}) = \frac{1}{2}\mathbf{a} + n\left(\mathbf{b} - \frac{1}{2}\mathbf{a}\right)$$

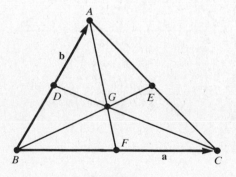

Figure 3-9

or, upon simplification and reorganization,

$$\frac{1}{2}(m + n - 1)\mathbf{a} + \left(\frac{m}{2} - n\right)\mathbf{b} = \mathbf{0}$$

Since **a** and **b** are linearly independent vectors,

$$\frac{1}{2}(m + n - 1) = 0, \qquad \frac{m}{2} - n = 0$$

Solving for m and n, we obtain $m = \frac{2}{3}$ and $n = \frac{1}{3}$. Thus $\overrightarrow{BG} = \frac{2}{3}\overrightarrow{BE}$ and $\overrightarrow{FG} = \frac{1}{3}\overrightarrow{FA}$, from which it follows that G is a point that divides AF and BE into a 2:1 ratio.

Similarly, we can show that the point of intersection G of the medians AF and CD divides them into a 2:1 ratio.

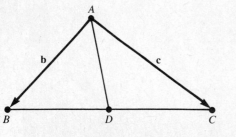

Figure 3-10

PROBLEM 3-2   Prove that the diagonals of a parallelogram bisect each other.

Solution: In the parallelogram $ABCD$ shown in Figure 3-10, let $\overrightarrow{BC} = \mathbf{a}$ and $\overrightarrow{BA} = \mathbf{b}$. Then $\overrightarrow{AD} = \overrightarrow{BC} = \mathbf{a}$ and $\overrightarrow{CD} = \overrightarrow{BA} = \mathbf{b}$. Hence

$$\overrightarrow{BD} = \overrightarrow{BC} + \overrightarrow{CD} = \mathbf{a} + \mathbf{b}, \qquad \overrightarrow{CA} = \overrightarrow{CB} + \overrightarrow{BA} = -\mathbf{a} + \mathbf{b} = \mathbf{b} - \mathbf{a}$$

If E is the midpoint of BD, then $\overrightarrow{BE} = \frac{1}{2}\overrightarrow{BD} = \frac{1}{2}(\mathbf{a} + \mathbf{b})$. Since $\overrightarrow{CA} = \mathbf{b} - \mathbf{a}$,

$$\overrightarrow{CE} = \overrightarrow{CB} + \overrightarrow{BE} = -\mathbf{a} + \frac{1}{2}(\mathbf{a} + \mathbf{b}) = \frac{1}{2}(\mathbf{b} - \mathbf{a}) = \frac{1}{2}\overrightarrow{CA}$$

So, E is the midpoint of both BD and CA, and the diagonals of the parallelogram bisect each other.

PROBLEM 3-3   Using vector methods, prove **Appolonius' theorem**: If in triangle ABC the segment AD is the median of the side BC, then

**APPOLONIUS'
THEOREM**
$$\overline{AB}^2 + \overline{AC}^2 = 2\overline{AD}^2 + \frac{1}{2}\overline{BC}^2$$

Solution: In the triangle ABC shown in Figure 3-11, let $\overrightarrow{AB} = \mathbf{b}$ and $\overrightarrow{AC} = \mathbf{c}$. Then $\overrightarrow{AD} = \frac{1}{2}(\mathbf{b} + \mathbf{c})$ and $\overrightarrow{BC} = \mathbf{c} - \mathbf{b}$. Hence

$$\overline{AD}^2 = \overrightarrow{AD} \cdot \overrightarrow{AD} = \frac{1}{4}(\mathbf{b} + \mathbf{c}) \cdot (\mathbf{b} + \mathbf{c}) = \frac{1}{4}(b^2 + c^2 + 2\mathbf{b} \cdot \mathbf{c})$$
$$\overline{BC}^2 = \overrightarrow{BC} \cdot \overrightarrow{BC} = (\mathbf{c} - \mathbf{b}) \cdot (\mathbf{c} - \mathbf{b}) = c^2 + b^2 - 2\mathbf{b} \cdot \mathbf{c}$$

Thus

$$2\overline{AD}^2 + \frac{1}{2}\overline{BC}^2 = \frac{1}{2}(b^2 + c^2 + 2\mathbf{b} \cdot \mathbf{c}) + \frac{1}{2}(c^2 + b^2 - 2\mathbf{b} \cdot \mathbf{c})$$
$$= b^2 + c^2$$
$$= \overline{AB}^2 + \overline{AC}^2$$

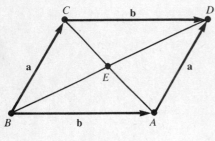

Figure 3-11

PROBLEM 3-4   In the tetrahedron $OABC$ shown in Figure 3-12 the edge OA is perpendicular to the edge BC and the edge OB is perpendicular to CA. Show that the edge OC is perpendicular to the edge AB.

Solution: In the tetrahedron $OABC$, if $\overrightarrow{OA} = \mathbf{a}$, $\overrightarrow{OB} = \mathbf{b}$, and $\overrightarrow{OC} = \mathbf{c}$, then $\overrightarrow{AB} = \mathbf{b} - \mathbf{a}$, $\overrightarrow{BC} = \mathbf{c} - \mathbf{b}$, and $\overrightarrow{CA} = \mathbf{a} - \mathbf{c}$. Since $\overrightarrow{OA} \perp \overrightarrow{BC}$, we have

$$\mathbf{a} \cdot (\mathbf{c} - \mathbf{b}) = 0 \quad \text{or} \quad \mathbf{a} \cdot \mathbf{c} = \mathbf{a} \cdot \mathbf{b}$$

Similarly, $\overrightarrow{OB} \perp \overrightarrow{CA}$ implies

$$\mathbf{b} \cdot (\mathbf{a} - \mathbf{c}) = 0 \quad \text{or} \quad \mathbf{b} \cdot \mathbf{a} = \mathbf{b} \cdot \mathbf{c}$$

Because of the commutativity of the scalar product, we know that $\mathbf{a} \cdot \mathbf{b} = \mathbf{b} \cdot \mathbf{a}$, and so

$$\mathbf{a} \cdot \mathbf{c} = \mathbf{b} \cdot \mathbf{c} \quad \text{or} \quad \mathbf{c} \cdot (\mathbf{b} - \mathbf{a}) = 0$$

which implies that $\overrightarrow{OC} \perp \overrightarrow{AB}$. [Compare this result to Example 3-2.]

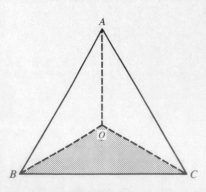

Figure 3-12

PROBLEM 3-5   Prove that an angle inscribed in a semicircle is a right angle.

Solution: Let $\angle ACB$ be any angle inscribed in a semicircle with center at O and radius $r = OA = OB = OC$. (See Figure 3-13.) If $\overrightarrow{OA} = \mathbf{a}$ and $\overrightarrow{OC} = \mathbf{c}$, then

$$\overrightarrow{OB} = -\mathbf{a}, \qquad \overrightarrow{BC} = \mathbf{a} + \mathbf{c}, \qquad \overrightarrow{AC} = -\mathbf{a} + \mathbf{c} = \mathbf{c} - \mathbf{a}$$

Since $a^2 = c^2 = r^2$,

$$\overrightarrow{AC} \cdot \overrightarrow{BC} = (\mathbf{c} - \mathbf{a}) \cdot (\mathbf{a} + \mathbf{c}) = \mathbf{c} \cdot \mathbf{c} - \mathbf{a} \cdot \mathbf{a} = c^2 - a^2 = 0$$

So $\overrightarrow{AC}$ and $\overrightarrow{BC}$ are perpendicular, which means that $\angle ACB$ is a right angle.

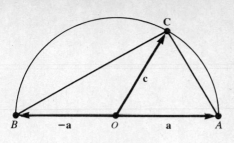

Figure 3-13

PROBLEM 3-6   Prove that the line joining the vertex of an isosceles triangle to the midpoint of its base is perpendicular to the base.

Solution: In the isosceles triangle ABC shown in Figure 3-14, let $\overrightarrow{OC} = \mathbf{a}$ and $\overrightarrow{OA} = \mathbf{c}$, where O is the midpoint of the base BC. Then,

$$\overrightarrow{OB} = -\mathbf{a}, \qquad \overrightarrow{AC} = \overrightarrow{AO} + \overrightarrow{OC} = \mathbf{a} - \mathbf{c}, \qquad \overrightarrow{BA} = \overrightarrow{BO} + \overrightarrow{OA} = \mathbf{a} + \mathbf{c}$$

Now

$$\overrightarrow{AC}^2 = \overrightarrow{AC} \cdot \overrightarrow{AC} = (\mathbf{a} - \mathbf{c}) \cdot (\mathbf{a} - \mathbf{c}) = a^2 + c^2 - 2\mathbf{a} \cdot \mathbf{c}$$
$$\overrightarrow{AB}^2 = \overrightarrow{AB} \cdot \overrightarrow{AB} = (\mathbf{a} + \mathbf{c}) \cdot (\mathbf{a} + \mathbf{c}) = a^2 + c^2 + 2\mathbf{a} \cdot \mathbf{c}$$

Since $\overrightarrow{AC} = \overrightarrow{AB}$, we have $\overrightarrow{AC}^2 = \overrightarrow{AB}^2$; that is,

$$a^2 + c^2 - 2\mathbf{a} \cdot \mathbf{c} = a^2 + c^2 + 2\mathbf{a} \cdot \mathbf{c}$$

Thus, on simplification,

$$4\mathbf{a} \cdot \mathbf{c} = 0 \quad \text{or} \quad \mathbf{a} \cdot \mathbf{c} = 0$$

And so $\mathbf{a} \perp \mathbf{c}$; that is, OA is perpendicular to BC.

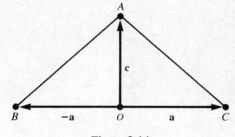

Figure 3-14

Vector Equation of a Straight Line

PROBLEM 3-7   Show that the equation of the line that is parallel to the vector **B** and that passes through a point whose position vector is **A** is

$$(\mathbf{r} - \mathbf{A}) \times \mathbf{B} = 0$$

where **r** is the position vector of any point on the line.

Solution: The equation of a line that is parallel to a nonzero vector **B** and that passes through a point whose position vector is **A** (Figure 3-15) is given by $\mathbf{r} - \mathbf{A} = t\mathbf{B}$ (3.1), for some parameter t. [See Example 3-3.]

   Since the cross product of a vector with a scalar multiple of itself is the zero vector, it follows from condition (1.42) that

$$(\mathbf{r} - \mathbf{A}) \times \mathbf{B} = t\mathbf{B} \times \mathbf{B} = 0$$

Conversely, if $(\mathbf{r} - \mathbf{A}) \times \mathbf{B} = 0$, then $\mathbf{r} - \mathbf{A}$ is parallel to **B**, so that $\mathbf{r} - \mathbf{A} = t\mathbf{B}$ for some parameter t.

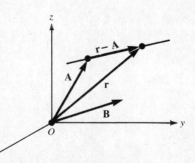

Figure 3-15

PROBLEM 3-8   If a line passes through two points whose position vectors are **A** and **B**, show that the equation of that line is

$$(\mathbf{r} - \mathbf{A}) \times (\mathbf{B} - \mathbf{A}) = 0$$

where **r** is the position vector of any point on the line.

Solution: As we can see from Figure 3-16, we have to find the equation of the line passing through the point that has the position vector **A** and is parallel to $\mathbf{B} - \mathbf{A}$. From the result of Problem 3-7, the required equation is

$$(\mathbf{r} - \mathbf{A}) \times (\mathbf{B} - \mathbf{A}) = 0$$

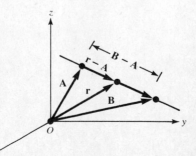

Figure 3-16

PROBLEM 3-9 Show that if the two straight lines represented by $(\mathbf{r} - \mathbf{A}) \times \mathbf{B} = \mathbf{0}$ and $(\mathbf{r} - \mathbf{C}) \times \mathbf{D} = \mathbf{0}$ intersect, then

$$(\mathbf{A} - \mathbf{C}) \cdot \mathbf{B} \times \mathbf{D} = 0$$

Solution: From definition (3.1) the general equations of the two straight lines can be written as $\mathbf{r} = \mathbf{A} + t\mathbf{B}$ and $\mathbf{r} = \mathbf{C} + s\mathbf{D}$ for some scalars s and t. Now if the straight lines intersect, there exist specific values of scalars s and t such that

$$\mathbf{A} + t\mathbf{B} = \mathbf{C} + s\mathbf{D}$$

Taking the dot product of both sides with $\mathbf{B} \times \mathbf{D}$, we obtain

$$(\mathbf{A} + t\mathbf{B}) \cdot (\mathbf{B} \times \mathbf{D}) = (\mathbf{C} + s\mathbf{D}) \cdot (\mathbf{B} \times \mathbf{D})$$

Since by the triple product condition (1.51)

$$\mathbf{B} \cdot \mathbf{B} \times \mathbf{D} = \mathbf{D} \cdot \mathbf{B} \times \mathbf{D} = 0$$

we obtain

$$\mathbf{A} \cdot \mathbf{B} \times \mathbf{D} = \mathbf{C} \cdot \mathbf{B} \times \mathbf{D}$$

Then, by transposing and factoring, we have the result

$$(\mathbf{A} - \mathbf{C}) \cdot \mathbf{B} \times \mathbf{D} = 0$$

PROBLEM 3-10 Find equations of the line passing through $(2, 0, 3)$ and parallel to $[3, 1, 2]$.

Solution: By the definition of the vector equation of a line (3.1), we have

$$\mathbf{r} = \mathbf{A} + t\mathbf{B} = [2, 0, 3] + t[3, 1, 2] = [2 + 3t, t, 3 + 2t]$$

By (3.2), the parametric equations are

$$x = 2 + 3t, \qquad y = t, \qquad z = 3 + 2t$$

For the nonparametric form we solve each of these for t and equate to obtain

$$\frac{x - 2}{3} = y = \frac{z - 3}{2}$$

Vector Equation of a Plane

PROBLEM 3-11 Show that the equation of a plane can be expressed as

$$ax + by + cz + k = 0$$

Solution: According to definition (3.7), the vector equation of the plane that is perpendicular to a nonzero vector $\mathbf{B}$ and passes through a point whose position vector is $\mathbf{A}$ is expressed as

$$(\mathbf{r} - \mathbf{A}) \cdot \mathbf{B} = 0$$

Expressing each term here in component form using

$$\mathbf{r} = [x, y, z], \qquad \mathbf{A} = [A_1, A_2, A_3], \qquad \mathbf{B} = [B_1, B_2, B_3],$$

the equation for the plane (3.7) becomes

$$(x - A_1)B_1 + (y - A_2)B_2 + (z - A_3)B_3 = 0$$

or

$$B_1 x + B_2 y + B_3 z - (A_1 B_1 + A_2 B_2 + A_3 B_3) = 0 \qquad \text{(a)}$$

If we let $B_1 = a$, $B_2 = b$, $B_3 = c$, and $-(A_1 B_1 + A_2 B_2 + A_3 B_3) = k$, eq. (a) reduces to

$$ax + by + cz + k = 0$$

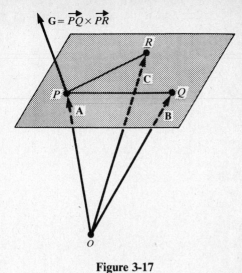

Figure 3-17

PROBLEM 3-12 Find a vector equation to represent the plane that passes through three given points whose position vectors are **A**, **B**, and **C**.

Solution: Let P, Q, and R be the three given points with position vectors $\overrightarrow{OP} = \mathbf{A}$, $\overrightarrow{OQ} = \mathbf{B}$, and $\overrightarrow{OR} = \mathbf{C}$ (see Figure 3-17). Then $\overrightarrow{PQ} = \mathbf{B} - \mathbf{A}$ and $\overrightarrow{PR} = \mathbf{C} - \mathbf{A}$. From the definition of the cross product (1.35), the vector

$$\mathbf{G} = \overrightarrow{PQ} \times \overrightarrow{PR} = (\mathbf{B} - \mathbf{A}) \times (\mathbf{C} - \mathbf{A}) = \mathbf{A} \times \mathbf{B} + \mathbf{B} \times \mathbf{C} + \mathbf{C} \times \mathbf{A} \qquad (a)$$

is perpendicular to the plane. Then from definition (3.7), the vector equation of the plane is

$$(\mathbf{r} - \mathbf{A}) \cdot \mathbf{G} = 0 \qquad (b)$$

where **r** is the position vector of any point on the plane. Since

$$\mathbf{A} \cdot \mathbf{G} = \mathbf{A} \cdot (\mathbf{A} \times \mathbf{B} + \mathbf{B} \times \mathbf{C} + \mathbf{C} \times \mathbf{A}) = \mathbf{A} \cdot \mathbf{B} \times \mathbf{C} = [\mathbf{ABC}]$$

we can rewrite eq. (b) to give the vector equation of the plane as

$$\mathbf{r} \cdot \mathbf{G} = [\mathbf{ABC}] \qquad (c)$$

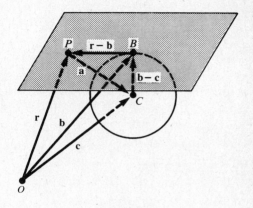

Figure 3-18

PROBLEM 3-13 The sphere shown in Figure 3-18 has radius a, and its center C is specified by the position vector **c** relative to O. An arbitrary point B on the sphere is specified by the position vector **b**. If **r** is the position vector of any point P on the plane tangent to the sphere at B, show that the vector equation of the tangent plane is

$$(\mathbf{r} - \mathbf{c}) \cdot (\mathbf{b} - \mathbf{c}) = a^2$$

Solution: As shown in Figure 3-18, vector $\overrightarrow{BP}$ is in the plane and is specified by $\mathbf{r} - \mathbf{b}$. Since the plane is tangent to the sphere at B, $\mathbf{r} - \mathbf{b}$ is perpendicular to the radius vector $\mathbf{b} - \mathbf{c}$ of the sphere. Thus

$$(\mathbf{r} - \mathbf{b}) \cdot (\mathbf{b} - \mathbf{c}) = 0 \qquad (a)$$

Since $\mathbf{b} = \mathbf{c} + (\mathbf{b} - \mathbf{c})$, we can rewrite eq. (a) as

$$[(\mathbf{r} - \mathbf{c}) - (\mathbf{b} - \mathbf{c})] \cdot (\mathbf{b} - \mathbf{c}) = 0 \qquad (b)$$

Simplifying, we obtain

$$(\mathbf{r} - \mathbf{c}) \cdot (\mathbf{b} - \mathbf{c}) = (\mathbf{b} - \mathbf{c}) \cdot (\mathbf{b} - \mathbf{c}) \qquad (c)$$

But $(\mathbf{b} - \mathbf{c}) \cdot (\mathbf{b} - \mathbf{c}) = a^2$, where a is the radius of the sphere. Thus the required equation is

$$(\mathbf{r} - \mathbf{c}) \cdot (\mathbf{b} - \mathbf{c}) = a^2$$

PROBLEM 3-14 At what point does the line that passes through the point $(0, 1, 1)$ and is parallel to the vector $[1, 1, -2]$ cross the plane whose equation is $x + 5y + 6z + 1 = 0$?

Solution: From their definition equations (3.2), the parametric equations of the line are

$$x = t, \qquad y = 1 + t, \qquad z = 1 - 2t$$

Substituting these coordinates of a general point into the equation of the given plane, we find that

$$t + 5(1 + t) + 6(1 - 2t) + 1 = 0$$

or

$$12 - 6t = 0$$

So $t = 2$, and the point has coordinates $(2, 3, -3)$.

Distance Formulas

PROBLEM 3-15   Find the distance d from the point $(1, -2, 1)$ to the straight line that joins the points $(1, 2, -1)$ and $(-1, -2, 1)$.

Solution: Let $\mathbf{r}_0 = [1, -2, 1]$, $\mathbf{A} = [1, 2, -1]$, and $\mathbf{B} = [-1, -2, 1]$. Then from the result of Problem 3-8 and the position vector $\mathbf{r} = [x, y, z]$, the equation of the straight line that joins the points $(1, 2, -1)$ and $(-1, -2, 1)$ is

$$(\mathbf{r} - \mathbf{A}) \times (\mathbf{B} - \mathbf{A}) = \mathbf{0}$$

Replacing $\mathbf{B}$ in the distance formula (3.8) by $\mathbf{B} - \mathbf{A}$, we get

$$d = \frac{|(\mathbf{r}_0 - \mathbf{A}) \times (\mathbf{B} - \mathbf{A})|}{|\mathbf{B} - \mathbf{A}|}$$

Now clearly $\mathbf{B} - \mathbf{A} = [-2, -4, 2]$, $|\mathbf{B} - \mathbf{A}| = 2\sqrt{6}$, and $\mathbf{r}_0 - \mathbf{A} = [0, -4, 2]$. So in determinant form

$$(\mathbf{r}_0 - \mathbf{A}) \times (\mathbf{B} - \mathbf{A}) = \begin{vmatrix} \mathbf{i} & \mathbf{j} & \mathbf{k} \\ 0 & -4 & 2 \\ -2 & -4 & 2 \end{vmatrix} = [0, 4, -8]$$

and the magnitude $|(\mathbf{r}_0 - \mathbf{A}) \times (\mathbf{B} - \mathbf{A})| = \sqrt{0^2 + 4^2 + (-8)^2} = \sqrt{80} = 4\sqrt{5}$. Thus the distance is

$$d = \frac{4\sqrt{5}}{2\sqrt{6}} = \frac{2\sqrt{5}}{\sqrt{6}}$$

PROBLEM 3-16   Find the shortest distance d between the line that joins the points $P(1, 2, -1)$ and $R(1, -1, 1)$ and the line that joins the points $Q(2, -2, 1)$ and $S(2, 0, -2)$.

Solution: Let the position vectors of P and Q be denoted by $\mathbf{A}$ and $\mathbf{C}$, and $\overrightarrow{PR} = \mathbf{B}$ and $\overrightarrow{QS} = \mathbf{D}$ as shown in Figure 3-19. Then we have the same problem as that in Example 3-7 (see also Figure 3-7), that is, finding the shortest distance between two lines that pass through two distinct points. Now

$$\mathbf{A} = [1, 2, -1]$$
$$\mathbf{B} = \overrightarrow{PR} = [1 - 1, -1 - 2, 1 - (-1)] = [0, -3, 2]$$
$$\mathbf{C} = [2, -2, 1]$$
$$\mathbf{D} = \overrightarrow{QS} = [2 - 2, 0 - (-2), -2 - 1] = [0, 2, -3]$$
$$\overrightarrow{PQ} = \mathbf{C} - \mathbf{A} = [2 - 1, -2 - 2, 1 - (-1)] = [1, -4, 2]$$

Since

$$\mathbf{B} \times \mathbf{D} = \begin{vmatrix} \mathbf{i} & \mathbf{j} & \mathbf{k} \\ 0 & -3 & 2 \\ 0 & 2 & -3 \end{vmatrix} = [5, 0, 0]$$

the magnitude $|\mathbf{B} \times \mathbf{D}| = 5$. So by distance formula (3.9), we obtain the required distance between the two lines:

$$d = \frac{|(\mathbf{C} - \mathbf{A}) \cdot (\mathbf{B} \times \mathbf{D})|}{|\mathbf{B} \times \mathbf{D}|} = \frac{5}{5} = 1$$

PROBLEM 3-17   Show that the distance d between an arbitrary point (x_0, y_0, z_0) and the plane $ax + by + cz + k = 0$ is

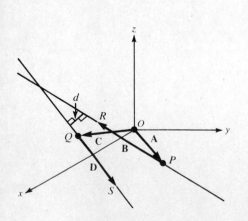

Figure 3-19

$$d - \frac{|ax_0 + by_0 + cz_0 + k|}{(a^2 + b^2 + c^2)^{1/2}}$$

Solution: As in the solution to Problem 3-11, if we let $\mathbf{B} = [a, b, c]$, $-\mathbf{A} \cdot \mathbf{B} = k$, and $\mathbf{r}_0 = [x_0, y_0, z_0]$, then

$$B = |\mathbf{B}| = (a^2 + b^2 + c^2)^{1/2}$$

Hence on taking the dot product,

$$(\mathbf{r}_0 - \mathbf{A}) \cdot \mathbf{B} = \mathbf{r}_0 \cdot \mathbf{B} - \mathbf{A} \cdot \mathbf{B} = ax_0 + by_0 + cz_0 + k$$

Then using the formula for the distance from a point to a plane (3.10), we obtain the required distance:

$$d = \frac{|(\mathbf{r}_0 - \mathbf{A}) \cdot \mathbf{B}|}{B} = \frac{|ax_0 + by_0 + cz_0 + k}{(a^2 + b^2 + c^2)^{1/2}}$$

Supplementary Exercises

PROBLEM 3-18   Prove that the sum of the squares of the lengths of the diagonals of a parallelogram is equal to the sum of the squares of the lengths of its sides.

PROBLEM 3-19   Prove that the midpoint of the hypotenuse of a right triangle is equidistant from the three vertices.

PROBLEM 3-20   Prove that the line that joins one vertex of a parallelogram to the midpoint of its opposite side divides the diagonal in a 2:1 ratio.

PROBLEM 3-21   If $ABCD$ is a quadrilateral, with M and N the midpoints of the diagonals AC and BD, respectively, prove that

$$\overline{AB}^2 + \overline{BC}^2 + \overline{CD}^2 + \overline{DA}^2 = \overline{AC}^2 + \overline{BD}^2 + 4\overline{MN}^2$$

PROBLEM 3-22   If the diagonals of a parallelogram are orthogonal, show that the parallelogram is a rhombus.

PROBLEM 3-23   Show that if two circles intersect, the line that joins their centers is the perpendicular bisector of the line segment that joins their points of intersection.

PROBLEM 3-24   Find the distance from the point $(2, 2, -5)$ to the straight line that joins the points $(1, 2, -5)$ and $(7, 5, -9)$.

Answers: $\dfrac{5}{\sqrt{61}}$

PROBLEM 3-25   Find the shortest distance between the line that joins the points $P(1, 2, 3)$ and $R(-1, 0, 2)$ and the line that joins the points $Q(0, 1, 7)$ and $S(2, 0, 5)$.

Answer: 3

PROBLEM 3-26   Find the shortest distance from the point $(1, -2, 1)$ to the plane determined by the three points $(2, 4, 1)$, $(-1, 0, 1)$, and $(-1, 4, 2)$.

Answer: 14/13

PROBLEM 3-27 Show that the equation of the plane that encompasses three given points whose position vectors are **A**, **B**, and **C** can be written in the symmetric form

$$(3\mathbf{r} - \mathbf{A} - \mathbf{B} - \mathbf{C}) \cdot (\mathbf{A} \times \mathbf{B} + \mathbf{B} \times \mathbf{C} + \mathbf{C} \times \mathbf{A}) = 0$$

where **r** is the position vector of any point on the plane.

PROBLEM 3-28 Show that the point of intersection of the straight line represented by $(\mathbf{r} - \mathbf{A}) \times \mathbf{B} = \mathbf{0}$ and the plane represented by $(\mathbf{r} - \mathbf{C}) \cdot \mathbf{D} = 0$ is the point whose position vector is given by

$$\mathbf{A} + \left[(\mathbf{C} - \mathbf{A}) \cdot \frac{\mathbf{D}}{(\mathbf{B} \cdot \mathbf{D})} \right] \mathbf{B}$$

4 DIFFERENTIATION OF VECTORS

THIS CHAPTER IS ABOUT

☑ **Vector Functions**
☑ **Ordinary Differentiation of Vectors**
☑ **Partial Derivatives of Vector Functions of More Than One Variable**
☑ **Space Curves**
☑ **Surfaces**
☑ **Tangent, Curvature, and Torsion**

4-1. Vector Functions

A. Scalar and vector functions

Definition: If t is a scalar variable, then a **scalar function** f assigns to each t in some interval a unique scalar $f(t)$ called the *value* of f at t. In general, the scalar variable represents either time or a set of coordinates.

Definition: A **vector function f** of single scalar variable t assigns to each t in some interval a unique vector $\mathbf{f}(t)$ called the *value* of **f** at t.

In a rectangular coordinate system, the vector function $\mathbf{f}(t)$ can be expressed in component form, namely,

VECTOR FUNCTION
$$\mathbf{f}(t) = [f_1(t), f_2(t), f_3(t)]$$
$$= f_1(t)\mathbf{i} + f_2(t)\mathbf{j} + f_3(t)\mathbf{k} \qquad (4.1)$$

where $f_1(t)$, $f_2(t)$, and $f_3(t)$ are scalar functions of t and are called the components of $\mathbf{f}(t)$.

B. Limit vector and continuity

Definition: Let a vector function $\mathbf{f}(t)$ be defined for all values of t in some neighborhood about a point t_0, except possibly for the value of t_0 itself. Then vector **a** is the **limit vector** of $\mathbf{f}(t)$ as t approaches t_0 and is written as

LIMIT VECTOR
$$\lim_{t \to t_0} \mathbf{f}(t) = \mathbf{a} \qquad (4.2)$$

if and only if, for any arbitrary real number $\varepsilon > 0$, there exists a real number $\delta > 0$ such that

$$|\mathbf{f}(t) - \mathbf{a}| < \varepsilon \quad \text{whenever} \quad 0 < |t - t_0| < \delta$$

This definition becomes the familiar definition for the limit of a scalar function if vector function $\mathbf{f}(t)$ is replaced by a scalar function f, and vector **a** by a scalar a.

Definition: A vector function $\mathbf{f}(t)$ is said to be *continuous* at $t = t_0$ if it is defined in some neighborhood of t_0 and

CONTINUITY OF A VECTOR FUNCTION
$$\lim_{t \to t_0} \mathbf{f}(t) = \mathbf{f}(t_0) \qquad (4.3)$$

Thus using the definition of the limit vector (4.2), $\mathbf{f}(t)$ is continuous at $t = t_0$ if and only if, for a given $\varepsilon > 0$, there is a $\delta > 0$ such that

$$|\mathbf{f}(t) - \mathbf{f}(t_0)| < \varepsilon \quad \text{whenever} \quad |t - t_0| < \delta$$

EXAMPLE 4-1: If $\mathbf{f} = f_1(t)\mathbf{i} + f_2(t)\mathbf{j} + f_3(t)\mathbf{k}$ and $\mathbf{a} = a_1\mathbf{i} + a_2\mathbf{j} + a_3\mathbf{k}$, show that

$$\lim_{t \to t_0} \mathbf{f}(t) = \mathbf{a} \tag{4.4}$$

if and only if

$$\lim_{t \to t_0} f_i(t) = a_i, \quad i = 1, 2, 3 \tag{4.5}$$

Solution: If $\lim_{t \to t_0} \mathbf{f}(t) = \mathbf{a}$, then for an arbitrary $\varepsilon > 0$, there exists $\delta > 0$ such that $|f(t) - \mathbf{a}| < \varepsilon$ whenever $0 < |t - t_0| < \delta$. But

$$\mathbf{f}(t) - \mathbf{a} = [f_1(t) - a_1]\mathbf{i} + [f_2(t) - a_2]\mathbf{j} + [f_3(t) - a_3]\mathbf{k}$$

and for $0 < |t - t_0| < \delta$,

$$|f_i(t) - a_i| \leqslant |\mathbf{f}(t) - \mathbf{a}| < \varepsilon$$

where $i = 1, 2, 3$. Hence

$$\lim_{t \to t_0} f_i(t) = a_i$$

Conversely, if $\lim_{t \to t_0} f_i(t) = a_i$ $(i = 1, 2, 3)$ for every $\varepsilon > 0$, then there exists $\delta > 0$ such that $|f_i(t) - a_i| < \varepsilon/3$ whenever $0 < |t - t_0| < \delta$. Then we use the triangle inequality for absolute values to write

$$\begin{aligned}
|\mathbf{f}(t) - \mathbf{a}| &= |[f_1(t) - a_1]\mathbf{i} + [f_2(t) - a_2]\mathbf{j} + [f_3(t) - a_3]\mathbf{k}| \\
&\leqslant |f_1(t) - a_1| + |f_2(t) - a_2| + |f_3(t) - a_1| \\
&< \varepsilon
\end{aligned}$$

which implies that

$$\lim_{t \to t_0} \mathbf{f}(t) = \mathbf{a}$$

EXAMPLE 4-2: Show that the limit of the sum of two vector functions is the sum of their limits; that is, if $\lim_{t \to t_0} \mathbf{f}(t) = \mathbf{a}$ and $\lim_{t \to t_0} \mathbf{g}(t) = \mathbf{b}$, then

$$\lim_{t \to t_0} [\mathbf{f}(t) + \mathbf{g}(t)] = \mathbf{a} + \mathbf{b} \tag{4.6}$$

Solution: If $\lim_{t \to t_0} \mathbf{f}(t) = \mathbf{a}$ and $\lim_{t \to t_0} \mathbf{g}(t) = \mathbf{b}$, then for every $\varepsilon > 0$, there exists $\delta > 0$ such that for $0 < |t - t_0| < \delta$,

$$|\mathbf{f}(t) - \mathbf{a}| < \tfrac{1}{2}\varepsilon \quad \text{and} \quad |\mathbf{g}(t) - \mathbf{b}| < \tfrac{1}{2}\varepsilon$$

Using the triangle inequality for absolute values, we write

$$\begin{aligned}
|[\mathbf{f}(t) + \mathbf{g}(t)] - (\mathbf{a} + \mathbf{b})| &= |[\mathbf{f}(t) - \mathbf{a}] + [\mathbf{g}(t) - \mathbf{b}]| \\
&\leqslant |\mathbf{f}(t) - \mathbf{a}| + |\mathbf{g}(t) - \mathbf{b}| \\
&< \varepsilon
\end{aligned}$$

So by the definition of the limit vector,

$$\lim_{t \to t_0} [\mathbf{f}(t) + \mathbf{g}(t)] = \mathbf{a} + \mathbf{b}$$

EXAMPLE 4-3: If $\lim_{t \to t_0} \mathbf{f}(t) = \mathbf{a}$ and $\lim_{t \to t_0} \phi(t) = c$, show that

$$\lim_{t \to t_0} \phi(t)\mathbf{f}(t) = c\mathbf{a} \qquad (4.7)$$

Solution: If $\lim_{t \to t_0} \mathbf{f}(t) = \mathbf{a}$ and $\lim_{t \to t_0} \phi(t) = c$, then for $0 < \varepsilon' < 1$ (ε' to be found later) there exists $\delta > 0$ such that for $|t - t_0| < \delta$, $|\mathbf{f}(t) - \mathbf{a}| < \varepsilon'$ and $|\phi(t) - c| < \varepsilon'$. But it is also true that

$$|\phi(t)| = |c + [\phi(t) - c]| < |c| + \varepsilon' < |c| + 1$$

Writing

$$\phi(t)\mathbf{f}(t) - c\mathbf{a} = \phi(t)\mathbf{f}(t) - \phi(t)\mathbf{a} + \phi(t)\mathbf{a} - c\mathbf{a}$$
$$= \phi(t)[\mathbf{f}(t) - \mathbf{a}] + [\phi(t) - c]\mathbf{a}$$

we have

$$|\phi(t)\mathbf{f}(t) - c\mathbf{a}| = |\phi(t)[\mathbf{f}(t) - \mathbf{a}] + [\phi(t) - c]\mathbf{a}|$$
$$\leqslant |\phi(t)|\,|\mathbf{f}(t) - \mathbf{a}| + |\phi(t) - c|\,|\mathbf{a}|$$
$$< (|c| + 1)\varepsilon' + \varepsilon'|\mathbf{a}|$$
$$= \varepsilon'(|c| + 1 + |\mathbf{a}|)$$

Thus if we choose $\varepsilon' < \varepsilon/(|c| + 1 + |\mathbf{a}|)$ and also $\varepsilon' < 1$ (as indicated earlier), we have

$$|\phi(t)\mathbf{f}(t) - c\mathbf{a}| < \varepsilon$$

So

$$\lim_{t \to t_0} \phi(t)\mathbf{f}(t) = c\mathbf{a}$$

Note: If $\mathbf{f}(t) = \mathbf{a}$ and $\lim_{t \to t_0} \phi(t) = c$, then (4.7) reduces to

$$\lim_{t \to t_0} \phi(t)\mathbf{a} = c\mathbf{a} \qquad (4.8)$$

EXAMPLE 4-4: Show that a vector function $\mathbf{f}(t) = f_1(t)\mathbf{i} + f_2(t)\mathbf{j} + f_3(t)\mathbf{k}$ is continuous at t_0 if each scalar function $f_i(t)$ $(i = 1, 2, 3)$ is continuous at t_0.

Solution: If $f_i(t)$ $(i = 1, 2, 3)$ is continuous at t_0, then we have

$$\lim_{t \to t_0} f_i(t) = f_i(t_0) \qquad (i = 1, 2, 3)$$

Now

$$\mathbf{f}(t) = f_1(t)\mathbf{i} + f_2(t)\mathbf{j} + f_3(t)\mathbf{k}$$

and so

$$\mathbf{f}(t_0) = f_1(t_0)\mathbf{i} + f_2(t_0)\mathbf{j} + f_3(t_0)\mathbf{k}$$

Using relation (4.6), we have

$$\lim_{t \to t_0} \mathbf{f}(t) = \lim_{t \to t_0} [f_1(t)\mathbf{i} + f_2(t)\mathbf{j} + f_3(t)\mathbf{k}]$$
$$= \lim_{t \to t_0} f_1(t)\mathbf{i} + \lim_{t \to t_0} f_2(t)\mathbf{j} + \lim_{t \to t_0} f_3(t)\mathbf{k}$$

Since $\mathbf{i}, \mathbf{j}, \mathbf{k}$ are constant vectors, from relation (4.8) we have

$$\lim_{t \to t_0} \mathbf{f}(t) = \mathbf{i} \lim_{t \to t_0} f_1(t) + \mathbf{j} \lim_{t \to t_0} f_2(t) + \mathbf{k} \lim_{t \to t_0} f_3(t)$$
$$= \mathbf{i} f_1(t_0) + \mathbf{j} f_2(t_0) + \mathbf{k} f_3(t_0)$$
$$= \mathbf{f}(t_0)$$

Hence by definition (4.3), $\mathbf{f}(t)$ is continuous at t_0.

4-2. Ordinary Differentiation of Vectors

A. Derivative of a vector function

The **derivative** $\mathbf{f}'(t)$ of a vector function $\mathbf{f}(t)$ is defined by

DERIVATIVE
$$\mathbf{f}'(t) = \frac{d\mathbf{f}(t)}{dt} = \lim_{\Delta t \to 0} \frac{\mathbf{f}(t + \Delta t) - \mathbf{f}(t)}{\Delta t} \tag{4.9}$$

The vector function $\mathbf{f}(t)$ is also said to be *differentiable*, and so the derivative $\mathbf{f}'(t)$ is also a vector function. Thus if the vector function is in component form, $\mathbf{f}(t) = f_1(t)\mathbf{i} + f_2(t)\mathbf{j} + f_3(t)\mathbf{k}$, then $\mathbf{f}'(t)$ exists if and only if the derivatives

$$f_i'(t) = \lim_{\Delta t \to 0} \frac{f_i(t + \Delta t) - f_i(t)}{\Delta t} \tag{4.10}$$

of scalar functions $f_i(t)$ $(i = 1, 2, 3)$ exist. Then we can write

$$\mathbf{f}'(t) = f_1'(t)\mathbf{i} + f_2'(t)\mathbf{j} + f_3'(t)\mathbf{k} \tag{4.11}$$

Higher derivatives of a vector function are defined in a manner similar to that for a scalar function of a single variable. Thus

$$\mathbf{f}''(t) = \frac{d^2\mathbf{f}(t)}{dt^2} = \frac{d}{dt}\left[\frac{d\mathbf{f}(t)}{dt}\right] \tag{4.12}$$

and so forth.

From elementary calculus we know that a differentiable function is necessarily continuous but that the converse is not always true. The same can be said about a vector function, in view of expression (4.11). In this book—unless we say otherwise—we'll consider only those functions that are differentiable to any order needed in a particular discussion.

B. Rules for differentiation of vector functions

The rules for differentiation of vector functions are similar to those for scalar functions—with one exception:

- To differentiate the vector product of vector functions, the order of factors must be preserved because the vector product is not a commutative operation.

If $\mathbf{A}(t)$, $\mathbf{B}(t)$, and $\mathbf{C}(t)$ are differentiable vector functions and $\phi(t)$ is a differentiable scalar function, then

$$\frac{d}{dt}[\mathbf{A}(t) + \mathbf{B}(t)] = \mathbf{A}'(t) + \mathbf{B}'(t) \tag{4.13}$$

$$\frac{d}{dt}[\phi(t)\mathbf{A}(t)] = \phi(t)\mathbf{A}'(t) + \phi'(t)\mathbf{A}(t) \tag{4.14}$$

$$\frac{d}{dt}[\mathbf{A}(t) \cdot \mathbf{B}(t)] = \mathbf{A}(t) \cdot \mathbf{B}'(t) + \mathbf{A}'(t) \cdot \mathbf{B}(t) \tag{4.15}$$

RULES FOR
DIFFERENTIATION
$$\frac{d}{dt}[\mathbf{A}(t) \times \mathbf{B}(t)] = \mathbf{A}(t) \times \mathbf{B}'(t) + \mathbf{A}'(t) \times \mathbf{B}(t) \tag{4.16}$$

$$\frac{d}{dt}[\mathbf{A}(t) \cdot \mathbf{B}(t) \times \mathbf{C}(t)] = \mathbf{A}'(t) \cdot \mathbf{B}(t) \times \mathbf{C}(t) + \mathbf{A}(t) \cdot \mathbf{B}'(t) \times \mathbf{C}(t)$$
$$+ \mathbf{A}(t) \cdot \mathbf{B}(t) \times \mathbf{C}'(t) \tag{4.17}$$

$$\frac{d}{dt}\{\mathbf{A}(t) \times [\mathbf{B}(t) \times \mathbf{C}(t)]\} = \mathbf{A}'(t) \times [\mathbf{B}(t) \times \mathbf{C}(t)] + \mathbf{A}(t) \times [\mathbf{B}'(t) \times \mathbf{C}(t)]$$
$$+ \mathbf{A}(t) \times [\mathbf{B}(t) \times \mathbf{C}'(t)] \tag{4.18}$$

If $t = g(s)$ is differentiable, then

$$\frac{d\mathbf{A}(t)}{ds} = \frac{d\mathbf{A}[g(s)]}{ds} = \frac{d\mathbf{A}(t)}{dt}\frac{dg(s)}{ds} \qquad \text{[Chain rule]} \qquad (4.19)$$

EXAMPLE 4-5: Verify rule (4.14) concerning the differentiation of a scalar function times a vector function.

Proof: If $\mathbf{f}(t) = \phi(t)\mathbf{A}(t)$, then we have $\mathbf{f}(t + \Delta t) = \phi(t + \Delta t)\mathbf{A}(t + \Delta t)$, and

$$\mathbf{f}(t + \Delta t) - \mathbf{f}(t) = \phi(t + \Delta t)\mathbf{A}(t + \Delta t) - \phi(t)\mathbf{A}(t)$$
$$= \phi(t + \Delta t)[\mathbf{A}(t + \Delta t) - \mathbf{A}(t)] + [\phi(t + \Delta t) - \phi(t)]\mathbf{A}(t)$$

Divide both sides by Δt and let Δt approach zero to obtain

$$\frac{d}{dt}[\phi(t)\mathbf{A}(t)] = \mathbf{f}'(t) = \lim_{\Delta t \to 0} \frac{\mathbf{f}(t + \Delta t) - \mathbf{f}(t)}{\Delta t}$$

$$= \lim_{\Delta t \to 0} \phi(t + \Delta t)\left[\frac{\mathbf{A}(t + \Delta t) - \mathbf{A}(t)}{\Delta t}\right]$$

$$+ \lim_{\Delta t \to 0} \left[\frac{\phi(t + \Delta t) - \phi(t)}{\Delta t}\right]\mathbf{A}(t)$$

$$= \phi(t)\mathbf{A}'(t) + \phi'(t)\mathbf{A}(t)$$

> *Note: Special cases of rule* (4.14):
> If $\phi(t) = k$ is a constant scalar, then

$$\frac{d}{dt}[k\mathbf{A}(t)] = k\mathbf{A}'(t) \qquad (4.20)$$

> If $\mathbf{A}(t) = \mathbf{a}$ is a constant vector, then

$$\frac{d}{dt}[\phi(t)\mathbf{a}] = \phi'(t)\mathbf{a} \qquad (4.21)$$

EXAMPLE 4-6: Verify rule (4.15) concerning the differentiation of the scalar product of two vector functions.

Proof: If $\psi(t) = \mathbf{A}(t) \cdot \mathbf{B}(t)$, then

$$\psi(t + \Delta t) - \psi(t) = \mathbf{A}(t + \Delta t) \cdot \mathbf{B}(t + \Delta t) - \mathbf{A}(t) \cdot \mathbf{B}(t)$$
$$= \mathbf{A}(t + \Delta t) \cdot [\mathbf{B}(t + \Delta t) - \mathbf{B}(t)] + [\mathbf{A}(t + \Delta t) - \mathbf{A}(t)] \cdot \mathbf{B}(t)$$

Divide both sides by Δt and let Δt approach zero to obtain

$$\psi'(t) = \frac{d}{dt}[\mathbf{A}(t) \cdot \mathbf{B}(t)] = \mathbf{A}(t) \cdot \mathbf{B}'(t) + \mathbf{A}'(t) \cdot \mathbf{B}(t)$$

Alternate Proof: Let $\mathbf{A}(t) = [A_1(t), A_2(t), A_3(t)]$ and $\mathbf{B}(t) = [B_1(t), B_2(t), B_3(t)]$. Then, from the definition of the dot product (2.24),

$$\frac{d}{dt}[\mathbf{A}(t) \cdot \mathbf{B}(t)] = \frac{d}{dt}[A_1(t)B_1(t) + A_2(t)B_2(t) + A_3(t)B_3(t)]$$

$$= A_1(t)B'_1(t) + A'_1(t)B_1(t) + A_2(t)B'_2(t) + A'_2(t)B_2(t)$$
$$+ A_3(t)B'_3(t) + A'_3(t)B_3(t)$$

$$= [A_1(t)B'_1(t) + A_2(t)B'_2(t) + A_3(t)B'_3(t)]$$
$$+ [A'_1(t)B_1(t) + A'_2(t)B_2(t) + A'_3(t)B_3(t)]$$

$$= \mathbf{A}(t) \cdot \mathbf{B}'(t) + \mathbf{A}'(t) \cdot \mathbf{B}(t)$$

EXAMPLE 4-7: Verify rule (4.16) concerning the differentiation of the vector product of two vector functions.

Solution: If $\mathbf{f}(t) = \mathbf{A}(t) \times \mathbf{B}(t)$, then

$$\mathbf{f}(t + \Delta t) - \mathbf{f}(t) = \mathbf{A}(t + \Delta t) \times \mathbf{B}(t + \Delta t) - \mathbf{A}(t) \times \mathbf{B}(t)$$

Adding and subtracting $\mathbf{A}(t + \Delta t) \times \mathbf{B}(t)$ and simplifying, we get

$$\mathbf{f}(t + \Delta t) - \mathbf{f}(t) = \mathbf{A}(t + \Delta t) \times [\mathbf{B}(t + \Delta t) - \mathbf{B}(t)] + [\mathbf{A}(t + \Delta t) - \mathbf{A}(t)] \times \mathbf{B}(t)$$

Dividing both sides by Δt and taking the limit as $\Delta t \to 0$,

$$\mathbf{f}'(t) = \frac{d}{dt}[\mathbf{A}(t) \times \mathbf{B}(t)] = \mathbf{A}(t) \times \mathbf{B}'(t) + \mathbf{A}'(t) \times \mathbf{B}(t)$$

Note: Don't forget that the order of factors in (4.16) must be preserved.

4-3. Partial Derivatives of Vector Functions of More Than One Variable

A. Vector functions of more than one variable

A vector function $\mathbf{f}$ of two variables u and v assigns to each point (u, v) in some region R a unique vector $\mathbf{f}(u, v)$ called the value of $\mathbf{f}$ at (u, v). Similarly, $\mathbf{f}(u, v, w)$ is a vector function of three variables u, v, and w.

In a rectangular coordinate system

$$\begin{aligned}\mathbf{f}(u, v, w) &= [f_1(u, v, w), f_2(u, v, w), f_3(u, v, w)] \\ &= f_1(u, v, w)\mathbf{i} + f_2(u, v, w)\mathbf{j} + f_3(u, v, w)\mathbf{k} \end{aligned} \quad (4.22)$$

where f_1, f_2, and f_3 are the scalar functions of u, v, and w and are called the components of $\mathbf{f}(u, v, w)$.

B. Partial derivatives of a vector function

The partial derivative $\mathbf{f}_u$ of the vector function $\mathbf{f}$ with respect to u is defined by

PARTIAL DERIVATIVE
$$\mathbf{f}_u = \frac{\partial \mathbf{f}}{\partial u} = \lim_{\Delta u \to 0} \frac{\mathbf{f}(u + \Delta u, v, w) - \mathbf{f}(u, v, w)}{\Delta u} \quad (4.23)$$

provided that this limit exists. Similarly, we define the partial derivatives of $\mathbf{f}$ with respect to v and w by

$$\mathbf{f}_v = \frac{\partial \mathbf{f}}{\partial v} = \lim_{\Delta v \to 0} \frac{\mathbf{f}(v, v + \Delta v, w) - \mathbf{f}(u, v, w)}{\Delta v} \quad (4.24)$$

$$\mathbf{f}_w = \frac{\partial \mathbf{f}}{\partial w} = \lim_{\Delta w \to 0} \frac{\mathbf{f}(u, v, w + \Delta w) - \mathbf{f}(u, v, w)}{\Delta w} \quad (4.25)$$

provided that these limits exist.

Higher-order partial derivatives can also be defined in the same way. For example,

$$\mathbf{f}_{uu} = \frac{\partial^2 \mathbf{f}}{\partial u^2} = \frac{\partial}{\partial u}\left(\frac{\partial \mathbf{f}}{\partial u}\right)$$

$$\mathbf{f}_{uv} = \frac{\partial^2 \mathbf{f}}{\partial v\, \partial u} = \frac{\partial}{\partial v}\left(\frac{\partial \mathbf{f}}{\partial u}\right)$$

$$\mathbf{f}_{vu} = \frac{\partial^2 \mathbf{f}}{\partial u\, \partial v} = \frac{\partial}{\partial u}\left(\frac{\partial \mathbf{f}}{\partial v}\right)$$

C. Rules for partial differentiation of vector functions

If $\mathbf{A}$ and $\mathbf{B}$ are differentiable vector functions of u, v, and w, and ϕ is a differentiable scalar function of u, v, and w, then (using u as an example)

$$\frac{\partial}{\partial u}(\mathbf{A} + \mathbf{B}) = \mathbf{A}_u + \mathbf{B}_u \tag{4.26}$$

RULES FOR PARTIAL DIFFERENTIATION

$$\frac{\partial}{\partial u}(\phi\mathbf{A}) = \phi\mathbf{A}_u + \phi_u\mathbf{A} \tag{4.27}$$

$$\frac{\partial}{\partial u}(\mathbf{A} \cdot \mathbf{B}) = \mathbf{A} \cdot \mathbf{B}_u + \mathbf{A}_u \cdot \mathbf{B} \tag{4.28}$$

$$\frac{\partial}{\partial u}(\mathbf{A} \times \mathbf{B}) = \mathbf{A} \times \mathbf{B}_u + \mathbf{A}_u \times \mathbf{B} \tag{4.29}$$

[Once again, note that the order of the factors that appear in (4.29) must be maintained.]

EXAMPLE 4-8: If $\mathbf{f}(u, v, w) = f_1(u, v, w)\mathbf{i} + f_2(u, v, w)\mathbf{j} + f_3(u, v, w)\mathbf{k}$, show that

$$\mathbf{f}_u = \frac{\partial \mathbf{f}}{\partial u} = \frac{\partial f_1}{\partial u}\mathbf{i} + \frac{\partial f_2}{\partial u}\mathbf{j} + \frac{\partial f_3}{\partial u}\mathbf{k} \tag{4.30}$$

Solution: From the definition of a partial derivative (4.23),

$$\mathbf{f}_u = \lim_{\Delta u \to 0} \frac{\mathbf{f}(u + \Delta u, v, w) - \mathbf{f}(u, v, w)}{\Delta u}$$

$$= \lim_{\Delta u \to 0} \left[\frac{f_1(u + \Delta u, v, w) - f_1(u, v, w)}{\Delta u}\mathbf{i} + \frac{f_2(u + \Delta u, v, w) - f_2(u, v, w)}{\Delta u}\mathbf{j} \right.$$

$$\left. + \frac{f_3(u + \Delta u, v, w) - f_3(u, v, w)}{\Delta u}\mathbf{k} \right]$$

$$= \frac{\partial f_1}{\partial u}\mathbf{i} + \frac{\partial f_2}{\partial u}\mathbf{j} + \frac{\partial f_3}{\partial u}\mathbf{k}$$

4-4. Space Curves

A. Representation of space curves

Let $\mathbf{f}(t)$ be a vector function of a scalar variable t; that is,

$$\mathbf{f}(t) = f_1(t)\mathbf{i} + f_2(t)\mathbf{j} + f_3(t)\mathbf{k} \tag{4.31}$$

where $f_1(t)$, $f_2(t)$, and $f_3(t)$ are scalar functions of t. Then for each value of t there is a position vector

$$\mathbf{r} = x\mathbf{i} + y\mathbf{j} + z\mathbf{k} \tag{4.32}$$

whose initial point is at the origin of a given coordinate system and whose terminal point specifies a point P in three-dimensional space. As t varies, P moves in a curved path C. Thus from the definition of the equality of vectors we get the *parametric equations*

$$x = f_1(t), \qquad y = f_2(t), \qquad z = f_3(t) \tag{4.33}$$

of the curve C in space, which is a function of $\mathbf{f}(t)$ with t as a parameter.

The terminal point of the vector $\mathbf{r}(a)$ is called the *initial point* of curve C, and the terminal point of the vector $\mathbf{r}(b)$ is called the *terminal point* of curve C. If $\mathbf{r}(a) = \mathbf{r}(b)$, then curve C is a *closed curve*. In other words, a curve is closed if its initial and terminal points coincide. Now if there exist t_1 and t_2 ($t_1 \neq t_2$ and $a < t_1$ (or t_2) $< b$) such that $\mathbf{r}(t_1) = \mathbf{r}(t_2) = \mathbf{r}_0$, then the point $\mathbf{r}_0$ is called a *multiple point*. Geometrically, a multipoint occurs where curve C crosses itself or is tangent to itself. We call curve C a *simple curve* if it has no multiple points. In particular, we call curve C that is both simple and closed a *simple closed curve*.

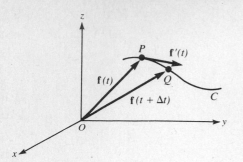

Figure 4-1
A tangent to a space curve.

Let P be a fixed point on a curve C for which $\mathbf{f} = \mathbf{f}(t)$, and let Q be the point that corresponds to $\mathbf{f}(t + \Delta t)$, as shown in Figure 4-1. Then

$$\lim_{\Delta t \to 0} \frac{\mathbf{f}(t + \Delta t) - \mathbf{f}(t)}{\Delta t} = \lim_{\Delta t \to 0} \frac{\overrightarrow{PQ}}{\Delta t} = \mathbf{f}'(t) \qquad (4.34)$$

is a vector that is tangent to the space curve C at P. This vector is called the **tangent vector** to C at P.

- A point $\mathbf{f}(t_i)$ on a space curve C is called a *singular point* of C if $|\mathbf{f}'(t_i)| = 0$. If $|\mathbf{f}'(t_i)| \neq 0$, it is called a *nonsingular point*.
- The *direction* of the space curve C at a nonsingular point P is the same as that of the tangent vector to C at P.
- A *smooth vector function* is a vector function that has a continuous derivative and no singular points.

EXAMPLE 4-9: Determine the curve C represented by the vector function

$$\mathbf{f}(t) = (a\cos t)\mathbf{i} + (a\sin t)\mathbf{j}, \qquad 0 \leqslant t \leqslant 2\pi$$

in space, and find its direction at the point $(0, a, 0)$.

Solution: The parametric equations of the curve C represented by $\mathbf{f}(t)$ are

$$x = a\cos t, \qquad y = a\sin t, \qquad z = 0$$

Since $z = 0$, we already know that the curve is in the xy-plane. So we get

$$x^2 + y^2 = a^2(\cos^2 t + \sin^2 t) = a^2$$

At $t = 0$,

$$x = a, \qquad y = 0, \qquad z = 0$$

and at $t = 2\pi$,

$$x = a, \qquad y = 0, \qquad z = 0$$

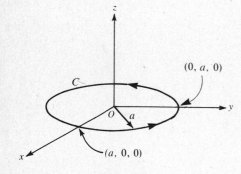

Figure 4-2

The curve C is a circle on the xy-plane centered at the origin and it has radius a; it begins and terminates at the point $(a, 0, 0)$ as t changes from 0 to 2π. (See Figure 4-2.)

The point $(0, a, 0)$ corresponds to that at $t = \pi/2$. Now by the formula for differentiation (4.11) we have

$$\mathbf{f}'(t) = -a\sin t\,\mathbf{i} + a\cos t\,\mathbf{j}$$

Thus,

$$\mathbf{f}'\left(\frac{\pi}{2}\right) = -a\sin\left(\frac{\pi}{2}\right)\mathbf{i} + a\cos\left(\frac{\pi}{2}\right)\mathbf{j}$$

$$= -a\mathbf{i}$$

Hence the direction of the curve at $(0, a, 0)$ is in the negative direction of the x-axis.

EXAMPLE 4-10: If a space curve C is represented by a smooth vector function $\mathbf{f}(t)$ for $a \leqslant t \leqslant b$, show that its length l is given by the integral

$$l = \int_a^b |\mathbf{f}'(t)|\,dt = \int_a^b [\mathbf{f}'(t) \cdot \mathbf{f}'(t)]^{1/2}\,dt \qquad (4.35)$$

Solution: If $\mathbf{f}(t) = f_1(t)\mathbf{i} + f_2(t)\mathbf{j} + f_3(t)\mathbf{k}$, then curve C can be described by the parametric equations $x = f_1(t)$, $y = f_2(t)$, and $z = f_3(t)$ for $a < t < b$.

Then we define the element of arc ds on C, from elementary calculus, as

$$ds^2 = dx^2 + dy^2 + dz^2 \qquad (4.36)$$

Using this representation and the parametric equations, we have

$$\frac{ds}{dt} = \left[\left(\frac{df_1}{dt} \right)^2 + \left(\frac{df_2}{dt} \right)^2 + \left(\frac{df_3}{dt} \right)^2 \right]^{1/2}$$
$$= \{ [f'_1(t)]^2 + [f'_2(t)]^2 + [f'_3(t)]^2 \}^{1/2}$$
$$= [\mathbf{f}'(t) \cdot \mathbf{f}'(t)]^{1/2}$$
$$= |\mathbf{f}'(t)| \tag{4.37}$$

So the element of arc ds can be represented as

$$ds = |\mathbf{f}'(t)| \, dt \tag{4.38}$$

But the total length l of the curve C is the integral of ds over C; that is,

$$l = \int_C ds = \int_a^b |\mathbf{f}'(t)| \, dt = \int_a^b [\mathbf{f}'(t) \cdot \mathbf{f}'(t)]^{1/2} \, dt$$

EXAMPLE 4-11: Find the length of the curve represented by the vector function

$$\mathbf{f}(t) = (a \cos t)\mathbf{i} + (a \sin t)\mathbf{j}, \qquad 0 \leqslant t \leqslant 2\pi$$

Solution: The derivative of $\mathbf{f}(t)$ is

$$\mathbf{f}'(t) = -a \sin t\mathbf{i} + a \cos t\mathbf{j}, \qquad 0 \leqslant t \leqslant 2\pi$$

and its magnitude is

$$|\mathbf{f}'(t)| = (a^2 \sin^2 t + a^2 \cos^2 t)^{1/2} = a$$

Thus from formula (4.35) for the length of the curve, we get

$$l = \int_0^{2\pi} |\mathbf{f}'(t)| \, dt = \int_0^{2\pi} a \, dt = 2\pi a$$

Note: We should have expected this answer as the circumference of the planar circle shown in Figure 4-2.

B. Arc length of a space curve

The **arc length** $s(t)$ of a space curve is a function of the scalar variable t from some fixed point a to t. It is defined as the length l of curve C (as given by the length formula (4.35), with the fixed upper limit b replaced by the variable t; that is,

ARC LENGTH OF A SPACE CURVE
$$s(t) = \int_a^t |\mathbf{f}'(t)| \, dt \tag{4.39}$$

The arc length s may serve as a parameter in the representations of space curves having no singular points. Differentiating the arc length formula (4.39) for a curve, we obtain for $a \leqslant t \leqslant b$

$$\frac{ds}{dt} = |\mathbf{f}'(t)| > 0 \tag{4.40}$$

So we can see that s is an increasing and continuous function in $a \leqslant t \leqslant b$. This implies that s has a unique inverse $t = q(s)$ for $0 \leqslant s \leqslant l$. Hence a smooth vector function

$$\mathbf{g}(s) = \mathbf{f}[q(s)] \tag{4.41}$$

with parameter s can represent the same space curve C that is represented by $\mathbf{f}(t)$.

By representing the position vector $\mathbf{r} = [x, y, z]$ as

$$\mathbf{r} = \mathbf{g}(s) = g_1(s)\mathbf{i} + g_2(s)\mathbf{j} + g_3(s)\mathbf{k} \tag{4.42}$$

we obtain the parametric equations of curve C with parameter s:

$$x = g_1(s), \qquad y = g_2(s), \qquad z = g_3(s) \tag{4.43}$$

So instead of using the vector functions $\mathbf{f}(t)$ or $\mathbf{g}(s)$ to represent the space curve C, we often use $\mathbf{r}(t)$ or $\mathbf{r}(s)$. Thus

$$\mathbf{f}(t) = \mathbf{r}(t) = x(t)\mathbf{i} + y(t)\mathbf{j} + z(t)\mathbf{k}, \qquad a \leqslant t \leqslant b \tag{4.44}$$

$$\mathbf{g}(s) = \mathbf{r}(s) = x(s)\mathbf{i} + y(s)\mathbf{j} + z(s)\mathbf{k}, \qquad 0 \leqslant s \leqslant l \tag{4.45}$$

Note: Although the same functional forms $\mathbf{r}(t)$ and $\mathbf{r}(s)$ are used, we have to recognize that $\mathbf{r}(s)$ is not obtained from $\mathbf{r}(t)$ simply by changing the parameter t to parameter s. Rather we obtain $\mathbf{r}(s)$ from $\mathbf{r}(t)$ by changing the parameter t to $q(s)$. See Example 4-12 for an illustration of this point.

EXAMPLE 4-12: A curve C is denoted by the vector function

$$\mathbf{f}(t) = a\cos t\mathbf{i} + a\sin t\mathbf{j}, \qquad 0 \leqslant t \leqslant 2\pi$$

Represent curve C by a vector function having the arc length s as the parameter.

Solution: From the definition of arc length (4.39) and Example 4-11, we have

$$s = \int_0^t |\mathbf{f}'(t)|\, dt = \int_0^t a\, dt = at$$

Hence $t = s/a$, and the desired vector function is

$$\mathbf{g}(s) = \mathbf{f}\!\left(\frac{s}{a}\right) = \left[a\cos\!\left(\frac{s}{a}\right)\right]\mathbf{i} + \left[a\sin\!\left(\frac{s}{a}\right)\right]\mathbf{j}, \qquad 0 \leqslant s \leqslant 2\pi a$$

EXAMPLE 4-13: When a curve C is represented by a vector function $\mathbf{g}(s)$ with arc length s as the parameter, show that

$$\frac{d\mathbf{g}(s)}{ds} = \mathbf{g}'(s) = \mathbf{T} \tag{4.46}$$

where $\mathbf{T}$ is the unit tangent vector to the curve C at any point.

Solution: From the definition formula (4.34) for the tangent vector, we know that the tangent vector to the curve C represented by the smooth vector function $\mathbf{f}(t)$ at any point is the vector $\mathbf{f}'(t)$. Hence for $|\mathbf{f}'(t)| \neq 0$ the unit tangent vector $\mathbf{T}$ to the curve C at any point is expressed by

$$\mathbf{T} = \frac{\mathbf{f}'(t)}{|\mathbf{f}'(t)|} \tag{4.47}$$

Now since $\mathbf{g}(s) = \mathbf{f}[q(s)]$ and $t = q(s)$, we have upon differentiating and using the chain rule (4.19),

$$\frac{d\mathbf{g}(s)}{ds} = \mathbf{g}'(s) = \mathbf{f}'(t)\frac{dt}{ds} = \frac{\mathbf{f}'(t)}{ds/dt} = \frac{\mathbf{f}'(t)}{|\mathbf{f}'(t)|} \tag{4.48}$$

Hence from (4.47) and (4.48) we obtain

$$\mathbf{g}'(s) = \frac{\mathbf{f}'(t)}{|\mathbf{f}'(t)|} = \mathbf{T}$$

EXAMPLE 4-14: Find the unit tangent vector to the curve C represented by the vector function

$$\mathbf{f}(t) = (a \cos t)\mathbf{i} + (a \sin t)\mathbf{j}, \qquad 0 \leqslant t \leqslant 2\pi$$

at $t = \pi/2$ (See Example 4-9.)

Solution: From Example 4-12 the curve C is also represented by

$$\mathbf{g}(s) = \left[a \cos\left(\frac{s}{a}\right) \right]\mathbf{i} + \left[a \sin\left(\frac{s}{a}\right) \right]\mathbf{j}, \qquad 0 \leqslant s \leqslant 2\pi a$$

From result (4.46) the unit tangent vector $\mathbf{T}$ to C at any point is

$$\mathbf{T} = \mathbf{g}'(s) = -\left[\sin\left(\frac{s}{a}\right) \right]\mathbf{i} + \left[\cos\left(\frac{s}{a}\right) \right]\mathbf{j}, \qquad 0 \leqslant s \leqslant 2\pi a$$

Now, from Example 4-12 we have $s = at \big|_{t=\pi/2} = \frac{1}{2}a\pi$ at $t = \pi/2$. Hence at point $t = \pi/2$ (or $s = a\pi/2$) the unit tangent vector is

$$\mathbf{T} = -\left[\sin\left(\frac{\pi}{2}\right) \right]\mathbf{i} + \left[\cos\left(\frac{\pi}{2}\right) \right]\mathbf{j} = -\mathbf{i}$$

since $\sin(\pi/2) = 1$ and $\cos(\pi/2) = 0$.

Alternate Solution: Using the results of Example 4-11 and from representation (4.47) for the unit tangent vector, we obtain

$$\mathbf{T} = \frac{\mathbf{f}'(t)}{|\mathbf{f}'(t)|} = \frac{1}{a}[-(a \sin t)\mathbf{i} + (a \cos t)\mathbf{j}]$$

$$= -(\sin t)\mathbf{i} + (\cos t)\mathbf{j}, \qquad 0 \leqslant t \leqslant 2\pi$$

Hence at $t = \pi/2$,

$$\mathbf{T} = -\left[\sin\left(\frac{\pi}{2}\right) \right]\mathbf{i} + \left[\cos\left(\frac{\pi}{2}\right) \right]\mathbf{j} = -\mathbf{i}$$

EXAMPLE 4-15: Show that the vector defined by $d\mathbf{T}/ds$ is normal, or perpendicular, to the unit tangent vector $\mathbf{T}$ at any point on the curve C.

Solution: Since $\mathbf{T}$ is the unit tangent vector, the dot product yields

$$\mathbf{T} \cdot \mathbf{T} = 1$$

Differentiating both sides with respect to s, we obtain

$$\frac{d\mathbf{T}}{ds} \cdot \mathbf{T} = 0 \qquad\qquad (4.49)$$

which implies that $d\mathbf{T}/ds$ is normal to $\mathbf{T}$.

4-5. Surfaces

A. Curvilinear coordinates and surfaces in space

In Section 4-4 vector equations of the type

$$\mathbf{r}(t) = x(t)\mathbf{i} + y(t)\mathbf{j} + z(t)\mathbf{k} \qquad\qquad (4.50)$$

in the single parameter t describe space curves. The parametric representation of the space curves is

$$x = x(t), \qquad y = y(t), \qquad z = z(t) \qquad\qquad (4.51)$$

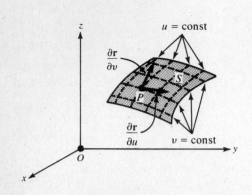

Figure 4-3
A surface in space.

Surfaces, in general, are described by the parametric equations of the type

SURFACE IN
SPACE $$x = x(u, v), \qquad y = y(u, v), \qquad z = z(u, v) \qquad (4.52)$$

where u and v are parameters. If v is fixed constant, that is, $v = c$, then the definition of a surface in (4.52) becomes a one-parameter expression describing a space curve along which u varies. This is the curve designated by the equation $v = c$. Thus for each v there exists a space curve. Similarly, v varies along the curve $u = k$, where k is a constant. The loci of all the curves $v = c$ and $u = k$ form a surface S. The parameters u and v are called the **curvilinear coordinates** of the points on the surfaces, and the u-curves and v-curves are called **parametric curves** (see Figure 4-3.)

If the terminal point of the position vector **r** generates the surface S, then definition (4.52) can be rewritten as

$$\mathbf{r}(u, v) = x(u, v)\mathbf{i} + y(u, v)\mathbf{j} + z(u, v)\mathbf{k} \qquad (4.53)$$

EXAMPLE 4-16: Show that the unit vector **n**, defined by

$$\mathbf{n} = \frac{\mathbf{r}_u \times \mathbf{r}_v}{|\mathbf{r}_u \times \mathbf{r}_v|} \qquad (4.54)$$

where $\mathbf{r}_u = \partial \mathbf{r}/\partial u$ and $\mathbf{r}_v$ $\partial \mathbf{r}/\partial v$ is normal to the surface S represented by $\mathbf{r}(u, v)$ if $\mathbf{r}_u \times \mathbf{r}_v \neq \mathbf{0}$.

Solution: According to definition equation (4.34), at any point P,

$$\mathbf{r}_u = \frac{\partial \mathbf{r}}{\partial u} = \frac{\partial x}{\partial u}\mathbf{i} + \frac{\partial y}{\partial u}\mathbf{j} + \frac{\partial z}{\partial u}\mathbf{k} \qquad (4.55)$$

is a tangent vector to a constant curve $v = C$ at P. Similarly,

$$\mathbf{r}_v = \frac{\partial \mathbf{r}}{\partial v} = \frac{\partial x}{\partial v}\mathbf{i} + \frac{\partial y}{\partial v}\mathbf{j} + \frac{\partial z}{\partial v}\mathbf{k} \qquad (4.56)$$

is a tangent vector to a constant curve $u = k$ at P. So it follows that at any point P the vector $\mathbf{r}_u \times \mathbf{r}_v$ is normal to the surface S at P. Then because $|\mathbf{r}_u \times \mathbf{r}_v|$ is the magnitude of this vector,

$$\mathbf{n} = \frac{\mathbf{r}_u \times \mathbf{r}_v}{|\mathbf{r}_u \times \mathbf{r}_v|}$$

is a unit normal vector to S at P.

A point $\mathbf{r}(u, v)$ on a surface S is called a **singular point** if $\mathbf{r}_u \times \mathbf{r}_v = \mathbf{0}$. If $\mathbf{r}_u \times \mathbf{r}_v \neq \mathbf{0}$, it is called a **nonsingular point**. If $\mathbf{r}_u$ and $\mathbf{r}_v$ are continuous, then the tangent planes exist only at the nonsingular points. Geometrically, the condition for $\mathbf{r}_u \times \mathbf{r}_v \neq \mathbf{0}$ is that the curves $u = k$ and $v = c$, where k and c are constant, are nonsingular and are not tangent to each other at their point of intersection.

B. Differential element of surface area

The **differential element of surface area** is a vector given by

DIFFERENTIAL
ELEMENT $$d\mathbf{S} = \frac{\partial \mathbf{r}}{\partial u}\, du \times \frac{\partial \mathbf{r}}{\partial v}\, dv = \mathbf{r}_u \times \mathbf{r}_v\, du\, dv \qquad (4.57)$$

From the definition of a unit normal vector (4.54), we see that $d\mathbf{S}$ is a vector normal to the surface represented by $\mathbf{r}(u, v)$ at any point P; its magnitude,

$$dS = |d\mathbf{S}| = |\mathbf{r}_u \times \mathbf{r}_v|\, du\, dv \qquad (4.58)$$

is approximately equal to the surface area S that is bound by four curves on S (see Figure 4-4).

From the unit normal vector $\mathbf{n}$ defined in (4.54) and the magnitude of $d\mathbf{S}$ we can rewrite the equation of the differential element (4.57) as

$$dS = \mathbf{n}\, dS \qquad (4.59)$$

We then obtain the surface area of S by integrating (4.59) over S:

$$S = \iint dS = \iint \mathbf{n} \cdot d\mathbf{S} \qquad (4.60)$$

where $\mathbf{n}$ is a unit normal vector of S at any point P.

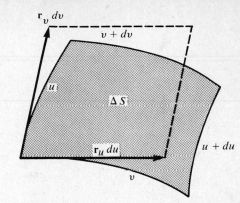

Figure 4-4
A differential element of surface area.

EXAMPLE 4-17: Find the surface area of the surface S represented by

$$x = a\sin\theta\cos\phi, \qquad y = a\sin\theta\sin\phi, \qquad z = a\cos\theta \qquad (4.61)$$

where $0 < \theta < \pi$ and $0 < \phi < 2\pi$.

Solution: In this representation the two parameters θ and ϕ are used instead of u and v. Hence

$$\mathbf{r}_\theta = \frac{\partial \mathbf{r}}{\partial \theta} = \frac{\partial x}{\partial \theta}\mathbf{i} + \frac{\partial y}{\partial \theta}\mathbf{j} + \frac{\partial z}{\partial \theta}\mathbf{k}$$

$$= a\cos\theta\cos\phi\,\mathbf{i} + a\cos\theta\sin\phi\,\mathbf{j} - a\sin\theta\,\mathbf{k}$$

$$\mathbf{r}_\phi = \frac{\partial \mathbf{r}}{\partial \phi} = \frac{\partial x}{\partial \phi}\mathbf{i} + \frac{\partial y}{\partial \phi}\mathbf{j} + \frac{\partial z}{\partial \phi}\mathbf{k}$$

$$= -a\sin\theta\sin\phi\,\mathbf{i} + a\sin\theta\cos\phi\,\mathbf{j}$$

So the vector product is

$$\mathbf{r}_\theta \times \mathbf{r}_\phi = \begin{vmatrix} \mathbf{i} & \mathbf{j} & \mathbf{k} \\ a\cos\theta\cos\phi & a\cos\theta\sin\phi & -a\sin\theta \\ -a\sin\theta\sin\phi & a\sin\theta\cos\phi & 0 \end{vmatrix}$$

$$= a^2\sin^2\theta\cos\phi\,\mathbf{i} + a^2\sin^2\theta\sin\phi\,\mathbf{j} + a^2\sin\theta\cos\theta\,\mathbf{k}$$

and its magnitude is

$$|\mathbf{r}_\theta \times \mathbf{r}_\phi| = a^2\sin\theta[\sin^2\theta\cos^2\phi + \sin^2\theta\sin^2\phi + \cos^2\theta]^{1/2}$$

$$= a^2\sin\theta[\sin^2\theta(\cos^2\phi + \sin^2\phi) + \cos^2\theta]^{1/2}$$

$$= a^2\sin\theta$$

Hence the magnitude of the differential element of the surface area (4.58) is

$$dS = a^2\sin\theta\, d\theta\, d\phi \qquad (4.62)$$

Then we obtain the surface area by integrating:

$$S = \iint dS = \int_0^{2\pi}\int_0^{\pi} a^2\sin\theta\, d\theta\, d\phi = 2\pi a^2 \int_0^{\pi} \sin\theta\, d\theta$$

$$= 2\pi a^2(-\cos\theta)\big|_0^{\pi}$$

$$= 2\pi a^2\{-[(-1) - 1]\}$$

$$= 4\pi a^2$$

Note: Representations (4.61) are the parametric equations for a sphere expressed by

$$x^2 + y^2 + z^2 = a^2 \qquad (4.63)$$

Using Figure 4-5, we can obtain (4.62) by observation. If we hold ϕ fixed and vary θ by $d\theta$, we obtain an arc of length $a\,d\theta$. Holding θ fixed and varying ϕ by $d\phi$, we obtain an arc of a circle of radius $a\sin\theta$ with the length $a\sin\theta\,d\phi$. Hence the element of differential area dS can be expressed as

$$dS = a\,d\theta\,a\sin\theta\,d\phi = a^2\sin\theta\,d\theta\,d\phi$$

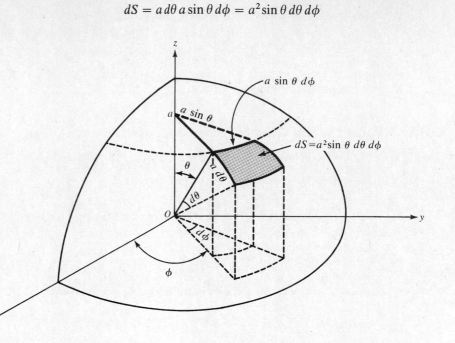

Figure 4-5
A differential element of surface area of a sphere.

4-6. Tangent, Curvature, and Torsion

In this section, we'll derive some formulas for space curves by vector methods.

A. Trihedral

We saw in Section 4-4 that the space curve C can be represented by the vector function $\mathbf{r}(t)$ with a scalar parameter t; that is, $\mathbf{r}(t) = x(t)\mathbf{i} + y(t)\mathbf{j} + z(t)\mathbf{k}$ (4.44). Then using (4.47), we can express the unit tangent vector $\mathbf{T}$ to curve C at any point on C as

$$\mathbf{T} = \frac{\mathbf{r}'(t)}{|\mathbf{r}'(t)|}$$

where $\mathbf{r}'(t) = d\mathbf{r}(t)/dt$ and $|\mathbf{r}'(t)| \neq 0$.

If the parameter t is replaced by arc length s, which measures the distance along C from some fixed point on the curve, then C can be represented by $\mathbf{r}(s) = x(s)\mathbf{i} + y(s)\mathbf{j} + z(s)\mathbf{k}$ (4.45). Then using (4.46), we can express the unit tangent vector to C by

$$\mathbf{T} = \mathbf{r}'(s) = \frac{d\mathbf{r}(s)}{ds}$$

in the direction of increasing s.

If $\mathbf{T}$ is the unit tangent vector to a curve C at any point on C, then its derivative $\mathbf{T}'(s) = d\mathbf{T}(s)/ds$ is normal to $\mathbf{T}$, as we saw in Example 4-15.

Definitions:

- The **unit principal normal vector** $\mathbf{N}$ is defined by

$$\frac{d\mathbf{T}}{ds} = \kappa\mathbf{N} \tag{4.64}$$

where

$$\kappa = \left| \frac{d\mathbf{T}}{ds} \right| \qquad (4.65)$$

is called the *curvature* of curve C at any point. The *radius of curvature* ρ of C at any point is defined by

$$\rho = \frac{1}{\kappa} \qquad (4.66)$$

where $\kappa \neq 0$.

- The **unit binormal vector B** to curve C at any point is defined by

$$\mathbf{B} = \mathbf{T} \times \mathbf{N} \qquad (4.67)$$

Its magnitude is

$$|\mathbf{B}| = |\mathbf{T} \times \mathbf{N}| = |\mathbf{T}||\mathbf{N}| \sin\frac{\pi}{2} = 1 \qquad (4.68)$$

since both $\mathbf{T}$ and $\mathbf{N}$ are unit vectors. The binormal vector $\mathbf{B}$ is perpendicular to the derivative of $\mathbf{B}$ with respect to s; that is

$$\mathbf{B} \perp \frac{d\mathbf{B}}{ds} \qquad (4.69)$$

- The **torsion** τ of a curve C at any point is defined by

$$\frac{d\mathbf{B}}{ds} = -\tau\mathbf{N} \qquad (4.70)$$

The reciprocal of τ, denoted by $\sigma = 1/\tau$, is called the *radius of torsion*.

- The unit tangent vector $\mathbf{T}$, the unit principal normal vector $\mathbf{N}$, and the unit binormal vector $\mathbf{B}$ are said to be **trihedral** at any point on C because they form a right-handed system of unit vectors; that is,

TRIHEDRAL $\qquad \mathbf{B} = \mathbf{T} \times \mathbf{N}, \qquad \mathbf{N} = \mathbf{B} \times \mathbf{T}, \qquad \mathbf{T} = \mathbf{N} \times \mathbf{B} \qquad (4.71)$

EXAMPLE 4-18: Verify that $\mathbf{B}$ and $d\mathbf{B}/ds$ are perpendicular (4.69).

Proof: Since $|\mathbf{B}| = 1$,

$$\mathbf{B} \cdot \mathbf{B} = B^2 = 1$$

Differentiating both sides with respect to s yields $(d\mathbf{B}/ds) \cdot \mathbf{B} = 0$. Since vectors whose scalar product is zero are perpendicular (see Example 1-3), we conclude that

$$\mathbf{B} \perp d\mathbf{B}/ds$$

EXAMPLE 4-19: Show that the trihedral equations (4.71) are consistent.

Solution: From the definition of the unit binormal vector (4.67) and the definition of cross product, we have

$$\mathbf{B} \cdot \mathbf{T} = 0 \quad \text{and} \quad \mathbf{B} \cdot \mathbf{N} = 0$$

Differentiating both sides of the first equation with respect to s,

$$\frac{d\mathbf{B}}{ds} \cdot \mathbf{T} + \mathbf{B} \cdot \frac{d\mathbf{T}}{ds} = 0$$

Thus, using the definition of the unit principal normal vector (4.64), we write

$$\frac{d\mathbf{B}}{ds} \cdot \mathbf{T} = -\mathbf{B} \cdot \frac{d\mathbf{T}}{ds} = -\mathbf{B} \cdot (\kappa\mathbf{N}) = -\kappa\mathbf{B} \cdot \mathbf{N} = 0$$

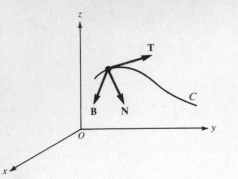

Figure 4-6
Trihedral.

which implies that $d\mathbf{B}/ds$ is also orthogonal to $\mathbf{T}$. Since $\mathbf{T}$ and $\mathbf{N}$ are perpendicular, $d\mathbf{B}/ds$ must be parallel to $\mathbf{N}$. So

$$\frac{d\mathbf{B}}{ds} = -\tau\mathbf{N}$$

From these results we conclude that $\mathbf{T}$, $\mathbf{N}$, and $\mathbf{B}$ form a right-handed system of unit vectors called a trihedral, as shown in Figure 4-6. So we have proved that

$$\mathbf{B} = \mathbf{T} \times \mathbf{N}, \qquad \mathbf{N} = \mathbf{B} \times \mathbf{T}, \qquad \mathbf{T} = \mathbf{N} \times \mathbf{B}$$

B. Frenet–Serret formulas

$$\frac{d\mathbf{T}}{ds} = \kappa\mathbf{N} \tag{4.72a}$$

$$\frac{d\mathbf{B}}{ds} = -\tau\mathbf{N} \tag{4.72b}$$

$$\frac{d\mathbf{N}}{ds} = \tau\mathbf{B} - \kappa\mathbf{T} \tag{4.72c}$$

These formulas are known collectively as the **Frenet–Serret formulas**. [Note that (4.72a) and (4.72b) are the equations that define the unit principal normal vector (4.64) and the torsion (4.70), respectively, of a curve.]

EXAMPLE 4-20: Verify the third Frenet–Serret formula (4.72c).

Proof: Using the definition of the trihedral (4.71),

$$\mathbf{N} = \mathbf{B} \times \mathbf{T}$$

we take its derivative

$$\frac{d\mathbf{N}}{ds} = \frac{d\mathbf{B}}{ds} \times \mathbf{T} + \mathbf{B} \times \frac{d\mathbf{T}}{ds}$$

Then using definition (4.70) for $d\mathbf{B}/ds$ and (4.64) for $d\mathbf{T}/ds$, we can write

$$\frac{d\mathbf{N}}{ds} = -\tau\mathbf{N} \times \mathbf{T} + \mathbf{B} \times (\kappa\mathbf{N})$$

$$= \tau(\mathbf{T} \times \mathbf{N}) - \kappa(\mathbf{N} \times \mathbf{B})$$

$$= \tau\mathbf{B} - \kappa\mathbf{T}$$

SUMMARY

1. The derivative $\mathbf{f}'(t)$ of a vector function $\mathbf{f}(t)$ is defined by

$$\mathbf{f}'(t) = \frac{d\mathbf{f}(t)}{dt} = \lim_{\Delta t \to 0} \frac{\mathbf{f}(t + \Delta t) - \mathbf{f}(t)}{\Delta t}$$

If $\mathbf{f}(t) = f_1(t)\mathbf{i} + f_2(t)\mathbf{j} + f_3(t)\mathbf{k}$, then

$$\mathbf{f}'(t) = f'_1(t)\mathbf{i} + f'_2(t)\mathbf{j} + f'_3(t)\mathbf{k}$$

Higher derivatives of $\mathbf{f}(t)$ are similarly defined by

$$\mathbf{f}''(t) = \frac{d^2\mathbf{f}(t)}{dt^2} = \frac{d}{dt}\left[\frac{d\mathbf{f}(t)}{dt}\right]$$

and so forth.

2. The partial derivative $\mathbf{f}_u$ of $\mathbf{f}(u, v, w)$ with respect to u is defined by

$$\mathbf{f}_u = \frac{\partial \mathbf{f}}{\partial u} = \lim_{\Delta u \to 0} \frac{\mathbf{f}(u + \Delta u, v, w) - \mathbf{f}(u, v, w)}{\Delta u}$$

The partial derivatives of $\mathbf{f}(u, v, w)$ with respect to v and w are defined similarly:

$$\mathbf{f}_v = \frac{\partial \mathbf{f}}{\partial v}, \qquad \mathbf{f}_w = \frac{\partial \mathbf{f}}{\partial w}$$

3. A space curve C can be represented by

$$\mathbf{r}(t) = x(t)\mathbf{i} + y(t)\mathbf{j} + z(t)\mathbf{k}, \qquad a \leqslant t \leqslant b$$

where t is a scalar variable and $\mathbf{r}$ is the position vector of a point on the curve.

4. The arc length $s(t)$ of a curve is given by

$$s = \int_0^t |\mathbf{r}'(t)| \, dt$$

5. Using the arc length s as a parameter, the space curve C can be represented by

$$\mathbf{r}(s) = x(s)\mathbf{i} + y(s)\mathbf{j} + z(s)\mathbf{k}, \qquad 0 \leqslant s \leqslant l$$

where l is the total length of the curve C.

6. A surface S can be represented by

$$\mathbf{r}(u, v) = x(u, v)\mathbf{i} + y(u, v)\mathbf{j} + z(u, v)\mathbf{k}$$

7. The unit vector $\mathbf{n}$ normal to the surface S is given by

$$\mathbf{n} = \frac{\mathbf{r}_u \times \mathbf{r}_v}{|\mathbf{r}_u \times \mathbf{r}_v|}$$

The differential element of surface area is expressed as

$$d\mathbf{S} = \mathbf{r}_u \times \mathbf{r}_v \, du \, dv = \mathbf{n} \, dS$$

8. If a space curve C is represented by $\mathbf{r}(s)$, where s is the arc length, then the unit tangent vector $\mathbf{T}$ to C is given by

$$\mathbf{T} = \mathbf{r}'(s) = \frac{d\mathbf{r}(s)}{ds}$$

9. The unit principal normal vector $\mathbf{N}$ to C is defined by

$$\frac{d\mathbf{T}}{ds} = \kappa \mathbf{N}$$

where κ is the curvature of C and $\rho = 1/\kappa$ is the radius of curvature of C.

10. The unit binormal vector $\mathbf{B}$ to curve C is defined by

$$\frac{d\mathbf{B}}{ds} = -\tau \mathbf{N}$$

and

$$\mathbf{B} = \mathbf{T} \times \mathbf{N}$$

where τ is the torsion of C and $\sigma = 1/\tau$ is the radius of torsion.

RAISE YOUR GRADES

Can you explain . . . ?

☑ how to define the derivative of a vector function
☑ what the rules are for differentiation of vector functions

☑ why the order of factors must be preserved when differentiating the vector product of vector functions
☑ how to define the partial derivative of a vector function
☑ how to define a curve in space
☑ how to represent a space curve using the arc length
☑ how to find the unit tangent vector to a space curve
☑ how to define a surface in space
☑ how to find a unit normal vector to a surface
☑ how to define the differential element of surface area
☑ how to find the radius of curvature of a space curve

SOLVED PROBLEMS

Ordinary Differentiation of Vectors

PROBLEM 4-1   Show that the derivative of a vector of constant magnitude is orthogonal to that vector. (This includes the possibility that the derivative is the zero vector.)

Solution: If $\mathbf{B}(t)$ is set equal to $\mathbf{A}(t)$ in rule (4.15), then

$$\frac{d}{dt}\left[\mathbf{A}(t) \cdot \mathbf{A}(t)\right] = 2\mathbf{A}(t) \cdot \mathbf{A}'(t) \tag{a}$$

But, by hypothesis,

$$\mathbf{A}(t) \cdot \mathbf{A}(t) = |\mathbf{A}(t)|^2 = \text{constant}$$

Since the derivative of a constant is zero,

$$\frac{d}{dt}\left[\mathbf{A}(t) \cdot \mathbf{A}(t)\right] = 0$$

Hence from eq. (a),

$$2\mathbf{A}(t) \cdot \mathbf{A}'(t) = 0$$

Therefore,

$$\mathbf{A}(t) \cdot \mathbf{A}'(t) = 0$$

that is, the vector $\mathbf{A}(t)$ is orthogonal to its derivative.

PROBLEM 4-2   Show that

$$\frac{d}{dt}\left[\mathbf{A}(t) \times \mathbf{A}'(t)\right] = \mathbf{A}(t) \times \mathbf{A}''(t) \tag{a}$$

Solution: If $\mathbf{B}(t)$ is set equal to $\mathbf{A}'(t)$ in rule (4.16), then

$$\frac{d}{dt}\left[\mathbf{A}(t) \times \mathbf{A}'(t)\right] = \mathbf{A}(t) \times \mathbf{A}''(t) + \mathbf{A}'(t) \times \mathbf{A}'(t)$$

Since the vector product of any vector with itself is zero, we have

$$\mathbf{A}'(t) \times \mathbf{A}'(t) = \mathbf{0}$$

and eq. (a) follows.

PROBLEM 4-3 Given $f(t) = e^{2t}\mathbf{a} + e^{3t}\mathbf{b}$, where $\mathbf{a}$ and $\mathbf{b}$ are constant vectors, show that

$$\mathbf{f}''(t) - 5\mathbf{f}'(t) + 6\mathbf{f}(t) = 0$$

Solution: Since $\mathbf{f}(t) = e^{2t}\mathbf{a} + e^{3t}\mathbf{b}$, then from differentiation rules (4.13) and (4.21)

$$\mathbf{f}'(t) = 2e^{2t}\mathbf{a} + 3e^{3t}\mathbf{b}$$
$$\mathbf{f}''(t) = 4e^{2t}\mathbf{a} + 9e^{3t}\mathbf{b}$$

Therefore,

$$\mathbf{f}''(t) - 5\mathbf{f}'(t) + 6\mathbf{f}(t) = 4e^{2t}\mathbf{a} + 9e^{3t}\mathbf{b} - 10e^{2t}\mathbf{a} - 15e^{3t}\mathbf{b} + 6e^{2t}\mathbf{a} + 6e^{3t}\mathbf{b}$$
$$= (4 - 10 + 6)e^{2t}\mathbf{a} + (9 - 15 + 6)e^{3t}\mathbf{b}$$
$$= 0$$

[*Note:* This is an example of a vector differential equation being satisfied by a vector function.]

Partial Derivatives of Vector Functions of More Than One Variable

PROBLEM 4-4 If $\mathbf{f}(u, v) = e^{uv}\mathbf{i} + (u - v)\mathbf{j} + u(\sin v)\mathbf{k}$, calculate (a) $\mathbf{f}_u$, (b) $\mathbf{f}_v$, (c) $\mathbf{f}_{uu}$, and (d) $\mathbf{f}_u \times \mathbf{f}_v$.

Solution: Using formula (4.30), we obtain

(a) $\mathbf{f}_u = \dfrac{\partial}{\partial u}(e^{uv})\mathbf{i} + \dfrac{\partial}{\partial u}(u - v)\mathbf{j} + \dfrac{\partial}{\partial u}(u \sin v)\mathbf{k}$

$\quad = ve^{uv}\mathbf{i} + \mathbf{j} + (\sin v)\mathbf{k}$

(b) $\mathbf{f}_v = \dfrac{\partial}{\partial v}(e^{uv})\mathbf{i} + \dfrac{\partial}{\partial v}(u - v)\mathbf{j} + \dfrac{\partial}{\partial v}(u \sin v)\mathbf{k}$

$\quad = ue^{uv}\mathbf{i} - \mathbf{j} + u(\cos v)\mathbf{k}$

(c) $\mathbf{f}_{uu} = \dfrac{\partial}{\partial u}\mathbf{f}_u = \dfrac{\partial}{\partial u}(ve^{uv})\mathbf{i} + \dfrac{\partial}{\partial u}(1)\mathbf{j} + \dfrac{\partial}{\partial u}(\sin v)\mathbf{k}$

$\quad = v^2 e^{uv}\mathbf{i}$

Using the definition of vector product (2.28) the results of parts (a) and (b),

(d) $\mathbf{f}_u \times \mathbf{f}_v = \begin{vmatrix} \mathbf{i} & \mathbf{j} & \mathbf{k} \\ ve^{uv} & 1 & \sin v \\ ue^{uv} & -1 & u\cos v \end{vmatrix}$

$\quad = (u\cos v + \sin v)\mathbf{i} + (u\sin v - v\cos v)e^{uv}\mathbf{j} - (v + u)e^{uv}\mathbf{k}$

PROBLEM 4-5 Given $\mathbf{f}(x, y, z) = x^2\mathbf{i} + y^2\mathbf{j} + z^2\mathbf{k}$, show that $\mathbf{f}_x, \mathbf{f}_y$, and $\mathbf{f}_z$ are perpendicular to one another.

Solution: Using formula (4.30), we obtain

$$\mathbf{f}_x = \frac{\partial \mathbf{f}}{\partial x} = \frac{\partial}{\partial x}(x^2)\mathbf{i} + \frac{\partial}{\partial x}(y^2)\mathbf{j} + \frac{\partial}{\partial x}(z^2)\mathbf{k} = 2x\mathbf{i}$$

$$\mathbf{f}_y = \frac{\partial \mathbf{f}}{\partial y} = \frac{\partial}{\partial y}(x^2)\mathbf{i} + \frac{\partial}{\partial y}(y^2)\mathbf{j} + \frac{\partial}{\partial y}(z^2)\mathbf{k} = 2y\mathbf{j}$$

$$\mathbf{f}_z = \frac{\partial \mathbf{f}}{\partial z} = \frac{\partial}{\partial z}(x^2)\mathbf{i} + \frac{\partial}{\partial z}(y^2)\mathbf{j} + \frac{\partial}{\partial z}(z^2)\mathbf{k} = 2z\mathbf{k}$$

Since $\mathbf{i}, \mathbf{j}$, and $\mathbf{k}$ are perpendicular to one another, $\mathbf{f}_x, \mathbf{f}_y$, and $\mathbf{f}_z$ are perpendicular to one another.

PROBLEM 4-6   Given $\mathbf{r}(t_1, t_2) = a \cos t_1 \mathbf{i} + a \sin t_1 \mathbf{j} + t_2 \mathbf{k}$, find $\partial \mathbf{r}/\partial t_1$ and $\partial \mathbf{r}/\partial t_2$.

Solution: From formula (4.30),

$$\frac{\partial \mathbf{r}}{\partial t_1} = \frac{\partial}{\partial t_1}(a \cos t_1)\mathbf{i} + \frac{\partial}{\partial t_1}(a \sin t_1)\mathbf{j} + \frac{\partial}{\partial t_1}(t_2)\mathbf{k}$$

$$= -a \sin t_1 \mathbf{i} + a \cos t_1 \mathbf{j}$$

$$\frac{\partial \mathbf{r}}{\partial t_2} = \frac{\partial}{\partial t_2}(a \cos t_1)\mathbf{i} + \frac{\partial}{\partial t_2}(a \sin t_1)\mathbf{j} + \frac{\partial}{\partial t_2}(t_2)\mathbf{k}$$

$$= \mathbf{k}$$

Space Curves

PROBLEM 4-7   Characterize the space curve represented by the vector function

$$\mathbf{r}(t) = a \cos t\mathbf{i} + b \sin t\mathbf{j}$$

Solution: The parametric equations of the curve represented by $\mathbf{r}(t)$ are

$$x = a \cos t, \qquad y = b \sin t, \qquad z = 0$$

Then,

$$\frac{x^2}{a^2} + \frac{y^2}{b^2} = \cos^2 t + \sin^2 t = 1, \qquad z = 0$$

Thus the curve represented by $\mathbf{r}(t)$ is an ellipse in the xy-plane; the center is at the origin, and the principal axes are in the direction of the x- and y-axes.

PROBLEM 4-8   Find the unit vector $\mathbf{T}$ tangent to the curve represented by

$$\mathbf{r}(t) = t\mathbf{i} + t^2\mathbf{j} + t^3\mathbf{k}$$

at the point $(1, 1, 1)$.

Solution: By the representation of the unit tangent vector in (4.47), we have

$$\mathbf{T} = \frac{\mathbf{r}'(t)}{|\mathbf{r}'(t)|} = \frac{\mathbf{i} + 2t\mathbf{j} + 3t^2\mathbf{k}}{\sqrt{1 + 4t^2 + 9t^4}}$$

When $t = 1$, we have

$$\mathbf{T} = \frac{1}{\sqrt{14}}(\mathbf{i} + 2\mathbf{j} + 3\mathbf{k})$$

PROBLEM 4-9   Find the arc length between the points $(1, 0, 0)$ and $(0, 1, \pi/2)$ on the curve represented by

$$\mathbf{r}(t) = \cos t\mathbf{i} + \sin t\mathbf{j} + t\mathbf{k}$$

Solution: For this curve

$$\mathbf{r}'(t) = -\sin t\mathbf{i} + \cos t\mathbf{j} + \mathbf{k}$$

From the definition of arc length (4.39), we have

$$s = \int_0^{\pi/2} |\mathbf{r}'(t)|\,dt$$

$$= \int_0^{\pi/2} \sqrt{\sin^2 t + \cos^2 t + 1}\,dt = \int_0^{\pi/2} \sqrt{2}\,dt = \sqrt{2}\,\frac{\pi}{2} = \frac{\pi}{\sqrt{2}}$$

[Note that the curve is a helix winding about the z-axis. It is on the cylindrical surface $x^2 + y^2 = 1$, as shown in Figure 4-7.]

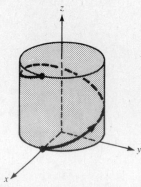

Figure 4-7
Helix.

Surfaces

PROBLEM 4-10 Find a unit normal vector **n** for a surface S represented by

$$x = x, \qquad y = y, \qquad z = z(x, y)$$

where x and y are parameters. (This would represent the "usual" surface in rectangular three-dimensional coordinates.)

Solution: From (4.53) we can represent the surface S by

$$\mathbf{r}(x, y) = x\mathbf{i} + y\mathbf{j} + z(x, y)\mathbf{k}$$

So, the partial derivatives are

$$\mathbf{r}_x = \frac{\partial \mathbf{r}}{\partial x} = \frac{\partial x}{\partial x}\mathbf{i} + \frac{\partial y}{\partial x}\mathbf{j} + \frac{\partial z}{\partial x}\mathbf{k} = \mathbf{i} + \frac{\partial z}{\partial x}\mathbf{k}$$

$$\mathbf{r}_y = \frac{\partial \mathbf{r}}{\partial y} = \frac{\partial x}{\partial y}\mathbf{i} + \frac{\partial y}{\partial y}\mathbf{j} + \frac{\partial z}{\partial y}\mathbf{k} = \mathbf{j} + \frac{\partial z}{\partial y}\mathbf{k}$$

The vector product of $\mathbf{r}_x$ and $\mathbf{r}_y$ is

$$\mathbf{r}_x \times \mathbf{r}_y = \begin{vmatrix} \mathbf{i} & \mathbf{j} & \mathbf{k} \\ 1 & 0 & \frac{\partial z}{\partial x} \\ 0 & 1 & \frac{\partial z}{\partial y} \end{vmatrix} = -\frac{\partial z}{\partial x}\mathbf{i} - \frac{\partial z}{\partial y}\mathbf{j} + \mathbf{k}$$

whose magnitude is

$$|\mathbf{r}_x \times \mathbf{r}_y| = \left[\left(\frac{\partial z}{\partial x}\right)^2 + \left(\frac{\partial z}{\partial y}\right)^2 + 1 \right]^{1/2}$$

So from definition (4.54), the unit normal vector is

$$\mathbf{n} = \frac{\mathbf{r}_x \times \mathbf{r}_y}{|\mathbf{r}_x \times \mathbf{r}_y|} = \frac{-\frac{\partial z}{\partial x}\mathbf{i} - \frac{\partial z}{\partial y}\mathbf{j} + \mathbf{k}}{\left[\left(\frac{\partial z}{\partial x}\right)^2 + \left(\frac{\partial z}{\partial y}\right)^2 + 1 \right]^{1/2}}$$

PROBLEM 4-11 Find a unit normal vector **n** to a surface S that is represented by

$$\phi(x, y, z) = 0 \qquad\qquad (a)$$

Solution: If we regard eq. (a) as implicitly defining z as a function of x and y, we can assume that $\phi_z \neq 0$. Then from elementary calculus, and using the results of Problem 4-10, we can find the partial derivatives:

$$\mathbf{r}_x = \mathbf{i} + \frac{\partial z}{\partial x}\mathbf{k} = \mathbf{i} - \frac{\frac{\partial \phi}{\partial x}}{\frac{\partial \phi}{\partial z}}\mathbf{k} = \mathbf{i} - \frac{\phi_x}{\phi_z}\mathbf{k}$$

$$\mathbf{r}_y = \mathbf{j} + \frac{\partial z}{\partial y}\mathbf{k} = \mathbf{j} - \frac{\frac{\partial \phi}{\partial y}}{\frac{\partial \phi}{\partial z}}\mathbf{k} = \mathbf{j} - \frac{\phi_y}{\phi_z}\mathbf{k}$$

Hence the vector product is

$$\mathbf{r}_x \times \mathbf{r}_y = \begin{vmatrix} \mathbf{i} & \mathbf{j} & \mathbf{k} \\ 1 & 0 & -\dfrac{\phi_x}{\phi_z} \\ 0 & 1 & -\dfrac{\phi_y}{\phi_z} \end{vmatrix}$$

$$= \frac{\phi_x}{\phi_z}\mathbf{i} + \frac{\phi_y}{\phi_z}\mathbf{j} + \mathbf{k}$$

$$= \frac{1}{\phi_z}(\phi_x\mathbf{i} + \phi_y\mathbf{j} + \phi_z\mathbf{k})$$

so that the unit normal vector is

$$\mathbf{n} = \frac{\mathbf{r}_x \times \mathbf{r}_y}{|\mathbf{r}_x \times \mathbf{r}_y|} = \frac{\phi_x\mathbf{i} + \phi_y\mathbf{j} + \phi_z\mathbf{k}}{[(\phi_x)^2 + (\phi_y)^2 + (\phi_z)^2]^{1/2}}$$

$$= \frac{\dfrac{\partial\phi}{\partial x}\mathbf{i} + \dfrac{\partial\phi}{\partial y}\mathbf{j} + \dfrac{\partial\phi}{\partial z}\mathbf{k}}{\left[\left(\dfrac{\partial\phi}{\partial x}\right)^2 + \left(\dfrac{\partial\phi}{\partial y}\right)^2 + \left(\dfrac{\partial\phi}{\partial z}\right)^2\right]^{1/2}}$$

[*Note:* If $\phi_x \neq 0$ or $\phi_y \neq 0$, a similar computation would yield the same result.]

PROBLEM 4-12   Find a unit normal vector $\mathbf{n}$ for a surface S that is represented by

$$x^2 + y^2 + z^2 = a^2$$

Solution: Note that the surface S is a sphere whose center is at the origin and whose radius is a. The sphere S can be represented by

$$\phi(x, y, z) = 0$$

where $\phi(x, y, z) = x^2 + y^2 + z^2 - a^2$. Now,

$$\frac{\partial\phi}{\partial x} = 2x, \qquad \frac{\partial\phi}{\partial y} = 2y, \qquad \frac{\partial\phi}{\partial z} = 2z$$

$$\left[\left(\frac{\partial\phi}{\partial x}\right)^2 + \left(\frac{\partial\phi}{\partial y}\right)^2 + \left(\frac{\partial\phi}{\partial z}\right)^2\right]^{1/2} = 2(x^2 + y^2 + z^2)^{1/2} = 2a$$

Then from the result of Problem 4-11 we have

$$\mathbf{n} = \frac{x}{a}\mathbf{i} + \frac{y}{a}\mathbf{j} + \frac{z}{a}\mathbf{k}$$

PROBLEM 4-13   Express the differential element of surface area dS for a surface S that is represented by $z = z(x, y)$.

Solution: From the result of Problem 4-10, we have

$$|\mathbf{r}_x \times \mathbf{r}_y| = \left[1 + \left(\frac{\partial z}{\partial x}\right)^2 + \left(\frac{\partial z}{\partial y}\right)^2\right]^{1/2}$$

So using result (4.58), we have

$$dS = |\mathbf{r}_x \times \mathbf{r}_y|\, dx\, dy = \left[1 + \left(\frac{\partial z}{\partial x}\right)^2 + \left(\frac{\partial z}{\partial y}\right)^2\right]^{1/2} dx\, dy$$

Tangent, Curvature, and Torsion

PROBLEM 4-14 Show that the Frenet–Serret formulas (4.72a, b, c) can be rewritten as follows:

$$\frac{d\mathbf{T}}{ds} = \boldsymbol{\omega} \times \mathbf{T}$$

$$\frac{d\mathbf{B}}{ds} = \boldsymbol{\omega} \times \mathbf{B}$$

$$\frac{d\mathbf{N}}{ds} = \boldsymbol{\omega} \times \mathbf{N}$$

where

$$\boldsymbol{\omega} = \tau\mathbf{T} + \kappa\mathbf{B}$$

This vector $\boldsymbol{\omega}$ is the **Darboux vector** of the curve C.

Solution: If $\boldsymbol{\omega} = \tau\mathbf{T} + \kappa\mathbf{B}$, then upon using the trihedral definition formulas (4.71), we can write

$$\boldsymbol{\omega} \times \mathbf{T} = (\tau\mathbf{T} + \kappa\mathbf{B}) \times \mathbf{T} = \tau\mathbf{T} \times \mathbf{T} + \kappa\mathbf{B} \times \mathbf{T} = \kappa\mathbf{N}$$

$$\boldsymbol{\omega} \times \mathbf{B} = (\tau\mathbf{T} + \kappa\mathbf{B}) \times \mathbf{B} = \tau\mathbf{T} \times \mathbf{B} + \kappa\mathbf{B} \times \mathbf{B} = -\tau\mathbf{N}$$

$$\boldsymbol{\omega} \times \mathbf{N} = (\tau\mathbf{T} + \kappa\mathbf{B}) \times \mathbf{N} = \tau\mathbf{T} \times \mathbf{N} + \kappa\mathbf{B} \times \mathbf{N} = \tau\mathbf{B} - \kappa\mathbf{T}$$

These now match the Frenet–Serret formulas.

PROBLEM 4-15 Show that

$$\frac{d\mathbf{T}}{ds} \cdot \frac{d\mathbf{B}}{ds} = -\kappa\tau$$

$$\mathbf{B} \cdot \frac{d\mathbf{N}}{ds} = \tau$$

$$\mathbf{T} \cdot \frac{d\mathbf{N}}{ds} = -\kappa$$

[Note that these may now be used to calculate curvature and torsion.]

Solution: Using the Frenet–Serret formulas (4.72a, b, c) and the fact that $\mathbf{N} \cdot \mathbf{N} = N^2 = 1$, we have

$$\frac{d\mathbf{T}}{ds} \cdot \frac{d\mathbf{B}}{ds} = \kappa\mathbf{N} \cdot (-\tau\mathbf{N}) = -\kappa\tau(\mathbf{N} \cdot \mathbf{N}) = -\kappa\tau$$

Since $\mathbf{B} \cdot \mathbf{B} = B^2 = 1$ and $\mathbf{B} \cdot \mathbf{T} = 0$, we obtain

$$\mathbf{B} \cdot \frac{d\mathbf{N}}{ds} = \mathbf{B} \cdot (\tau\mathbf{B} - \kappa\mathbf{T}) = \tau(\mathbf{B} \cdot \mathbf{B}) - \kappa(\mathbf{B} \cdot \mathbf{T}) = \tau$$

Since $\mathbf{T} \cdot \mathbf{T} = T^2 = 1$ and $\mathbf{T} \cdot \mathbf{B} = 0$, we have

$$\mathbf{T} \cdot \frac{d\mathbf{N}}{ds} = \mathbf{T} \cdot (\tau\mathbf{B} - \kappa\mathbf{T}) = \tau(\mathbf{T} \cdot \mathbf{B}) - \kappa(\mathbf{T} \cdot \mathbf{T}) = -\kappa$$

PROBLEM 4-16 Show that a necessary and sufficient condition for a curve C to be a straight line is that its curvature κ be zero.

Solution: If C is a straight line, then the unit tangent vector $\mathbf{T}$ is a constant vector; that is,

$$\frac{d\mathbf{T}}{ds} = \kappa\mathbf{N} = \mathbf{0}$$

This yields $\kappa = 0$, since $\mathbf{N} \neq \mathbf{0}$.

Conversely, if $\kappa = 0$, then we have $d\mathbf{T}/ds = \mathbf{0}$; consequently, $\mathbf{T}$ is a constant vector and C is a straight line.

PROBLEM 4-17   A curve that lies in a space plane is called a **plane curve**. A curve that doesn't lie in a space plane is called a **twisted curve**. Show that a necessary and sufficient condition for a curve C to be a plane curve is that the torsion τ of C be zero. (Assume that $\kappa \neq 0$.)

Solution: If C lies in a plane, we can choose the origin of this plane so that both vectors $\mathbf{r}'(s) = d\mathbf{r}(s)/ds$ and $\mathbf{r}''(s) = d^2\mathbf{r}(s)/ds^2$ also lie in this plane. Hence $\mathbf{T} = \mathbf{r}'(s)$ and

$$\mathbf{N} = \frac{1}{\kappa}\frac{d\mathbf{T}}{ds} = \frac{1}{\kappa}\mathbf{r}''(s)$$

will also lie in the plane. Consequently, $\mathbf{B} = \mathbf{T} \times \mathbf{N}$ is a constant vector normal to the plane. Thus

$$\frac{d\mathbf{B}}{ds} = -\tau\mathbf{N} = \mathbf{0}$$

which implies that $\tau = 0$.

Conversely, if $\tau = 0$, then $d\mathbf{B}/ds = -\tau\mathbf{N} = \mathbf{0}$; hence, $\mathbf{B}$ is a constant vector. Let the curve C be represented by $\mathbf{r}(s)$. Then, since $\mathbf{T} \perp \mathbf{B}$ and $\tau = 0$,

$$\frac{d}{ds}(\mathbf{r}\cdot\mathbf{B}) = \frac{d\mathbf{r}}{ds}\cdot\mathbf{B} + \mathbf{r}\cdot\frac{d\mathbf{B}}{ds} = \mathbf{T}\cdot\mathbf{B} + -\tau\mathbf{r}\cdot\mathbf{N} = 0$$

Since $\mathbf{r}\cdot\mathbf{B} = 0$ is the equation of a plane normal to $\mathbf{B}$ (see (3.7)), it follows that the curve is a plane curve.

PROBLEM 4-18   Show that if the curve C is represented by $r(t)$, which is a twice-differentiable function, then the curvature κ of C at any point is

$$\kappa = \frac{|\mathbf{r}'(t) \times \mathbf{r}''(t)|}{|\mathbf{r}'(t)|^3} \tag{a}$$

Solution: Let $\mathbf{r}(t) = \mathbf{r}[s(t)]$, where s is the arc length. The derivative of $\mathbf{r}(t)$ is

$$\mathbf{r}'(t) = \frac{d\mathbf{r}(t)}{dt} = \frac{d\mathbf{r}}{ds}\frac{ds}{dt} = \frac{ds}{dt}\mathbf{T} \tag{b}$$

and because $|\mathbf{T}| = 1$, the magnitude of $\mathbf{r}'(t)$ is

$$|\mathbf{r}'(t)| = \left|\frac{ds}{dt}\right| \tag{c}$$

Next, using the Frenet–Serret formulas (4.72a, b, c), we have

$$\frac{d\mathbf{T}}{dt} = \frac{d\mathbf{T}}{ds}\frac{ds}{dt} = \kappa\mathbf{N}\frac{ds}{dt} \tag{d}$$

Now differentiating eq. (b) using eq. (d), we obtain

$$\mathbf{r}''(t) = \frac{d}{dt}\mathbf{r}'(t) = \frac{d^2s}{dt^2}\mathbf{T} + \frac{ds}{dt}\frac{d\mathbf{T}}{dt} = \frac{d^2s}{dt^2}\mathbf{T} + \left(\frac{ds}{dt}\right)^2\kappa\mathbf{N} \tag{e}$$

Combining the results from eqs. (b) and (e), we find

$$\mathbf{r}'(t) \times \mathbf{r}''(t) = \left(\frac{ds}{dt}\mathbf{T}\right) \times \left[\frac{d^2s}{dt^2}\mathbf{T} + \left(\frac{ds}{dt}\right)^2\kappa\mathbf{N}\right]$$

$$= \left(\frac{ds}{dt}\right)^3\kappa\mathbf{T} \times \mathbf{N}$$

$$= \left(\frac{ds}{dt}\right)^3\kappa\mathbf{B} \tag{f}$$

and because $|\mathbf{B}| = 1$, the magnitude is

$$|\mathbf{r}'(t) \times \mathbf{r}''(t)| = \left|\frac{ds}{dt}\right|^3 \kappa = |\mathbf{r}'(t)|^3 \kappa \qquad \text{(g)}$$

Hence, the curvature is

$$\kappa = \frac{|\mathbf{r}'(t) \times \mathbf{r}''(t)|}{|\mathbf{r}'(t)|^3}$$

PROBLEM 4-19 Show that if the curve C is represented by $\mathbf{r}(t)$, which is thrice-differentiable, then the torsion τ of C at any point is

$$\tau = \frac{[\mathbf{r}'(t)\mathbf{r}''(t)\mathbf{r}'''(t)]}{|\mathbf{r}'(t) \times \mathbf{r}''(t)|^2} \qquad \text{(a)}$$

Solution: Differentiating the second derivative found in eq. (e) of Problem 4-18, we obtain

$$\mathbf{r}'''(t) = \frac{d\mathbf{r}''(t)}{dt}$$

$$= \frac{d}{dt}\left[\frac{d^2 s}{dt^2}\mathbf{T} + \left(\frac{ds}{dt}\right)^2 \kappa\mathbf{N}\right]$$

$$= \frac{d^3 s}{dt^3}\mathbf{T} + \frac{d^2 s}{dt^2}\frac{d\mathbf{T}}{dt} + 2\left(\frac{ds}{dt}\right)\frac{d^2 s}{dt^2}\kappa\mathbf{N} + \left(\frac{ds}{dt}\right)^2 \frac{d\kappa}{dt}\mathbf{N} + \left(\frac{ds}{dt}\right)^2 \kappa\frac{d\mathbf{N}}{dt} \qquad \text{(b)}$$

From eq. (d) of Problem 4-18, we have

$$\frac{d\mathbf{T}}{dt} = \kappa\frac{ds}{dt}\mathbf{N}$$

and from the Frenet–Serret formula (4.72c),

$$\frac{d\mathbf{N}}{dt} = \frac{d\mathbf{N}}{ds}\frac{ds}{dt} = \frac{ds}{dt}(\tau\mathbf{B} - \kappa\mathbf{T})$$

Substituting these results into eq. (b), and rearranging the terms, we obtain

$$\mathbf{r}'''(t) = \left[\frac{d^3 s}{dt^3} - \left(\frac{ds}{dt}\right)^3 \kappa^2\right]\mathbf{T} + \left[3\left(\frac{ds}{dt}\right)\left(\frac{d^2 s}{dt^2}\right)\kappa + \left(\frac{ds}{dt}\right)^2\left(\frac{d\kappa}{dt}\right)\right]\mathbf{N} + \left(\frac{ds}{dt}\right)^3 \kappa\tau\mathbf{B}$$

$$\text{(c)}$$

Now, combining eq. (f) of Problem 4-18 and the preceding eq. (c), we have the triple scalar product:

$$[\mathbf{r}'(t)\mathbf{r}''(t)\mathbf{r}'''(t)] = [\mathbf{r}'''(t)\mathbf{r}'(t)\mathbf{r}''(t)]$$

$$= \mathbf{r}'''(t) \cdot [\mathbf{r}'(t) \times \mathbf{r}''(t)]$$

$$= \left(\frac{ds}{dt}\right)^6 \kappa^2 \tau$$

Solving for τ and substituting from the equation for magnitude in eq. (g) of Problem 4-18, the torsion is

$$\tau = \frac{[\mathbf{r}'(t)\mathbf{r}''(t)\mathbf{r}'''(t)]}{\left[\left(\frac{ds}{dt}\right)^3 \kappa\right]^2} = \frac{[\mathbf{r}'(t)\mathbf{r}''(t)\mathbf{r}'''(t)]}{|\mathbf{r}'(t) \times \mathbf{r}''(t)|^2}$$

PROBLEM 4-20 Show that if a curve C is represented by $\mathbf{r}(t)$, which is a twice-differentiable function, then

$$\mathbf{T} = \frac{\mathbf{r}'(t)}{|\mathbf{r}'(t)|} \qquad \text{(a)}$$

$$B = \frac{r'(t) \times r''(t)}{|r'(t) \times r''(t)|} \qquad (b)$$

$$N = \frac{[r'(t) \times r''(t)] \times r'(t)}{|[r'(t) \times r''(t)] \times r'(t)|} \qquad (c)$$

Solution: Equation (a) is already derived in (4.47). To verify eq. (b), we use the formulas for the cross product and its magnitude from eqs. (f) and (g) of Problem 4-18:

$$B = \frac{1}{\left(\frac{ds}{dt}\right)^3 \kappa} r'(t) \times r''(t) = \frac{r'(t) \times r''(t)}{|r'(t) \times r''(t)|}$$

To verify eq. (c), substitute eqs. (a) and (b) into the second equation $N = B \times T$ of (4.71). Then observing that N is a unit vector, we obtain

$$N = B \times T = \frac{[r'(t) \times r''(t)] \times r'(t)}{|[r'(t) \times r''(t)] \times r'(t)|}$$

PROBLEM 4-21 Find T, κ, N, B, and τ for the circle of radius a represented by

$$r(s) = a \cos\left(\frac{s}{a}\right) i + a \sin\left(\frac{s}{a}\right) j \qquad (a)$$

Solution: We already know from definition (4.46) that the tangent vector to the circle (a) is

$$T = \frac{dr(s)}{ds} = -\sin\left(\frac{s}{a}\right) i + \cos\left(\frac{s}{a}\right) j \qquad (b)$$

To find the unit principal normal N, we take the derivative of T (4.64), getting

$$\frac{dT}{ds} = -\frac{1}{a} \cos\left(\frac{s}{a}\right) i - \frac{1}{a} \sin\left(\frac{s}{a}\right) j = \kappa N$$

Thus the curvature (4.65) is

$$\kappa = \left|\frac{dT}{ds}\right| = \left\{\left[\frac{1}{a} \cos\left(\frac{s}{a}\right)\right]^2 + \left[\frac{1}{a} \sin\left(\frac{s}{a}\right)\right]^2\right\}^{1/2}$$

$$= \frac{1}{a} \left[\cos^2\left(\frac{s}{a}\right) + \sin^2\left(\frac{s}{a}\right)\right]^{1/2}$$

$$= \frac{1}{a} \qquad (c)$$

and, the unit principal normal is

$$N = \frac{1}{\kappa} \frac{dT}{ds} = -\cos\left(\frac{s}{a}\right) i - \sin\left(\frac{s}{a}\right) j = -\frac{1}{a} r(s) \qquad (d)$$

From the definition of the binormal vector B (4.67), we have

$$B = T \times N = \begin{vmatrix} i & j & k \\ -\sin\left(\frac{s}{a}\right) & \cos\left(\frac{s}{a}\right) & 0 \\ -\cos\left(\frac{s}{a}\right) & -\sin\left(\frac{s}{a}\right) & 0 \end{vmatrix} = k\left[\sin^2\left(\frac{s}{a}\right) + \cos^2\left(\frac{s}{a}\right)\right] = k \qquad (e)$$

Using the definition of torsion (4.70), we find that $dB/ds = 0 = -\tau N$; hence the torsion is

$$\tau = 0$$

The value $\kappa = 1/a$ shows that the circle has constant curvature equal to the reciprocal of its radius, and $\tau = 0$ indicates that the circle is a plane curve.

PROBLEM 4-22 Find **T**, **B**, **N**, κ, and τ for a circular helix represented by

$$\mathbf{r}(t) = (a\cos t)\mathbf{i} + (a\sin t)\mathbf{j} + bt\mathbf{k} \qquad (a)$$

Solution: Differentiating the given equation,

$$\mathbf{r}'(t) = -(a\sin t)\mathbf{i} + (a\cos t)\mathbf{j} + b\mathbf{k}$$
$$\mathbf{r}''(t) = -(a\cos t)\mathbf{i} - (a\sin t)\mathbf{j}$$
$$\mathbf{r}'''(t) = (a\sin t)\mathbf{i} - (a\cos t)\mathbf{k}$$

Accordingly, we obtain

$$\mathbf{r}'(t) \times \mathbf{r}''(t) = \begin{vmatrix} \mathbf{i} & \mathbf{j} & \mathbf{k} \\ -a\sin t & a\cos t & b \\ -a\cos t & -a\sin t & 0 \end{vmatrix}$$

$$= (ab\sin t)\mathbf{i} - (ab\cos t)\mathbf{j} + a^2\mathbf{k}$$

$$[\mathbf{r}'(t) \times \mathbf{r}''(t)] \times \mathbf{r}'(t) = \begin{vmatrix} \mathbf{i} & \mathbf{j} & \mathbf{k} \\ ab\sin t & -ab\cos t & a^2 \\ -a\sin t & a\cos t & b \end{vmatrix}$$

$$= (-ab^2 - a^3)\cos t\,\mathbf{i} - (ab^2 - a^3)\sin t\,\mathbf{j}$$
$$= -a(a^2 + b^2)\cos t\,\mathbf{i} - a(a^2 + b^2)\sin t\,\mathbf{j}$$

$$[\mathbf{r}'(t)\mathbf{r}''(t)\mathbf{r}'''(t)] = \begin{vmatrix} -a\sin t & a\cos t & b \\ -a\cos t & -a\sin t & 0 \\ a\sin t & -a\cos t & 0 \end{vmatrix}$$

$$= b\begin{vmatrix} -a\cos t & -a\sin t \\ a\sin t & -a\cos t \end{vmatrix}$$

$$= ba^2(\cos^2 t + \sin^2 t)$$
$$= a^2 b$$

So the magnitudes are

$$|\mathbf{r}'(t)| = (a^2\sin^2 t + a^2\cos^2 t + b^2)^{1/2} = (a^2 + b^2)^{1/2}$$

$$|\mathbf{r}'(t) \times \mathbf{r}''(t)|$$
$$= (a^2b^2\sin^2 t + a^2b^2\cos^2 t + a^4)^{1/2} = [a^2(a^2 + b^2)]^{1/2} = a(a^2 + b^2)^{1/2}$$

$$|[\mathbf{r}'(t) \times \mathbf{r}''(t)] \times \mathbf{r}'(t)| = [a^2(a^2 + b^2)^2(\cos^2 t + \sin^2 t)]^{1/2} = a(a^2 + b^2)$$

Hence from eqs. (a), (b), and (c) of Problem 4-19, eq. (a) of Problem 4-18, and eq. (a) of Problem 4-19, we obtain

$$\mathbf{T} = \frac{\mathbf{r}'(t)}{|\mathbf{r}'(t)|} = -\left[\frac{a}{(a^2 + b^2)^{1/2}}\sin t\right]\mathbf{i} + \left[\frac{a}{(a^2 + b^2)^{1/2}}\cos t\right]\mathbf{j} + \left[\frac{b}{(a^2 + b^2)^{1/2}}\right]\mathbf{k}$$

$$\mathbf{B} = \frac{\mathbf{r}'(t) \times \mathbf{r}''(t)}{|\mathbf{r}'(t) \times \mathbf{r}''(t)|}$$

$$= \left[\frac{b}{(a^2 + b^2)^{1/2}}\sin t\right]\mathbf{i} - \left[\frac{b}{(a^2 + b^2)^{1/2}}\cos t\right]\mathbf{j} + \left[\frac{a}{(a^2 + b^2)^{1/2}}\right]\mathbf{k}$$

$$\mathbf{N} = \frac{[\mathbf{r}'(t) \times \mathbf{r}''(t)] \times \mathbf{r}'(t)}{|[\mathbf{r}'(t) \times \mathbf{r}''(t)] \times \mathbf{r}'(t)|} = -(\cos t)\mathbf{i} - (\sin t)\mathbf{j}$$

$$\kappa = \frac{|\mathbf{r}'(t) \times \mathbf{r}''(t)|}{|\mathbf{r}'(t)|^3} = \frac{a(a^2 + b^2)^{1/2}}{(a^2 + b^2)^{3/2}} = \frac{a}{a^2 + b^2}$$

$$\tau = \frac{[\mathbf{r}'(t)\mathbf{r}''(t)\mathbf{r}'''(t)]}{|\mathbf{r}'(t) \times \mathbf{r}''(t)|^2} = \frac{a^2 b}{a^2(a^2 + b^2)} = \frac{b}{a^2 + b^2}$$

Note:

$$\mathbf{T} \cdot \mathbf{k} = \frac{b}{(a^2 + b^2)^{1/2}} = \text{constant}$$

Thus, a helix is a space curve whose tangent makes a constant angle with a fixed line called its axis—in this case, the z-axis.

Supplementary Exercises

PROBLEM 4-23   Let $\mathbf{A}(t) = t\mathbf{i} + t^2\mathbf{j} + t^3\mathbf{k}$ and $\mathbf{B}(t) = \mathbf{i} + t\mathbf{j} + (1 - t)\mathbf{k}$. At $t = 1$, evaluate (a) $\mathbf{A}'(t)$, (b) $(d/dt)[\mathbf{A}(t) \cdot \mathbf{B}(t)]$, and (c) $(d/dt)[\mathbf{A}(t) \times \mathbf{B}(t)]$.

Answer: (a) $[1, 2, 3]$ (b) 3   (c) $[-5, 4, 0]$

PROBLEM 4-24   If $\mathbf{f} = uvw\mathbf{i} + uw^2\mathbf{j} - v^3\mathbf{k}$ and $\mathbf{g} = u^3\mathbf{i} - uvw\mathbf{j} + u^2w\mathbf{k}$, calculate

(a) $\dfrac{\partial^2 \mathbf{f}}{\partial u\, \partial v}$ at the origin

(b) $\dfrac{\partial^2 \mathbf{f}}{\partial v^2} \times \dfrac{\partial^2 \mathbf{g}}{\partial u^2}$ at the point $(1, 1, 0)$

Answer: (a) $\mathbf{0}$ (b) $[0, -36, 0]$

PROBLEM 4-25   Identify the curves represented by the following vector functions:

(a) $\mathbf{r}(t) = \cos t\mathbf{i} + \sin t\mathbf{j} + \sin t\mathbf{k}$, $0 \le t \le 2\pi$
(b) $\mathbf{r}(t) = a\cos t\mathbf{i} + a\sin t\mathbf{j} + bt\mathbf{k}$, $b > 0$ and $0 \le t < \infty$

Answer: (a) an ellipse   (b) a cylindrical helix

PROBLEM 4-26   Find the length of the curves represented by the following vector functions:

(a) $\mathbf{r}(t) = \cos t\mathbf{i} + \sin t\mathbf{j} + \sin t\mathbf{k}$, $0 \le t \le 2\pi$
(b) $\mathbf{r}(t) = t\mathbf{i} + \sin(2\pi t)\mathbf{j} + \cos(2\pi t)\mathbf{k}$, $0 \le t \le 1$
(c) $\mathbf{r}(t) = \cos 3t\mathbf{i} + \sin 3t\mathbf{j}$, $0 \le t \le 2\pi$

Answer: (a) 3π (b) $\sqrt{1 + 4\pi^2}$ (c) 6π

PROBLEM 4-27   Find the unit tangent vector $\mathbf{T}$ to the curves represented by the following vector functions at the points specified:

(a) $\mathbf{r}(t) = \cos t\mathbf{i} + \sin t\mathbf{j} + t\mathbf{k}$,   at $t = \pi$
(b) $\mathbf{r}(t) = \left(t - \dfrac{t^3}{3}\right)\mathbf{i} + t^2\mathbf{j} + \left(t + \dfrac{t^3}{3}\right)\mathbf{k}$,   at $t = 1$
(c) $\mathbf{r}(t) = a(t - \sin t)\mathbf{i} + a(1 - \cos t)\mathbf{j}$,   at any point t

Answer: (a) $\left[0, -\dfrac{1}{\sqrt{2}}, \dfrac{1}{\sqrt{2}}\right]$ (b) $\left[0, \dfrac{1}{\sqrt{2}}, \dfrac{1}{\sqrt{2}}\right]$ (c) $\left(\sin\dfrac{t}{2}\right)\mathbf{i} + \left(\cos\dfrac{t}{2}\right)\mathbf{j}$

PROBLEM 4-28 Find a unit normal vector **n** at $(0,0,0)$ for the surface S represented by $z = 3x^2 + 4y^2$.

Answer: $[0,0,-1]$

PROBLEM 4-29 Find a unit normal vector **n** at $(1,1,1)$ for the surface S represented by $x^2 + y^2 - z - 1 = 0$.

Answer: $[\frac{2}{3}, \frac{2}{3}, -\frac{1}{3}]$

PROBLEM 4-30 If the curve C is represented by $y = y(x)$ in the xy-plane, show that the radius of the curvature is

$$\kappa = \frac{|y''|}{(1 + y'^2)^{3/2}}$$

where the number of primes denotes the number of derivatives taken.

PROBLEM 4-31 Find **T**, **B**, **N**, κ, and τ at $t = 0$ for the curve C represented by

$$\mathbf{r}(t) = (3t \cos t)\mathbf{i} + (3t \sin t)\mathbf{j} + 4t\mathbf{k}$$

Answer: $\mathbf{T} = [\frac{3}{5}, 0, \frac{4}{5}]$; $\mathbf{B} = [-\frac{4}{5}, 0, \frac{3}{5}]$; $\mathbf{N} = [0,1,0]$; $\kappa = \tau = \frac{6}{25}$

PROBLEM 4-32 Find κ and τ for the curve C represented by

$$\mathbf{r}(t) = a(3t - t^3)\mathbf{i} + 3at^2\mathbf{j} + a(3t + t^3)\mathbf{k}$$

Answer: $\kappa = \tau = \dfrac{1}{3a(1 + t^2)^2}$

PROBLEM 4-33 If the curve C is represented by $\mathbf{r}(s)$, where s is the arc length, then show that

$$\kappa = [\mathbf{r}''(s) \cdot \mathbf{r}''(s)]^{1/2}, \qquad \tau = \frac{[\mathbf{r}'(s)\mathbf{r}''(s)\mathbf{r}'''(s)]}{\mathbf{r}''(s) \cdot \mathbf{r}''(s)}$$

PROBLEM 4-34 Show that $\mathbf{r}'(t) \times \mathbf{r}''(t) = \mathbf{0}$ at every point on a curve is a necessary and sufficient condition that the curve be a straight line.

PROBLEM 4-35 Show that $\mathbf{T}'(s) \times \mathbf{T}''(s) = \kappa^2\boldsymbol{\omega}$, where $\boldsymbol{\omega}$ is the Darboux vector.

PROBLEM 4-36 Prove that $\mathbf{r}'''(s) = \kappa'\mathbf{N} - \kappa^2\mathbf{T} + \kappa\tau\mathbf{B}$ where $\kappa' = d\kappa/ds$.

5 GRADIENT, DIVERGENCE, AND CURL

5-1. Directional Derivative and the Gradient of a Scalar Function

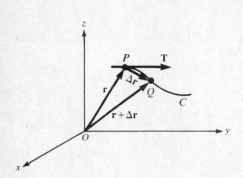

Figure 5-1
The directional derivative.

A. Directional derivative

A **scalar field** $\phi(x, y, z)$ is the totality of scalars $\phi(x, y, z)$ assigned to each point (x, y, z) of a region R in space. A **vector field** $\mathbf{f}(x, y, z)$ is the totality of vectors $\mathbf{f}(x, y, z)$ assigned to each point (x, y, z) of a region R in space.

Let $P(x, y, z)$ be a nonsingular point on a curve C in space and $Q(x + \Delta x, y + \Delta y, z + \Delta z)$ be any other point on C. (See Figure 5-1.) Then the position vectors $\mathbf{r}$ and $\mathbf{r} + \Delta \mathbf{r}$ of P and Q are

$$\mathbf{r} = x\mathbf{i} + y\mathbf{j} + z\mathbf{k}$$
$$\mathbf{r} + \Delta\mathbf{r} = (x + \Delta x)\mathbf{i} + (y + \Delta y)\mathbf{j} + (z + \Delta z)\mathbf{k}$$

Let $\phi(x, y, z)$ be a continuous and differentiable scalar function in a region R that contains the arc of curve C from P to Q. Then the **directional derivative** of $\phi(x, y, z)$ at P in the direction of the unit tangent vector $\mathbf{T}$ to curve C at P is defined as

DIRECTIONAL DERIVATIVE
$$\frac{\partial \phi}{\partial s} = \lim_{\Delta s \to 0} \frac{\phi(x + \Delta x, y + \Delta y, z + \Delta z) - \phi(x, y, z)}{\Delta s} \qquad (5.1)$$

where Δs is the arc length of C from P to Q.

B. Gradient

The **gradient** of the scalar function $\phi(x, y, z)$, written as grad ϕ, is a vector defined by

GRADIENT
$$\operatorname{grad} \phi = \frac{\partial \phi}{\partial x}\mathbf{i} + \frac{\partial \phi}{\partial y}\mathbf{j} + \frac{\partial \phi}{\partial z}\mathbf{k} \qquad (5.2)$$

Using the vector differential operator ∇ (read "*nabla*" or "*del*"),

$$\nabla = \frac{\partial}{\partial x}\mathbf{i} + \frac{\partial}{\partial y}\mathbf{j} + \frac{\partial}{\partial z}\mathbf{k} \qquad (5.3)$$

we can write the gradient of ϕ as

$$\operatorname{grad} \phi = \nabla \phi = \left(\frac{\partial}{\partial x}\mathbf{i} + \frac{\partial}{\partial y}\mathbf{j} + \frac{\partial}{\partial z}\mathbf{k} \right)\phi = \frac{\partial \phi}{\partial x}\mathbf{i} + \frac{\partial \phi}{\partial y}\mathbf{j} + \frac{\partial \phi}{\partial z}\mathbf{k} \qquad (5.4)$$

EXAMPLE 5-1: Show that the directional derivative of scalar function $\phi(x, y, z)$ in the direction of a curve C can be expressed as

$$\frac{\partial \phi}{\partial s} = \text{grad } \phi \cdot \mathbf{T} = \nabla \phi \cdot \mathbf{T} \tag{5.5}$$

where $\mathbf{T}$ is the unit tangent vector to C.

Solution: We recall from (4.46) that the unit tangent vector $\mathbf{T}$ to C at any point is given by

$$\mathbf{T} = \frac{d\mathbf{r}}{ds} = \frac{dx}{ds}\mathbf{i} + \frac{dy}{ds}\mathbf{j} + \frac{dz}{ds}\mathbf{k} \tag{5.6}$$

Now from elementary calculus, if ϕ is a function of x, y, and z, then

$$\frac{\partial \phi}{\partial s} = \frac{\partial \phi}{\partial x}\frac{dx}{ds} + \frac{\partial \phi}{\partial y}\frac{dy}{ds} + \frac{\partial \phi}{\partial z}\frac{dz}{ds} \tag{5.7}$$

which is the scalar product of $\nabla\phi$ and $\mathbf{T}$. Hence

$$\frac{\partial \phi}{\partial s} = \text{grad } \phi \cdot \mathbf{T} = \nabla \phi \cdot \mathbf{T}$$

EXAMPLE 5-2: Show that the magnitude and direction of $\nabla\phi = \text{grad } \phi$ is independent of the coordinate system. That is, show that the magnitude of grad ϕ, denoted $|\nabla\phi|$, is equal to the maximum value of the directional derivative of $\phi(x, y, z)$ and its direction is that of the maximum rate of increase of the function $\phi(x, y, z)$.

Solution: From the expression for the directional derivative (5.5) and the definition of scalar product (1.22), the directional derivative of ϕ at a point (x, y, z) is given by

$$\frac{\partial \phi}{\partial s} = \nabla\phi \cdot \mathbf{T} = |\nabla\phi||\mathbf{T}|\cos\theta = |\nabla\phi|\cos\theta \tag{5.8}$$

where θ is the angle between $\nabla\phi$ and the unit tangent vector $\mathbf{T}$ to the curve C. Since $-1 \leqslant \cos\theta \leqslant 1$, $\partial\phi/\partial s$ is a maximum when $\theta = 0$, that is, when the direction of $\mathbf{T}$ is the direction of $\nabla\phi$ and

$$\left.\frac{\partial \phi}{\partial s}\right|_{\text{max}} = |\nabla\phi| \tag{5.9}$$

Thus the magnitude of the gradient is $\partial\phi/\partial s|_{\text{max}}$, and its direction is in the direction of the maximum rate of increase of the function ϕ.

EXAMPLE 5-3: If $\phi(x, y, z)$ is a scalar function and $\nabla\phi \neq \mathbf{0}$ at a point P in space, show that $\nabla\phi$ is perpendicular to the surface S defined by

$$\phi(x, y, z) = c \tag{5.10}$$

where c is a constant.

Solution: If we assume a space curve C that lies on a surface S is represented by $\mathbf{r}(t) = x(t)\mathbf{i} + y(t)\mathbf{j} + z(t)\mathbf{k}$, then from (5.10), we have for constant c

$$\phi[x(t), y(t), z(t)] = c$$

By differentiating this with respect to t, we get

$$\frac{\partial \phi}{\partial x}x'(t) + \frac{\partial \phi}{\partial y}y'(t) + \frac{\partial \phi}{\partial z}z'(t) = \nabla\phi \cdot \mathbf{r}'(t) = 0 \tag{5.11}$$

where, from (4.34), the vector tangent to curve C at point P is

$$\mathbf{r}'(t) = x'(t)\mathbf{i} + y'(t)\mathbf{j} + z'(t)\mathbf{k}$$

Now (5.11) implies that $\nabla\phi \perp \mathbf{r}'(t)$; that is, $\nabla\phi$ is perpendicular to curve C at P.

This reasoning can be applied to any smooth curve on the surface S that passes through P. Hence $\nabla\phi$ is perpendicular to every such curve at P, which can be the case only if $\nabla\phi$ is perpendicular to the surface S.

C. Properties of the gradient

$$\nabla(c\phi) = c\,\nabla\phi \tag{5.12}$$

$$\nabla(\phi + \psi) = \nabla\phi + \nabla\psi \tag{5.13}$$

$$\nabla(\phi\psi) = \phi\,\nabla\psi + \psi\,\nabla\phi \tag{5.14}$$

where ϕ and ψ are differentiable scalar functions in some region in space and c is a constant.

EXAMPLE 5-4: Verify the property for taking the gradient of the product of two scalar functions (5.14).

Proof: From the definition of the gradient (5.4),

$$\nabla(\phi\psi) = \frac{\partial}{\partial x}(\phi\psi)\mathbf{i} + \frac{\partial}{\partial y}(\phi\psi)\mathbf{j} + \frac{\partial}{\partial z}(\phi\psi)\mathbf{k}$$

$$= \left(\phi\frac{\partial\psi}{\partial x} + \psi\frac{\partial\phi}{\partial x}\right)\mathbf{i} + \left(\phi\frac{\partial\psi}{\partial y} + \psi\frac{\partial\phi}{\partial y}\right)\mathbf{j} + \left(\phi\frac{\partial\psi}{\partial z} + \psi\frac{\partial\phi}{\partial z}\right)\mathbf{k}$$

$$= \phi\left(\frac{\partial\psi}{\partial x}\mathbf{i} + \frac{\partial\psi}{\partial y}\mathbf{j} + \frac{\partial\psi}{\partial z}\mathbf{k}\right) + \psi\left(\frac{\partial\phi}{\partial x}\mathbf{i} + \frac{\partial\phi}{\partial y}\mathbf{j} + \frac{\partial\phi}{\partial z}\mathbf{k}\right)$$

$$= \phi\,\nabla\psi + \psi\,\nabla\phi$$

EXAMPLE 5-5: If $\phi = \phi(u)$, where $u = u(x, y, z)$, then show that

$$\nabla\phi = \nabla\phi(u) = \phi'(u)\,\nabla u \tag{5.15}$$

Solution: From definition (5.4), the gradient is

$$\nabla\phi = \nabla\phi(u) = \frac{\partial\phi}{\partial x}\mathbf{i} + \frac{\partial\phi}{\partial y}\mathbf{j} + \frac{\partial\phi}{\partial z}\mathbf{k}$$

$$= \phi'(u)\frac{\partial u}{\partial x}\mathbf{i} + \phi'(u)\frac{\partial u}{\partial y}\mathbf{j} + \phi'(u)\frac{\partial u}{\partial z}\mathbf{k}$$

$$= \phi'(u)\left(\frac{\partial u}{\partial x}\mathbf{i} + \frac{\partial u}{\partial y}\mathbf{j} + \frac{\partial u}{\partial z}\mathbf{k}\right)$$

$$= \phi'(u)\,\nabla u$$

5-2. The Operator ∇

The **vector differential operator ∇**

$$\nabla = \frac{\partial}{\partial x}\mathbf{i} + \frac{\partial}{\partial y}\mathbf{j} + \frac{\partial}{\partial z}\mathbf{k}$$

is not a vector, but an operator: It may be considered as a *symbolic* vector. Thus if ϕ is a scalar field, then $\phi\nabla$ is still an operator, whereas $\nabla\phi$ yields the important vector function called the gradient. Similarly, if $\mathbf{f}$ is a differentiable vector function, then $\mathbf{f} \cdot \mathbf{V}$ and $\mathbf{f} \times \mathbf{V}$ are still operators, whereas $\mathbf{V} \cdot \mathbf{f}$ and $\mathbf{V} \times \mathbf{f}$ yield important scalar and vector functions, respectively. (See Sections 5-3 and 5-4.)

- Multiplying $\mathbf{V}$ on the left yields *operators*
- Multiplying $\mathbf{V}$ on the right yields important *scalar and vector functions*

In other words, $\mathbf{V}$ operates only on what follows it.

5-3. Divergence of a Vector Function

A. Divergence

If a vector function $\mathbf{f} = [f_1, f_2, f_3]$, where f_1, f_2, and f_3 are scalar functions, then its scalar or dot product with the symbolic vector $\mathbf{V}$ is

$$\mathbf{V} \cdot \mathbf{f} = \left(\frac{\partial}{\partial x} \mathbf{i} + \frac{\partial}{\partial y} \mathbf{j} + \frac{\partial}{\partial z} \mathbf{k} \right) \cdot (f_1 \mathbf{i} + f_2 \mathbf{j} + f_3 \mathbf{k})$$

$$= \frac{\partial f_1}{\partial x} + \frac{\partial f_2}{\partial y} + \frac{\partial f_3}{\partial z}$$

Hence a vector function $\mathbf{f}$ is transformed into a scalar function when operated on from the left by $\mathbf{V}$. This scalar function is called the **divergence** of the vector function $\mathbf{f}$ and is written as div $\mathbf{f}$:

DIVERGENCE $$\operatorname{div} \mathbf{f} = \mathbf{V} \cdot \mathbf{f} = \frac{\partial f_1}{\partial x} + \frac{\partial f_2}{\partial y} + \frac{\partial f_3}{\partial z} \tag{5.16}$$

Although the definition of divergence (5.16) does not give any physical or geometric meaning to the concept of divergence, it does give an easy computational form. (Other definitions of divergence, as well as its physical and geometric interpretations, are given in Chapter 6.)

EXAMPLE 5-6: If $\mathbf{r}$ is the position vector, find the divergence $\mathbf{V} \cdot \mathbf{r}$.

Solution: If $\mathbf{r} = x\mathbf{i} + y\mathbf{j} + z\mathbf{k}$, then from the divergence definition (5.16),

$$\mathbf{V} \cdot \mathbf{r} = \frac{\partial x}{\partial x} + \frac{\partial y}{\partial y} + \frac{\partial z}{\partial z} = 3 \tag{5.17}$$

EXAMPLE 5-7: If $\mathbf{f}$ and $\mathbf{g}$ are two vector functions, show that

$$\mathbf{V} \cdot (\mathbf{f} + \mathbf{g}) = \mathbf{V} \cdot \mathbf{f} + \mathbf{V} \cdot \mathbf{g} \tag{5.18}$$

Solution: Let $\mathbf{f} = [f_1, f_2, f_3]$ and $\mathbf{g} = [g_1, g_2, g_3]$. Then

$$\mathbf{f} + \mathbf{g} = (f_1 + g_1)\mathbf{i} + (f_2 + g_2)\mathbf{j} + (f_3 + g_3)\mathbf{k}$$

Now, again using definition (5.16), we get

$$\mathbf{V} \cdot (\mathbf{f} + \mathbf{g}) = \frac{\partial}{\partial x}(f_1 + g_1) + \frac{\partial}{\partial y}(f_2 + g_2) + \frac{\partial}{\partial z}(f_3 + g_3)$$

$$= \left(\frac{\partial f_1}{\partial x} + \frac{\partial f_2}{\partial y} + \frac{\partial f_3}{\partial z} \right) + \left(\frac{\partial g_1}{\partial x} + \frac{\partial g_2}{\partial y} + \frac{\partial g_3}{\partial z} \right)$$

$$= \mathbf{V} \cdot \mathbf{f} + \mathbf{V} \cdot \mathbf{g}$$

B. Laplacian

If ϕ is a scalar function, then the divergence of the gradient of ϕ is

$$\operatorname{div}(\operatorname{grad} \phi) = \frac{\partial^2 \phi}{\partial x^2} + \frac{\partial^2 \phi}{\partial y^2} + \frac{\partial^2 \phi}{\partial z^2} \tag{5.19}$$

EXAMPLE 5-8: Verify (5.19).

Solution: The gradient of ϕ is

$$\nabla\phi = \frac{\partial\phi}{\partial x}\mathbf{i} + \frac{\partial\phi}{\partial y}\mathbf{j} + \frac{\partial\phi}{\partial z}\mathbf{k}$$

Hence by (5.16)

$$\text{div(grad }\phi) = \nabla\cdot(\nabla\phi) = \frac{\partial}{\partial x}\left(\frac{\partial\phi}{\partial x}\right) + \frac{\partial}{\partial y}\left(\frac{\partial\phi}{\partial y}\right) + \frac{\partial}{\partial z}\left(\frac{\partial\phi}{\partial z}\right)$$

$$= \frac{\partial^2\phi}{\partial x^2} + \frac{\partial^2\phi}{\partial y^2} + \frac{\partial^2\phi}{\partial z^2}$$

Since the divergence of the gradient $\nabla\cdot\nabla$ is written as ∇^2, then $\nabla\cdot(\nabla\phi)$ is written as $\nabla\cdot\nabla\phi = \nabla^2\phi$. The operator ∇^2 is called the **Laplacian**:

$$\text{Laplacian} = \nabla^2 = \frac{\partial^2}{\partial x^2} + \frac{\partial^2}{\partial y^2} + \frac{\partial^2}{\partial z^2} \tag{5.20}$$

LAPLACIAN

$$\nabla^2\phi = \left(\frac{\partial^2}{\partial x^2} + \frac{\partial^2}{\partial y^2} + \frac{\partial^2}{\partial z^2}\right)\phi = \frac{\partial^2\phi}{\partial x^2} + \frac{\partial^2\phi}{\partial y^2} + \frac{\partial^2\phi}{\partial z^2} \tag{5.21}$$

A scalar function ϕ is said to be *harmonic* if it is continuous, has continuous second partial derivatives and satisfies **Laplace's equation**:

LAPLACE'S EQUATION $$\nabla^2\phi = 0 \tag{5.22}$$

EXAMPLE 5-9: Show that the function $1/r$, where $r = |\mathbf{r}| = (x^2 + y^2 + z^2)^{1/2}$, is a harmonic function, provided $r \neq 0$.

Solution: Clearly, $1/r$ for $r \neq 0$ is continuous because x^2, y^2, z^2, and r are continuous. Then from definition (5.21) we have

$$\nabla^2\left(\frac{1}{r}\right) = \left(\frac{\partial^2}{\partial x^2} + \frac{\partial^2}{\partial y^2} + \frac{\partial^2}{\partial z^2}\right)(x^2 + y^2 + z^2)^{-1/2}$$

Now the first and second partial derivatives of $1/r$ with respect to x are

$$\frac{\partial}{\partial x}(x^2 + y^2 + z^2)^{-1/2} = -x(x^2 + y^2 + z^2)^{-3/2}$$

$$\frac{\partial^2}{\partial x^2}(x^2 + y^2 + z^2)^{-1/2} = \frac{\partial}{\partial x}\left[-x(x^2 + y^2 + z^2)^{-3/2}\right]$$

$$= 3x^2(x^2 + y^2 + z^2)^{-5/2} - (x^2 + y^2 + z^2)^{-3/2}$$

$$= 3x^2r^{-5} - r^{-3} \tag{5.23}$$

Similarly, the second partial derivatives of $1/r$ with respect to y and z are

$$\frac{\partial^2}{\partial y^2}(x^2 + y^2 + z^2)^{-1/2} = 3y^2r^{-5} - r^{-3} \tag{5.24}$$

$$\frac{\partial^2}{\partial z^2}(x^2 + y^2 + z^2)^{-1/2} = 3z^2r^{-5} - r^{-3} \tag{5.25}$$

Since x, y, z, and r are continuous, the second partial derivatives are also continuous if $r \neq 0$.

Adding (5.23), (5.24), and (5.25), we obtain

$$\nabla^2\left(\frac{1}{r}\right) = 3r^{-5}(x^2 + y^2 + z^2) - 3r^{-3} = 3r^{-5}r^2 - 3r^{-3} = 3r^{-3} - 3r^{-3} = 0 \tag{5.26}$$

provided $r \neq 0$. Thus the function $1/r$ satisfies Laplace's equation and is therefore harmonic.

5-4. Curl of a Vector Function

A. Curl

If a given vector function $\mathbf{f} = [f_1, f_2, f_3]$, where f_1, f_2, and f_3 are scalar functions with continuous first derivatives, then its vector or cross product with the symbolic vector ∇ is called the **curl** or **rotation** of the vector function $\mathbf{f}$:

$$\text{curl or rot } \mathbf{f} = \nabla \times \mathbf{f} \qquad (5.27)$$

$$\nabla \times \mathbf{f} = \left(\frac{\partial}{\partial x} \mathbf{i} + \frac{\partial}{\partial y} \mathbf{j} + \frac{\partial}{\partial z} \mathbf{k} \right) \times (f_1 \mathbf{i} + f_2 \mathbf{j} + f_3 \mathbf{k})$$

CURL

$$= \begin{vmatrix} \mathbf{i} & \mathbf{j} & \mathbf{k} \\ \dfrac{\partial}{\partial x} & \dfrac{\partial}{\partial y} & \dfrac{\partial}{\partial z} \\ f_1 & f_2 & f_3 \end{vmatrix}$$

$$= \left(\frac{\partial f_3}{\partial y} - \frac{\partial f_2}{\partial z} \right) \mathbf{i} + \left(\frac{\partial f_1}{\partial z} - \frac{\partial f_3}{\partial x} \right) \mathbf{j} + \left(\frac{\partial f_2}{\partial x} - \frac{\partial f_1}{\partial y} \right) \mathbf{k} \qquad (5.28)$$

Note: $\nabla \times \mathbf{f}$ is not necessarily perpendicular to $\mathbf{f}$, although we treat ∇ as a symbolic vector. (The physical meaning of the curl of $\mathbf{f}$ is discussed in Chapter 6.)

EXAMPLE 5-10: If $\mathbf{r}$ is the position vector, find $\nabla \times \mathbf{r}$.

Solution: Since $\mathbf{r} = x\mathbf{i} + y\mathbf{j} + z\mathbf{k}$, from definition (5.28), the curl $\mathbf{r}$ is

$$\nabla \times \mathbf{r} = \begin{vmatrix} \mathbf{i} & \mathbf{j} & \mathbf{k} \\ \dfrac{\partial}{\partial x} & \dfrac{\partial}{\partial y} & \dfrac{\partial}{\partial z} \\ x & y & z \end{vmatrix} = \mathbf{0} \qquad (5.29)$$

EXAMPLE 5-11: If $\mathbf{f}$ and $\mathbf{g}$ are two vector functions, show that

$$\nabla \times (\mathbf{f} + \mathbf{g}) = \nabla \times \mathbf{f} + \nabla \times \mathbf{g} \qquad (5.30)$$

Solution: Let $\mathbf{f} = [f_1, f_2, f_3]$ and $\mathbf{g} = [g_1, g_2, g_3]$. Then we can write

$$\mathbf{f} + \mathbf{g} = (f_1 + g_1)\mathbf{i} + (f_2 + g_2)\mathbf{j} + (f_3 + g_3)\mathbf{k}$$

Hence from the definition of the curl (5.28), we find

$$\nabla \times (\mathbf{f} + \mathbf{g}) = \begin{vmatrix} \mathbf{i} & \mathbf{j} & \mathbf{k} \\ \dfrac{\partial}{\partial x} & \dfrac{\partial}{\partial y} & \dfrac{\partial}{\partial z} \\ f_1 + g_1 & f_2 + g_2 & f_3 + g_3 \end{vmatrix}$$

$$= \begin{vmatrix} \mathbf{i} & \mathbf{j} & \mathbf{k} \\ \dfrac{\partial}{\partial x} & \dfrac{\partial}{\partial y} & \dfrac{\partial}{\partial z} \\ f_1 & f_2 & f_3 \end{vmatrix} + \begin{vmatrix} \mathbf{i} & \mathbf{j} & \mathbf{k} \\ \dfrac{\partial}{\partial x} & \dfrac{\partial}{\partial y} & \dfrac{\partial}{\partial z} \\ g_1 & g_2 & g_3 \end{vmatrix}$$

$$= \nabla \times \mathbf{f} + \nabla \times \mathbf{g}$$

B. Curl of the gradient and divergence of the curl

If ϕ is a scalar function with continuous second partial derivatives, then the curl of the gradient of ϕ is the zero vector:

$$\nabla \times (\nabla \phi) = \mathbf{0} \tag{5.31}$$

If the vector function $\mathbf{f} = [f_1, f_2, f_3]$ has scalar components f_1, f_2, and f_3 that have continuous second derivatives, then the divergence of the curl of $\mathbf{f}$ is zero:

$$\nabla \cdot (\nabla \times \mathbf{f}) = 0 \tag{5.32}$$

EXAMPLE 5-12: Verify that the curl of the gradient of scalar function ϕ is zero: $\nabla \times (\nabla \phi) = \mathbf{0}$ (5.31).

Proof: From definition (5.4), the gradient of ϕ is

$$\nabla \phi = \frac{\partial \phi}{\partial x} \mathbf{i} + \frac{\partial \phi}{\partial y} \mathbf{j} + \frac{\partial \phi}{\partial z} \mathbf{k}$$

Hence, assuming that ϕ has continuous second partial derivatives, from definition (5.28),

$$\nabla \times (\nabla \phi) = \begin{vmatrix} \mathbf{i} & \mathbf{j} & \mathbf{k} \\ \dfrac{\partial}{\partial x} & \dfrac{\partial}{\partial y} & \dfrac{\partial}{\partial z} \\ \dfrac{\partial \phi}{\partial x} & \dfrac{\partial \phi}{\partial y} & \dfrac{\partial \phi}{\partial z} \end{vmatrix}$$

$$= \left(\frac{\partial^2 \phi}{\partial y\, \partial z} - \frac{\partial^2 \phi}{\partial z\, \partial y} \right) \mathbf{i} + \left(\frac{\partial^2 \phi}{\partial z\, \partial x} - \frac{\partial^2 \phi}{\partial x\, \partial z} \right) \mathbf{j} + \left(\frac{\partial^2 \phi}{\partial x\, \partial y} - \frac{\partial^2 \phi}{\partial y\, \partial x} \right) \mathbf{k}$$

$$= \mathbf{0}$$

since the second partial derivatives are continuous, and hence

$$\frac{\partial^2 \phi}{\partial y\, \partial z} = \frac{\partial^2 \phi}{\partial z\, \partial y}, \qquad \frac{\partial^2 \phi}{\partial z\, \partial x} = \frac{\partial^2 \phi}{\partial x\, \partial z}, \qquad \frac{\partial^2 \phi}{\partial x\, \partial y} = \frac{\partial^2 \phi}{\partial y\, \partial x}$$

EXAMPLE 5-13: Verify that the divergence of the curl of the vector function $\mathbf{f}$ is zero: $\nabla \cdot (\nabla \times \mathbf{f}) = 0$, (5.32).

Proof: If $\mathbf{f} = f_1 \mathbf{i} + f_2 \mathbf{j} + f_3 \mathbf{k}$, then from definition (5.28) its curl is

$$\nabla \times \mathbf{f} = \left(\frac{\partial f_3}{\partial y} - \frac{\partial f_2}{\partial z} \right) \mathbf{i} + \left(\frac{\partial f_1}{\partial z} - \frac{\partial f_3}{\partial x} \right) \mathbf{j} + \left(\frac{\partial f_2}{\partial x} - \frac{\partial f_1}{\partial y} \right) \mathbf{k}$$

Assuming that $\mathbf{f}$ has continuous second partial derivatives, from (5.16) the divergence of the curl is

$$\nabla \cdot (\nabla \times \mathbf{f}) = \frac{\partial}{\partial x} \left(\frac{\partial f_3}{\partial y} - \frac{\partial f_2}{\partial z} \right) + \frac{\partial}{\partial y} \left(\frac{\partial f_1}{\partial z} - \frac{\partial f_3}{\partial x} \right) + \frac{\partial}{\partial z} \left(\frac{\partial f_2}{\partial x} - \frac{\partial f_1}{\partial y} \right)$$

$$= \frac{\partial^2 f_3}{\partial x\, \partial y} - \frac{\partial^2 f_2}{\partial x\, \partial z} + \frac{\partial^2 f_1}{\partial y\, \partial z} - \frac{\partial^2 f_3}{\partial y\, \partial x} + \frac{\partial^2 f_2}{\partial z\, \partial x} - \frac{\partial^2 f_1}{\partial z\, \partial y}$$

$$= 0$$

because the second partial derivatives are continuous, and

$$\frac{\partial^2 f_3}{\partial x\, \partial y} = \frac{\partial^2 f_3}{\partial y\, \partial x}, \qquad \frac{\partial^2 f_2}{\partial x\, \partial z} = \frac{\partial^2 f_2}{\partial z\, \partial x}, \qquad \frac{\partial^2 f_1}{\partial y\, \partial z} = \frac{\partial^2 f_1}{\partial z\, \partial y}$$

EXAMPLE 5-14: Show that if $\mathbf{B} = \nabla \times \mathbf{A}$, then $\mathbf{A}$ is not uniquely determined by $\mathbf{B}$.

Solution: If ψ is any arbitrary scalar function, let

$$\mathbf{A}' = \mathbf{A} + \nabla\psi \tag{5.33}$$

Then from (5.30) the curl of $\mathbf{A}'$ is

$$\begin{aligned}
\nabla \times \mathbf{A}' &= \nabla \times (\mathbf{A} + \nabla\psi) \\
&= \nabla \times \mathbf{A} + \nabla \times (\nabla\psi) \\
&= \nabla \times \mathbf{A} \tag{5.34}
\end{aligned}$$

since $\nabla \times (\nabla\psi) = 0$ from (5.31). Thus $\mathbf{B}$ can be expressed as $\mathbf{B} = \nabla \times \mathbf{A}'$, which means that because ψ is arbitrary, $\mathbf{A}$ is not uniquely determined from $\mathbf{B}$.

5-5. Operations with ∇ and Some Vector Identities

A. Operations with ∇

Using the operator ∇, we have defined gradient, divergence, and curl to obtain both vector and scalar quantities. In this section we'll consider the various combinations of the operator ∇ with vector and scalar functions.

Note: Remember that the order in which the symbols appear in expressions using ∇ is very important: The operator ∇ operates only on what follows it. For example

$$(\mathbf{f} \cdot \nabla)\mathbf{g} \neq \mathbf{g}(\mathbf{f} \cdot \nabla)$$

The left side of this expression is a vector field, while the right side is an operator.

EXAMPLE 5-15: Determine the meaning that can be assigned to $(\mathbf{f} \cdot \nabla)\phi$ and $(\mathbf{f} \cdot \nabla)\mathbf{g}$.

Solution: Since ∇ is a differential operator with respect to the space coordinates, it operates only on what follows it. Thus we interpret $(\mathbf{f} \cdot \nabla)\phi$ as the dot product of the vector $\mathbf{f}$ and the gradient $\nabla\phi$:

$$(\mathbf{f} \cdot \nabla)\phi = \mathbf{f} \cdot (\nabla\phi) \tag{5.35}$$

Similar results can be obtained by interpreting ∇ as a symbolic vector; that is,

$$\nabla = \frac{\partial}{\partial x}\mathbf{i} + \frac{\partial}{\partial y}\mathbf{j} + \frac{\partial}{\partial z}\mathbf{k}$$

Taking its left scalar product with $\mathbf{f}$,

$$\mathbf{f} \cdot \nabla = f_1 \frac{\partial}{\partial x} + f_2 \frac{\partial}{\partial y} + f_3 \frac{\partial}{\partial z} \tag{5.36}$$

Thus

$$\begin{aligned}
(\mathbf{f} \cdot \nabla)\phi &= \left(f_1 \frac{\partial}{\partial x} + f_2 \frac{\partial}{\partial y} + f_3 \frac{\partial}{\partial z} \right)\phi \\
&= f_1 \frac{\partial\phi}{\partial x} + f_2 \frac{\partial\phi}{\partial y} + f_3 \frac{\partial\phi}{\partial z} \\
&= \mathbf{f} \cdot (\nabla\phi)
\end{aligned}$$

Since $\nabla\mathbf{g}$ is not defined, we can use the differential operator (5.36) that operates on $\mathbf{g}$ to obtain

$$(\mathbf{f} \cdot \nabla)\mathbf{g} = \left(f_1 \frac{\partial}{\partial x} + f_2 \frac{\partial}{\partial y} + f_3 \frac{\partial}{\partial z} \right)\mathbf{g}$$

$$= f_1 \frac{\partial \mathbf{g}}{\partial x} + f_2 \frac{\partial \mathbf{g}}{\partial y} + f_3 \frac{\partial \mathbf{g}}{\partial z} \qquad (5.37)$$

EXAMPLE 5-16: Determine the meaning that can be given to $(\mathbf{f} \times \nabla)\phi$ and $(\mathbf{f} \times \nabla)\mathbf{g}$.

Solution: Since the operator ∇ operates only on what follows it, we can interpret $(\mathbf{f} \times \nabla)\phi$ as the vector product of $\mathbf{f}$ and the gradient $\nabla\phi$:

$$(\mathbf{f} \times \nabla)\phi = \mathbf{f} \times (\nabla\phi) \qquad (5.38)$$

Similar results can be obtained by interpreting ∇ as a symbolic vector:

$$\nabla = \frac{\partial}{\partial x}\mathbf{i} + \frac{\partial}{\partial y}\mathbf{j} + \frac{\partial}{\partial z}\mathbf{k}$$

Since $\mathbf{f} \times \nabla$ is an operator,

$$\mathbf{f} \times \nabla = \begin{vmatrix} \mathbf{i} & \mathbf{j} & \mathbf{k} \\ f_1 & f_2 & f_3 \\ \dfrac{\partial}{\partial x} & \dfrac{\partial}{\partial y} & \dfrac{\partial}{\partial z} \end{vmatrix}$$

$$= \left(f_2 \frac{\partial}{\partial z} - f_3 \frac{\partial}{\partial y} \right)\mathbf{i} + \left(f_3 \frac{\partial}{\partial x} - f_1 \frac{\partial}{\partial z} \right)\mathbf{j} + \left(f_1 \frac{\partial}{\partial y} - f_2 \frac{\partial}{\partial x} \right)\mathbf{k} \quad (5.39)$$

Thus combining $\mathbf{f} \times \nabla$ and ϕ, we get

$$(\mathbf{f} \times \nabla)\phi = \left[\left(f_2 \frac{\partial}{\partial z} - f_3 \frac{\partial}{\partial y} \right)\mathbf{i} + \left(f_3 \frac{\partial}{\partial x} - f_1 \frac{\partial}{\partial z} \right)\mathbf{j} + \left(f_1 \frac{\partial}{\partial y} - f_2 \frac{\partial}{\partial x} \right)\mathbf{k} \right]\phi$$

$$= \left(f_2 \frac{\partial \phi}{\partial z} - f_3 \frac{\partial \phi}{\partial y} \right)\mathbf{i} + \left(f_3 \frac{\partial \phi}{\partial x} - f_1 \frac{\partial \phi}{\partial z} \right)\mathbf{j} + \left(f_1 \frac{\partial \phi}{\partial y} - f_2 \frac{\partial \phi}{\partial x} \right)\mathbf{k}$$

$$= \mathbf{f} \times (\nabla\phi)$$

No definition or meaning can be assigned to $(\mathbf{f} \times \nabla)\mathbf{g}$, because it is a kind of differential operator with vector quantities.

EXAMPLE 5-17: If $\mathbf{f}$ and $\mathbf{g}$ are two vector functions, show that

$$(\mathbf{f} \times \nabla) \cdot \mathbf{g} = \mathbf{f} \cdot (\nabla \times \mathbf{g}) \qquad (5.40)$$

Solution: Using result (5.39) in Example 5-16, we find

$$(\mathbf{f} \times \nabla) \cdot \mathbf{g} = \left(f_2 \frac{\partial}{\partial z} - f_3 \frac{\partial}{\partial y} \right)(\mathbf{i} \cdot \mathbf{g}) + \left(f_3 \frac{\partial}{\partial x} - f_1 \frac{\partial}{\partial z} \right)(\mathbf{j} \cdot \mathbf{g})$$

$$+ \left(f_1 \frac{\partial}{\partial y} - f_2 \frac{\partial}{\partial x} \right)(\mathbf{k} \cdot \mathbf{g})$$

$$= \left(f_2 \frac{\partial g_1}{\partial z} - f_3 \frac{\partial g_1}{\partial y} \right) + \left(f_3 \frac{\partial g_2}{\partial x} - f_1 \frac{\partial g_2}{\partial z} \right) + \left(f_1 \frac{\partial g_3}{\partial y} - f_2 \frac{\partial g_3}{\partial x} \right)$$

$$= f_1 \left(\frac{\partial g_3}{\partial y} - \frac{\partial g_2}{\partial z} \right) + f_2 \left(\frac{\partial g_1}{\partial z} - \frac{\partial g_3}{\partial x} \right) + f_3 \left(\frac{\partial g_2}{\partial x} - \frac{\partial g_1}{\partial y} \right)$$

$$= \mathbf{f} \cdot (\nabla \times \mathbf{g})$$

Since $\mathbf{i} \cdot \mathbf{g} = g_1$, $\mathbf{j} \cdot \mathbf{g} = g_2$, and $\mathbf{k} \cdot \mathbf{g} = g_3$.

B. Vector identities

We'll consider some vector identities that involve the operator ∇. Although these identities can all be verified by direct expansion using the components of the vector function, we'll establish these identities heuristically by treating ∇ as both a symbolic vector and a differential operator. We'll then manipulate the expressions according to the appropriate formulas from vector algebra, and finally, we'll give ∇ its operational meaning.

EXAMPLE 5-18: Prove that

$$\nabla \cdot (\phi\mathbf{f}) = \phi\nabla \cdot \mathbf{f} + \mathbf{f} \cdot \nabla\phi \qquad (5.41a)$$

Solution: Since ∇ is a differential operator, we can express the rule for the differentiation of a product as

$$\nabla \cdot (\phi\mathbf{f}) = \nabla_\phi \cdot (\phi\mathbf{f}) + \nabla_\mathbf{f} \cdot (\phi\mathbf{f}) \qquad (5.41b)$$

The dot product $\nabla_\phi \cdot (\phi\mathbf{f})$ means that $\mathbf{f}$ is fixed and ∇ operates on ϕ. Similarly, $\nabla_\mathbf{f} \cdot (\phi\mathbf{f})$ means that ϕ is fixed and ∇ operates on $\mathbf{f}$. Since $\nabla \cdot \phi$ is not a defined quantity,

$$\nabla_\phi \cdot (\phi\mathbf{f}) = \mathbf{f} \cdot \nabla\phi$$
$$\nabla_\mathbf{f} \cdot (\phi\mathbf{f}) = \phi\nabla \cdot \mathbf{f}$$

Adding these two equations and using (5.41b), we get

$$\nabla \cdot (\phi\mathbf{f}) = \mathbf{f} \cdot \nabla\phi + \phi \cdot \nabla\mathbf{f}$$

Alternate Solution: By using components, we reach the same conclusion:

$$\nabla \cdot (\phi\mathbf{f}) = \frac{\partial}{\partial x}(\phi f_1) + \frac{\partial}{\partial y}(\phi f_2) + \frac{\partial}{\partial z}(\phi f_3)$$

$$= \phi\frac{\partial f_1}{\partial x} + f_1\frac{\partial \phi}{\partial x} + \phi\frac{\partial f_2}{\partial y} + f_2\frac{\partial \phi}{\partial y} + \phi\frac{\partial f_3}{\partial z} + f_3\frac{\partial \phi}{\partial z}$$

$$= \phi\left(\frac{\partial f_1}{\partial x} + \frac{\partial f_2}{\partial y} + \frac{\partial f_3}{\partial z}\right) + \left(f_1\frac{\partial \phi}{\partial x} + f_2\frac{\partial \phi}{\partial y} + f_3\frac{\partial \phi}{\partial z}\right)$$

$$= \phi\nabla \cdot \mathbf{f} + \mathbf{f} \cdot \nabla\phi$$

EXAMPLE 5-19: Prove that

$$\nabla \times (\phi\mathbf{f}) = \phi\nabla \times \mathbf{f} + (\nabla\phi) \times \mathbf{f} = \phi\nabla \times \mathbf{f} - \mathbf{f} \times \nabla\phi \qquad (5.42)$$

Solution: Since $\nabla \times (\phi\mathbf{f}) = \nabla_\phi \times (\phi\mathbf{f}) + \nabla_\mathbf{f} \times (\phi\mathbf{f})$ and since $\nabla \times \phi$ is not defined, we can write

$$\nabla_\phi \times (\phi\mathbf{f}) = (\nabla\phi) \times \mathbf{f} = -\mathbf{f} \times \nabla\phi$$
$$\nabla_\mathbf{f} \times (\phi\mathbf{f}) = \phi\nabla \times \mathbf{f}$$

Adding these two equations, we obtain

$$\nabla \times (\phi\mathbf{f}) = (\nabla\phi) \times \mathbf{f} + \phi\nabla \times \mathbf{f} = \phi\nabla \times \mathbf{f} - \mathbf{f} \times \nabla\phi$$

EXAMPLE 5-20: Prove that

$$\nabla \cdot (\mathbf{f} \times \mathbf{g}) = \mathbf{g} \cdot (\nabla \times \mathbf{f}) - \mathbf{f} \cdot (\nabla \times \mathbf{g}) \qquad (5.43)$$

Solution: We can write

$$\nabla \cdot (\mathbf{f} \times \mathbf{g}) = \nabla_\mathbf{f} \cdot (\mathbf{f} \times \mathbf{g}) + \nabla_\mathbf{g} \cdot (\mathbf{f} \times \mathbf{g})$$

where $\nabla_f \cdot (f \times g)$ means that g is fixed and ∇ operates on f, and $\nabla_g \cdot (f \times g)$ means that f is fixed and ∇ operates on g. From the results of Problem 1-16, we use the property of cyclic rotation of the triple scalar product to remove the subscript notation and obtain

$$\nabla_f \cdot (f \times g) = g \cdot \nabla \times f$$
$$\nabla_g \cdot (f \times g) = -f \cdot \nabla \times g$$

So we get

$$\nabla \cdot (f \times g) = g \cdot \nabla \times f - f \cdot \nabla \times g$$

EXAMPLE 5-21: If f is a vector function with continuous second derivatives, show that the curl of the curl of f is

$$\text{curl}(\text{curl}\,f) = \nabla \times (\nabla \times f) = \nabla(\nabla \cdot f) - \nabla^2 f$$
$$= \text{grad}(\text{div}\,f) - \nabla^2 f \qquad (5.44)$$

Solution: From the formula for the triple vector product (1.59), we have

$$A \times (B \times C) = (A \cdot C)B - (A \cdot B)C$$

Letting $A = B = \nabla$ and $C = f$, we obtain

$$\nabla \times (\nabla \times f) = \nabla(\nabla \cdot f) - (\nabla \cdot \nabla)f$$
$$= \nabla(\nabla \cdot f) - \nabla^2 f$$

Notice that rather than writing $(\nabla \cdot f)\nabla$, we must write $\nabla(\nabla \cdot f)$ to make sure that ∇ operates on f.

Note: From expression (5.44), we have a formula for evaluating $\nabla^2 f$:

$$\nabla^2 f = \nabla(\nabla \cdot f) - \nabla \times (\nabla \times f) \qquad (5.45)$$

SUMMARY

1. The directional derivative of a scalar function $\phi(x, y, z)$ in the direction of the unit tangent vector T to a curve C is defined as $\partial\phi/\partial s$, where s is the arc length of C.

2. The gradient of a scalar function $\phi(x, y, z)$, denoted by $\nabla\phi$, is a vector defined by

$$\text{grad}\,\phi = \nabla\phi = \frac{\partial\phi}{\partial x}\,i + \frac{\partial\phi}{\partial y}\,j + \frac{\partial\phi}{\partial z}\,k$$

where

$$\nabla = \frac{\partial}{\partial x}\,i + \frac{\partial}{\partial y}\,j + \frac{\partial}{\partial z}\,k$$

is an operator.

3. The divergence of a vector function $f = [f_1, f_2, f_3]$, denoted by $\nabla \cdot f$ is a scalar defined by

$$\text{div}\,f = \nabla \cdot f = \frac{\partial f_1}{\partial x} + \frac{\partial f_2}{\partial y} + \frac{\partial f_3}{\partial z}$$

4. The Laplacian of a scalar function ϕ, denoted by $\nabla^2\phi$ is defined as

$$\nabla^2\phi = \text{div}(\text{grad}\,\phi) = \nabla \cdot (\nabla\phi) = \frac{\partial^2\phi}{\partial x^2} + \frac{\partial^2\phi}{\partial y^2} + \frac{\partial^2\phi}{\partial z^2}$$

5. The curl of a vector function $\mathbf{f} = [f_1, f_2, f_3]$, denoted by $\nabla \times \mathbf{f}$ is a vector defined by

$$\nabla \times \mathbf{f} = \begin{vmatrix} \mathbf{i} & \mathbf{j} & \mathbf{k} \\ \dfrac{\partial}{\partial x} & \dfrac{\partial}{\partial y} & \dfrac{\partial}{\partial z} \\ f_1 & f_2 & f_3 \end{vmatrix} = \left(\dfrac{\partial f_3}{\partial y} - \dfrac{\partial f_2}{\partial z}\right)\mathbf{i} + \left(\dfrac{\partial f_1}{\partial z} - \dfrac{\partial f_3}{\partial x}\right)\mathbf{j} + \left(\dfrac{\partial f_2}{\partial x} - \dfrac{\partial f_1}{\partial y}\right)\mathbf{k}$$

6. Some vector identities involving the ∇ operator are

$$\nabla \cdot (\phi \mathbf{f}) = \phi \nabla \cdot \mathbf{f} + \mathbf{f} \cdot \nabla \phi$$
$$\nabla \times (\phi \mathbf{f}) = (\nabla \phi) \times \mathbf{f} + \phi \nabla \times \mathbf{f} = \phi \nabla \times \mathbf{f} - \mathbf{f} \times \nabla \phi$$
$$\nabla \cdot (\mathbf{f} \times \mathbf{g}) = \mathbf{g} \cdot (\nabla \times \mathbf{f}) - \mathbf{f} \cdot (\nabla \times \mathbf{g})$$
$$\nabla \times (\mathbf{f} \times \mathbf{g}) = \mathbf{f}(\nabla \cdot \mathbf{g}) - \mathbf{g}(\nabla \cdot \mathbf{f}) + (\mathbf{g} \cdot \nabla)\mathbf{f} - (\mathbf{f} \cdot \nabla)\mathbf{g}$$
$$\nabla^2 \mathbf{f} = \nabla(\nabla \cdot \mathbf{f}) - \nabla \times (\nabla \times \mathbf{f})$$

RAISE YOUR GRADES

Can you explain . . . ?

☑ how to define the gradient of a scalar function
☑ why $\nabla \phi$ is perpendicular to the surface S defined by $\phi(x, y, z) = \text{const}$
☑ how to define the divergence of a vector function
☑ how to define the curl of a vector function
☑ how to prove vector identities involving the ∇ operator

SOLVED PROBLEMS

Directional Derivative and Gradient of a Scalar Function

PROBLEM 5-1 Find (a) the directional derivative of $\phi(x, y, z) = x^2 + y^2 + z^2$ in the direction from $P(1, 1, 0)$ to $Q(2, 1, 1)$ and (b) its maximum value and direction at $(1, 1, 0)$.

Solution:
(a) Let the position vectors $\mathbf{r}_1 = [1, 1, 0]$ and $\mathbf{r}_2 = [2, 1, 1]$ represent the points $P(1, 1, 0)$ and $Q(2, 1, 1)$, and let $\mathbf{T}$ be the unit vector in the direction from P to Q. Then from the definition of a unit vector (2.35) and the representation of a vector (2.19),

$$\mathbf{T} = \frac{\mathbf{r}_2 - \mathbf{r}_1}{|\mathbf{r}_2 - \mathbf{r}_1|} = \frac{(2-1)\mathbf{i} + (1-1)\mathbf{j} + (1-0)\mathbf{k}}{[(2-1)^2 + (1-1)^2 + (1-0)^2]^{1/2}} = \frac{\mathbf{i} + \mathbf{k}}{\sqrt{2}}$$

$$= \frac{1}{\sqrt{2}}\mathbf{i} + \frac{1}{\sqrt{2}}\mathbf{k} = \left[\frac{1}{\sqrt{2}}, 0, \frac{1}{\sqrt{2}}\right]$$

Now, combining the definition of the gradient (5.4) with the given $\phi(x, y, z) = x^2 + y^2 + z^2$, we obtain

$$\nabla\phi = 2x\mathbf{i} + 2y\mathbf{j} + 2z\mathbf{k}$$

Thus at $(1, 1, 0)$ the gradient is

$$\nabla\phi = 2\mathbf{i} + 2\mathbf{j} + 0\mathbf{k} = [2, 2, 0]$$

and its magnitude is

$$|\nabla\phi| = (2^2 + 2^2 + 0)^{1/2} = 2\sqrt{2}$$

Hence by (5.5) the directional derivative of ϕ at $(1, 1, 0)$ is

$$\frac{\partial\phi}{\partial s} = \nabla\phi \cdot \mathbf{T} = (2)\left(\frac{1}{\sqrt{2}}\right) + (2)(0) + (0)\left(\frac{1}{\sqrt{2}}\right) = \sqrt{2}$$

This means that at the point $P(1, 1, 0)$ the value of ϕ increases by $\sqrt{2}$ per unit distance as it proceeds from $P(1, 1, 0)$ to $Q(2, 1, 1)$. Since the gradient value is computed by the coordinates of each point, this rate of increase will change as we move along the curve.

(b) The maximum values of $\partial\phi/\partial s$ at $(1, 1, 0)$ is $|\nabla\phi| = 2\sqrt{2}$, and its direction is that of $\nabla\phi = [2, 2, 0]$.

PROBLEM 5-2 Find a unit normal vector $\mathbf{n}$ to the surface given by $z = x^2 + y^2$ at the point $(1, 0, 1)$.

Solution: Since $z = x^2 + y^2$, the surface is defined as

$$\phi(x, y, z) \equiv x^2 + y^2 - z = 0$$

and from definition (5.4) the gradient is

$$\nabla\phi = 2x\mathbf{i} + 2y\mathbf{j} - \mathbf{k}$$

Thus $\nabla\phi = 2\mathbf{i} - \mathbf{k}$ at $(1, 0, 1)$. Hence the unit vector is

$$\mathbf{n} = \frac{\nabla\phi}{|\nabla\phi|} = \frac{2\mathbf{i} - \mathbf{k}}{[2^2 + 0 + (-1)^2]^{1/2}} = \frac{2\mathbf{i} - \mathbf{k}}{\sqrt{5}} = \left[\frac{2}{\sqrt{5}}, 0, -\frac{1}{\sqrt{5}}\right]$$

PROBLEM 5-3 For an arbitrary constant vector $\mathbf{a}$, show that

$$\nabla(\mathbf{a} \cdot \mathbf{r}) = \mathbf{a}$$

where $\mathbf{r}$ is the position vector.

Solution: Let $\mathbf{a} = [a_1, a_2, a_3]$ and $\mathbf{r} = [x, y, z]$. Then using definition of dot product (2.20), we have

$$\mathbf{a} \cdot \mathbf{r} = a_1 x + a_2 y + a_3 z$$

So

$$\nabla(\mathbf{a} \cdot \mathbf{r}) = \frac{\partial}{\partial x}(a_1 x + a_2 y + a_3 z)\mathbf{i} + \frac{\partial}{\partial y}(a_1 x + a_2 y + a_3 z)\mathbf{j}$$

$$+ \frac{\partial}{\partial z}(a_1 x + a_2 y + a_3 z)\mathbf{k}$$

$$= a_1\mathbf{i} + a_2\mathbf{j} + a_3\mathbf{k}$$

$$= \mathbf{a}$$

PROBLEM 5-4 If $\phi = \phi(x, y, z)$, show that

$$\nabla\phi \cdot d\mathbf{r} = d\phi$$

Solution: If $\mathbf{r} = x\mathbf{i} + y\mathbf{j} + z\mathbf{k}$, then its differential is

$$d\mathbf{r} = (dx)\mathbf{i} + (dy)\mathbf{j} + (dz)\mathbf{k}$$

From the definition of the gradient (5.4),

$$\nabla\phi = \frac{\partial\phi}{\partial x}\mathbf{i} + \frac{\partial\phi}{\partial y}\mathbf{j} + \frac{\partial\phi}{\partial z}\mathbf{k}$$

So when we take the scalar product of these two vectors, we obtain

$$\nabla\phi \cdot d\mathbf{r} = \frac{\partial\phi}{\partial x}dx + \frac{\partial\phi}{\partial y}dy + \frac{\partial\phi}{\partial z}dz = d\phi$$

PROBLEM 5-5   If $\phi = \phi(x, y, z, t)$, show that

$$d\phi = d\mathbf{r} \cdot \nabla\phi + \frac{\partial\phi}{\partial t}dt$$

Solution: If $\phi = \phi(x, y, z, t)$, from calculus, its differential is

$$d\phi = \frac{\partial\phi}{\partial x}dx + \frac{\partial\phi}{\partial y}dy + \frac{\partial\phi}{\partial z}dz + \frac{\partial\phi}{\partial t}dt$$

Then from the results of Problem 5-4, we have

$$d\phi = (\nabla\phi) \cdot d\mathbf{r} + \frac{\partial\phi}{\partial t}dt$$

$$= d\mathbf{r} \cdot \nabla\phi + \frac{\partial\phi}{\partial t}dt$$

PROBLEM 5-6   Find $\nabla\phi$ if $\phi = r = |\mathbf{r}| = (x^2 + y^2 + z^2)^{1/2}$.

Solution: The surface defined by $\phi = r = (x^2 + y^2 + z^2)^{1/2} = $ const is a sphere with the center at the origin $(0, 0, 0)$. Hence according to the result of Example 5-3, ∇r is normal to the sphere, and therefore it is parallel to the position vector $\mathbf{r} = [x, y, z]$. Thus we can write $\nabla r = k\mathbf{r}$. Now from the result of Problem 5-4,

$$dr = \nabla r \cdot d\mathbf{r} = k\mathbf{r} \cdot d\mathbf{r} = kr\,dr$$

Thus

$$k = \frac{1}{r}$$

and hence

$$\nabla r = \frac{1}{r}\mathbf{r} = \mathbf{e}_r$$

where $\mathbf{e}_r$ is the unit vector in the direction of the position vector $\mathbf{r}$.

Alternate Solution: From definition (5.4) the gradient is

$$\nabla_r = \nabla(x^2 + y^2 + z^2)^{1/2}$$

$$= \frac{\partial}{\partial x}(x^2 + y^2 + z^2)^{1/2}\mathbf{i} + \frac{\partial}{\partial y}(x^2 + y^2 + z^2)^{1/2}\mathbf{j} + \frac{\partial}{\partial z}(x^2 + y^2 + z^2)^{1/2}\mathbf{k}$$

$$= \frac{1}{2}\frac{2x}{(x^2 + y^2 + z^2)^{1/2}}\mathbf{i} + \frac{1}{2}\frac{2y}{(x^2 + y^2 + z^2)^{1/2}}\mathbf{j} + \frac{1}{2}\frac{2z}{(x^2 + y^2 + z^2)^{1/2}}\mathbf{k}$$

$$= \frac{x}{r}\mathbf{i} + \frac{y}{r}\mathbf{j} + \frac{z}{r}\mathbf{k}$$

$$= \frac{1}{r}\mathbf{r}$$

$$= \mathbf{e}_r$$

PROBLEM 5-7   If $r = |\mathbf{r}| = (x^2 + y^2 + z^2)^{1/2}$, find ∇r^n and $\nabla(1/r)$, where n is any real number.

Solution: Let $\phi = \phi(r) = r^n$. Then combining $\nabla\phi = \nabla\phi(u) = \phi'(u)\,\nabla u$ (5.15) and the result of Problem 5-6, we have

$$\nabla(r^n) = \frac{d}{dr}\,(r^n)\,\nabla r = nr^{n-1}\,\nabla r$$

$$= nr^{n-1}\,\mathbf{e}_r$$

$$= nr^{n-1}\left(\frac{\mathbf{r}}{|\mathbf{r}|}\right)$$

$$= nr^{n-2}\,\mathbf{r}$$

Setting $n = -1$, we obtain

$$\nabla\left(\frac{1}{r}\right) = -\frac{1}{r^3}\,\mathbf{r} = -\frac{1}{r^2}\,\mathbf{e}_r$$

PROBLEM 5-8   Verify that for arbitrary functions $f(x, y, z)$ and $g(x, y, z)$, where $g(x, y, z) \neq 0$,

$$\nabla\left(\frac{f}{g}\right) = \frac{g\,\nabla f - f\,\nabla g}{g^2}$$

Solution: By definition (5.4) we have

$$\nabla\left(\frac{f}{g}\right) = \frac{\partial}{\partial x}\left(\frac{f}{g}\right)\mathbf{i} + \frac{\partial}{\partial y}\left(\frac{f}{g}\right)\mathbf{j} + \frac{\partial}{\partial z}\left(\frac{f}{g}\right)\mathbf{k}$$

Now from elementary calculus

$$\frac{\partial}{\partial x}\left(\frac{f}{g}\right) = \frac{g\,\dfrac{\partial f}{\partial x} - f\,\dfrac{\partial g}{\partial x}}{g^2}$$

$$\frac{\partial}{\partial y}\left(\frac{f}{g}\right) = \frac{g\,\dfrac{\partial f}{\partial y} - f\,\dfrac{\partial g}{\partial y}}{g^2}$$

$$\frac{\partial}{\partial z}\left(\frac{f}{g}\right) = \frac{g\,\dfrac{\partial f}{\partial z} - f\,\dfrac{\partial g}{\partial z}}{g^2}$$

Therefore

$$\nabla\left(\frac{f}{g}\right) = \frac{1}{g^2}\left[g\left(\frac{\partial f}{\partial x}\mathbf{i} + \frac{\partial f}{\partial y}\mathbf{j} + \frac{\partial f}{\partial z}\,\mathbf{k}\right) - f\left(\frac{\partial g}{\partial x}\mathbf{i} + \frac{\partial g}{\partial y}\mathbf{j} + \frac{\partial g}{\partial z}\,\mathbf{k}\right)\right]$$

$$= \frac{g\,\nabla f - f\,\nabla g}{g^2}$$

Divergence of a Vector Function

PROBLEM 5-9   Find the divergence of $\mathbf{f} = xyz\mathbf{i} + x^2y^2z\mathbf{j} + yz^3\mathbf{k}$.

Solution: From definition (5.16), the divergence of $\mathbf{f}$ is

$$\operatorname{div}\mathbf{f} = \nabla\cdot\mathbf{f} = \frac{\partial f_1}{\partial x} + \frac{\partial f_2}{\partial y} + \frac{\partial f_3}{\partial z}$$

$$= \frac{\partial}{\partial x}\,(xyz) + \frac{\partial}{\partial y}\,(x^2y^2z) + \frac{\partial}{\partial z}\,(yz^3)$$

$$= yz + 2x^2yz + 3yz^2$$

PROBLEM 5-10   If $\mathbf{f} = [f_1, f_2, f_3]$, show that

$$\nabla \cdot \mathbf{f} = \nabla f_1 \cdot \mathbf{i} + \nabla f_2 \cdot \mathbf{j} + \nabla f_3 \cdot \mathbf{k}$$

Solution: Using the definition of the gradient (5.4), we have

$$\nabla f_1 = \frac{\partial f_1}{\partial x}\mathbf{i} + \frac{\partial f_1}{\partial y}\mathbf{j} + \frac{\partial f_1}{\partial z}\mathbf{k}$$

Taking the scalar product of ∇f_1 with the vector $\mathbf{i}$, we obtain

$$\nabla f_1 \cdot \mathbf{i} = \frac{\partial f_1}{\partial x}$$

and continuing the process with ∇f_2 and ∇f_3,

$$\nabla f_2 \cdot \mathbf{j} = \frac{\partial f_2}{\partial y}, \qquad \nabla f_3 \cdot \mathbf{k} = \frac{\partial f_3}{\partial z}$$

Finally, using the divergence definition (5.16),

$$\nabla \cdot \mathbf{f} = \frac{\partial f_1}{\partial x} + \frac{\partial f_2}{\partial y} + \frac{\partial f_3}{\partial z}$$

$$= \nabla f_1 \cdot \mathbf{i} + \nabla f_2 \cdot \mathbf{j} + \nabla f_3 \cdot \mathbf{k}$$

Curl of a Vector Function

PROBLEM 5-11   Find the curl of $\mathbf{f} = xyz\mathbf{i} + x^2y^2z\mathbf{j} + yz^3\mathbf{k}$.

Solution: From definition (5.28), the curl of $\mathbf{f}$ is

$$\nabla \times \mathbf{f} = \begin{vmatrix} \mathbf{i} & \mathbf{j} & \mathbf{k} \\ \dfrac{\partial}{\partial x} & \dfrac{\partial}{\partial y} & \dfrac{\partial}{\partial z} \\ xyz & x^2y^2z & yz^3 \end{vmatrix}$$

$$= \left[\frac{\partial}{\partial y}(yz^3) - \frac{\partial}{\partial z}(x^2y^2z)\right]\mathbf{i} + \left[\frac{\partial}{\partial z}(xyz) - \frac{\partial}{\partial x}(yz^3)\right]\mathbf{j}$$

$$+ \left[\frac{\partial}{\partial x}(x^2y^2z) - \frac{\partial}{\partial y}(xyz)\right]\mathbf{k}$$

$$= (z^3 - x^2y^2)\mathbf{i} + (xy)\mathbf{j} + (2xy^2z - xz)\mathbf{k}$$

PROBLEM 5-12   If $\mathbf{f} = [f_1, f_2, f_3]$, show that

$$\nabla \times \mathbf{f} = \nabla f_1 \times \mathbf{i} + \nabla f_2 \times \mathbf{j} + \nabla f_3 \times \mathbf{k}$$

Solution: Using the definition of grad f from (5.4), we obtain

$$\nabla f_1 = \frac{\partial f_1}{\partial x}\mathbf{i} + \frac{\partial f_1}{\partial y}\mathbf{j} + \frac{\partial f_1}{\partial z}\mathbf{k}$$

Hence by the definition of the vector product of ∇f_1 with unit vector $\mathbf{i}$ (2.29) we have

$$\nabla f_1 \times \mathbf{i} = -\frac{\partial f_1}{\partial y}\mathbf{k} + \frac{\partial f_1}{\partial z}\mathbf{j}$$

Similarly,

$$\nabla f_2 \times \mathbf{j} = \frac{\partial f_2}{\partial x}\mathbf{k} - \frac{\partial f_2}{\partial z}\mathbf{i}$$

$$\nabla f_3 \times \mathbf{k} = -\frac{\partial f_3}{\partial x}\mathbf{j} + \frac{\partial f_3}{\partial y}\mathbf{i}$$

Combining these and using definition (5.28), we obtain

$$\nabla \times \mathbf{f} = \left(\frac{\partial f_3}{\partial y} - \frac{\partial f_2}{\partial z} \right) \mathbf{i} + \left(\frac{\partial f_1}{\partial z} - \frac{\partial f_3}{\partial x} \right) \mathbf{j} + \left(\frac{\partial f_2}{\partial x} - \frac{\partial f_1}{\partial y} \right) \mathbf{k}$$

$$= \nabla f_1 \times \mathbf{i} + \nabla f_2 \times \mathbf{j} + \nabla f_3 \times \mathbf{k}$$

Operations with ∇ and Some Vector Identities

PROBLEM 5-13 Find $(\mathbf{f} \cdot \nabla)\phi$ and $(\mathbf{f} \cdot \nabla)\mathbf{g}$ at $(1, 1, 1)$ if $\mathbf{f} = -y\mathbf{i} + x\mathbf{j} + z\mathbf{k}$, $\mathbf{g} = 3xyz^2\mathbf{i} + 2xy^3\mathbf{j} - x^2yz\mathbf{k}$, and $\phi = xyz$.

Solution: Since

$$\nabla \phi = \frac{\partial}{\partial x}(xyz)\mathbf{i} + \frac{\partial}{\partial y}(xyz)\mathbf{j} + \frac{\partial}{\partial z}(xyz)\mathbf{k} = yz\mathbf{i} + xz\mathbf{j} + xy\mathbf{k}$$

we have

$$(\mathbf{f} \cdot \nabla)\phi = \mathbf{f} \cdot \nabla \phi = (-y)(yz) + (x)(xz) + (z)(xy)$$

$$= -y^2z + x^2z + xyz$$

Then at $(1, 1, 1)$

$$(\mathbf{f} \cdot \nabla)\phi = -1 + 1 + 1 = 1$$

Multiplying $\mathbf{f} \cdot \nabla$ by $\mathbf{g}$ on the right (5.37) yields

$$(\mathbf{f} \cdot \nabla)\mathbf{g} = (-y)\frac{\partial}{\partial x}(3xyz^2\mathbf{i} + 2xy^3\mathbf{j} - x^2yz\mathbf{k})$$

$$+ (x)\frac{\partial}{\partial y}(3xyz^2\mathbf{i} + 2xy^3\mathbf{j} - x^2yz\mathbf{k})$$

$$+ (z)\frac{\partial}{\partial z}(3xyz^2\mathbf{i} + 2xy^3\mathbf{j} - x^2yz\mathbf{k})$$

$$= (-y)(3yz^2\mathbf{i} + 2y^3\mathbf{j} - 2xyz\mathbf{k}) + (x)(3xz^2\mathbf{i} + 6xy^2\mathbf{j} - x^2z\mathbf{k})$$

$$+ (z)(6xyz\mathbf{i} - x^2y\mathbf{k})$$

$$= (-3y^2z^2 + 3x^2z^2 + 6xyz^2)\mathbf{i} + (-2y^4 + 6x^2y^2)\mathbf{j}$$

$$+ (2xy^2z - x^3z - x^2yz)\mathbf{k}$$

Evaluating this at $(1, 1, 1)$, we find

$$(\mathbf{f} \cdot \nabla)\mathbf{g} = (-3 + 3 + 6)\mathbf{i} + (-2 + 6)\mathbf{j} + (2 - 1 - 1)\mathbf{k} = 4\mathbf{j} = [0, 4, 0]$$

PROBLEM 5-14 If $\mathbf{r}$ is the position vector, show that $(\mathbf{f} \cdot \nabla)\mathbf{r} = \mathbf{f}$.

Solution: If $\mathbf{r} = x\mathbf{i} + y\mathbf{j} + z\mathbf{k}$, then

$$(\mathbf{f} \cdot \nabla)\mathbf{r} = f_1 \frac{\partial \mathbf{r}}{\partial x} + f_2 \frac{\partial \mathbf{r}}{\partial y} + f_3 \frac{\partial \mathbf{r}}{\partial z}$$

$$= f_1\mathbf{i} + f_2\mathbf{j} + f_3\mathbf{k}$$

$$= \mathbf{f}$$

PROBLEM 5-15 Show that $(d\mathbf{r} \cdot \nabla)\mathbf{f} = d\mathbf{f}$.

Solution: From the formula for the meaning of $\mathbf{f} \cdot \nabla$ (5.36), we obtain

$$d\mathbf{r} \cdot \nabla = dx \frac{\partial}{\partial x} + dy \frac{\partial}{\partial y} + dz \frac{\partial}{\partial z}$$

So

$$(d\mathbf{r} \cdot \nabla)\mathbf{f} = dx \frac{\partial \mathbf{f}}{\partial x} + dy \frac{\partial \mathbf{f}}{\partial y} + dz \frac{\partial \mathbf{f}}{\partial z}$$

$$= \frac{\partial \mathbf{f}}{\partial x} dx + \frac{\partial \mathbf{f}}{\partial y} dy + \frac{\partial \mathbf{f}}{\partial z} dz$$

$$= d\mathbf{f}$$

PROBLEM 5-16   If $\mathbf{f} = \mathbf{f}(x, y, z, t)$, show that

$$d\mathbf{f} = (d\mathbf{r} \cdot \nabla)\mathbf{f} + \frac{\partial \mathbf{f}}{\partial t} dt$$

Solution: Since $\mathbf{f} = \mathbf{f}(x, y, z, t)$, from calculus and the definition of differentials, we have

$$d\mathbf{f} = \frac{\partial \mathbf{f}}{\partial x} dx + \frac{\partial \mathbf{f}}{\partial y} dy + \frac{\partial \mathbf{f}}{\partial z} dz + \frac{\partial \mathbf{f}}{\partial t} dt$$

$$= dx \frac{\partial \mathbf{f}}{\partial x} + dy \frac{\partial \mathbf{f}}{\partial y} + dz \frac{\partial \mathbf{f}}{\partial z} + \frac{\partial \mathbf{f}}{\partial t} dt$$

$$= (d\mathbf{r} \cdot \nabla)\mathbf{f} + \frac{\partial \mathbf{f}}{\partial t} dt$$

PROBLEM 5-17   Calculate $\nabla \cdot [f(r)\mathbf{r}]$ if $\mathbf{r}$ is the position vector and $r = |\mathbf{r}|$.

Solution: Using the rule for the differentiation of a product (5.41a), we have

$$\nabla \cdot [f(r)\mathbf{r}] = f(r)\nabla \cdot \mathbf{r} + \mathbf{r} \cdot \nabla[f(r)]$$

and from the result of Example 5-6 (5.17), we have

$$\nabla \cdot \mathbf{r} = 3$$

Then using $\nabla \phi = \nabla \phi(u) = \phi'(u)\nabla u$ (5.15) and the result of Problem 5-6, we obtain

$$\nabla[f(r)] = f'(r)\nabla r = f'(r)\frac{\mathbf{r}}{r} = f'(r)\mathbf{e}_r \qquad \text{(a)}$$

By combining these equations with $\mathbf{r} \cdot \mathbf{r} = r^2$, we obtain

$$\nabla \cdot [f(r)\mathbf{r}] = 3f(r) + \frac{f'(r)}{r}\mathbf{r} \cdot \mathbf{r} = 3f(r) + rf'(r)$$

PROBLEM 5-18   Calculate $\nabla \cdot (r^{n-1}\mathbf{r})$.

Solution: Setting $f(r) = r^{n-1}$ in the result of Problem 5-17, we have

$$\nabla \cdot (r^{n-1}\mathbf{r}) = 3r^{n-1} + (n-1)r^{n-1}$$
$$= (n+2)r^{n-1}$$

Note: Setting $n = -2$, we obtain for $r \neq 0$

$$\nabla \cdot \left(\frac{\mathbf{r}}{r^3}\right) = 0$$

Now substituting $\nabla(1/r) = (-1/r^3)\mathbf{r}$ into $\nabla \cdot (\mathbf{r}/r^3) = 0$, we obtain for $r \neq 0$

$$\nabla \cdot \nabla\left(\frac{1}{r}\right) = \nabla^2\left(\frac{1}{r}\right) = 0$$

which is a verification of result (5.26).

PROBLEM 5-19 Determine the value of $\mathbf{V} \times [f(r)\mathbf{r}]$.

Solution: Replacing $(\phi\mathbf{f})$ in $\mathbf{V} \times (\phi\mathbf{f}) = \phi\mathbf{V} \times \mathbf{f} + (\mathbf{V}\phi) \times \mathbf{f}$ (5.42) with $[f(r)\mathbf{r}]$, you get

$$\mathbf{V} \times [f(r)\mathbf{r}] = f(r)\mathbf{V} \times \mathbf{r} + [\mathbf{V}f(r)] \times \mathbf{r}$$

Now since $\mathbf{V} \times \mathbf{r} = \mathbf{0}$ (5.29), $\mathbf{V}[f(r)] = f'(r)\mathbf{r}/r$ from eq. (a) of Problem 5-17, and $\mathbf{r} \times \mathbf{r} = \mathbf{0}$ (1.42), we get

$$\mathbf{V} \times [f(r)\mathbf{r}] = \frac{f'(r)}{r}\,\mathbf{r} \times \mathbf{r} = \mathbf{0}$$

PROBLEM 5-20 Calculate $\mathbf{V} \cdot \mathbf{V}[f(r)] = \mathbf{V}^2 f(r)$.

Solution: From the result of Problem 5-17, we have

$$\mathbf{V}[f(r)] = \frac{f'(r)}{r}\,\mathbf{r}$$

Thus, from (5.41a),

$$\mathbf{V} \cdot \mathbf{V}[f(r)] = \mathbf{V} \cdot \left[\frac{f'(r)}{r}\,\mathbf{r}\right]$$

$$= \frac{f'(r)}{r}\,\mathbf{V} \cdot \mathbf{r} + \mathbf{r} \cdot \mathbf{V}\left[\frac{f'(r)}{r}\right]$$

Working on the second term of the right-hand side, we use eq. (a) of Problem 5-17 to obtain

$$\mathbf{V}\left[\frac{f'(r)}{r}\right] = \frac{1}{r}\frac{d}{dr}\left[\frac{f'(r)}{r}\right]\mathbf{r} = \frac{1}{r}\left[\frac{1}{r}f''(r) - \frac{1}{r^2}f'(r)\right]\mathbf{r}$$

Thus, because $\mathbf{V} \cdot \mathbf{r} = 3$ (5.17) and $\mathbf{r} \cdot \mathbf{r} = r^2$, we obtain

$$\mathbf{V} \cdot \mathbf{V}[f(r)] = \frac{3}{r}\,f'(r) + \frac{1}{r}\left[\frac{1}{r}\,f''(r) - \frac{1}{r^2}\,f'(r)\right]\mathbf{r} \cdot \mathbf{r}$$

$$= \frac{2}{r}\,f'(r) + f''(r) \tag{a}$$

PROBLEM 5-21 Calculate $\mathbf{V} \cdot \mathbf{V}r^n = \mathbf{V}^2 r^n$

Solution: Setting $f(r) = r^n$ in eq. (a) of Problem 5-20, we obtain

$$\mathbf{V} \cdot \mathbf{V}r^n = \mathbf{V}^2 r^n = \frac{2}{r}\frac{d}{dr}\,r^n + \frac{d^2}{dr^2}\,r^n$$

$$= 2nr^{n-2} + n(n-1)r^{n-2}$$

$$= n(n+1)r^{n-2}$$

[*Note:* Setting $n = -1$ for $r \neq 0$, we have $\mathbf{V}^2(1/r) = 0$, which is a verification of (5.26).]

PROBLEM 5-22 Show that for arbitrary functions $\phi(x, y, z)$ and $\psi(x, y, z)$

$$\mathbf{V} \cdot (\phi\,\mathbf{V}\psi - \psi\,\mathbf{V}\phi) = \phi\,\mathbf{V}^2\psi - \psi\,\mathbf{V}^2\phi$$

Solution: Setting $\mathbf{f} = \mathbf{V}\psi$ in the rule for the differentiation of a product (5.41a), we have

$$\mathbf{V} \cdot (\phi\,\mathbf{V}\psi) = \phi\mathbf{V} \cdot \mathbf{V}\psi + \mathbf{V}\psi \cdot \mathbf{V}\phi = \phi\,\mathbf{V}^2\psi + \mathbf{V}\psi \cdot \mathbf{V}\phi \tag{a}$$

By interchanging ϕ and ψ in eq. (a)

$$\mathbf{V} \cdot (\psi\,\mathbf{V}\phi) = \psi\,\mathbf{V}^2\phi + \mathbf{V}\phi \cdot \mathbf{V}\psi \tag{b}$$

Finally, since $\nabla\phi \cdot \nabla\psi = \nabla\psi \cdot \nabla\phi$, subtracting eq. (b) from eq. (a) yields

$$\nabla \cdot (\phi\,\nabla\psi - \psi\,\nabla\phi) = \phi\,\nabla^2\psi - \psi\,\nabla^2\phi$$

PROBLEM 5-23   Verify the identity

$$\nabla \times (\mathbf{f} \times \mathbf{g}) = \mathbf{f}(\nabla \cdot \mathbf{g}) - \mathbf{g}(\nabla \cdot \mathbf{f}) + (\mathbf{g} \cdot \nabla)\mathbf{f} - (\mathbf{f} \cdot \nabla)\mathbf{g}$$

Solution: Since $\nabla \times (\mathbf{f} \times \mathbf{g}) = \nabla_\mathbf{f} \times (\mathbf{f} \times \mathbf{g}) + \nabla_\mathbf{g} \times (\mathbf{f} \times \mathbf{g})$, applying formulas (1.59) and (1.60), we have

$$\nabla_\mathbf{f} \times (\mathbf{f} \times \mathbf{g}) = (\mathbf{g} \cdot \nabla)\mathbf{f} - \mathbf{g}(\nabla \cdot \mathbf{f})$$
$$\nabla_\mathbf{g} \times (\mathbf{f} \times \mathbf{g}) = \mathbf{f}(\nabla \cdot \mathbf{g}) - (\mathbf{f} \cdot \nabla)\mathbf{g}$$

Hence

$$\nabla \times (\mathbf{f} \times \mathbf{g}) = \mathbf{f}(\nabla \cdot \mathbf{g}) - \mathbf{g}(\nabla \cdot \mathbf{f}) + (\mathbf{g} \cdot \nabla)\mathbf{f} - (\mathbf{f} \cdot \nabla)\mathbf{g}$$

[*Note:* The quantities $(\mathbf{g} \cdot \nabla)\mathbf{f}$ and $(\mathbf{f} \cdot \nabla)\mathbf{g}$ are defined in (5.37).]

PROBLEM 5-24   Prove that

$$\nabla(\mathbf{f} \cdot \mathbf{g}) = \mathbf{f} \times (\nabla \times \mathbf{g}) + \mathbf{g} \times (\nabla \times \mathbf{f}) + (\mathbf{f} \cdot \nabla)\mathbf{g} + (\mathbf{g} \cdot \nabla)\mathbf{f}$$

Solution: If we apply (1.59) to $\mathbf{f} \times (\nabla \times \mathbf{g})$, where the vector function $\mathbf{f}$ is a constant, we have

$$\mathbf{f} \times (\nabla \times \mathbf{g}) = \nabla_\mathbf{g}(\mathbf{f} \cdot \mathbf{g}) - (\mathbf{f} \cdot \nabla)\mathbf{g}$$

Hence

$$\nabla_\mathbf{g}(\mathbf{f} \cdot \mathbf{g}) = \mathbf{f} \times (\nabla \times \mathbf{g}) + (\mathbf{f} \cdot \nabla)\mathbf{g}$$

By interchanging $\mathbf{f}$ and $\mathbf{g}$, we obtain

$$\nabla_\mathbf{f}(\mathbf{g} \cdot \mathbf{f}) = \nabla_\mathbf{f}(\mathbf{f} \cdot \mathbf{g}) = \mathbf{g} \times (\nabla \times \mathbf{f}) + (\mathbf{g} \cdot \nabla)\mathbf{f}$$

Combining $\nabla_\mathbf{f}$ and $\nabla_\mathbf{g}$, we obtain

$$\nabla(\mathbf{f} \cdot \mathbf{g}) = \nabla_\mathbf{f}(\mathbf{f} \cdot \mathbf{g}) + \nabla_\mathbf{g}(\mathbf{f} \cdot \mathbf{g})$$
$$= \mathbf{g} \times (\nabla \times \mathbf{f}) + \mathbf{f} \times (\nabla \times \mathbf{g}) + (\mathbf{g} \cdot \nabla)\mathbf{f} + (\mathbf{f} \cdot \nabla)\mathbf{g}$$

Supplementary Exercises

PROBLEM 5-25   Find the directional derivative of

$$\phi(x, y, z) = z^2 y + y^2 z + z^2 x$$

at $(1, 1, 1)$ in the direction of curve C represented by $\mathbf{r}(t) = t\mathbf{i} + t^2\mathbf{j} + t^3\mathbf{k}$.

Answer: $22/\sqrt{14}$

PROBLEM 5-26   If $\phi(x, y, z) = xy + yz + zx$, find

(a) $\nabla\phi$ at $(1, 1, 3)$

(b) $\dfrac{\partial\phi}{\partial s}$ at $(1, 1, 3)$ in the direction of the vector $[1, 1, 1]$

(c) the normal derivative $\partial\phi/\partial n = \nabla\phi \cdot \mathbf{n}$ at $(1, 1, 3)$, where $\mathbf{n}$ is a unit vector normal to the surface S defined by a constant $\phi(x, y, z)$.

Answer: (a) $[4, 4, 2]$ (b) $10/\sqrt{3}$ (c) 6

PROBLEM 5-27 Find the divergence and curl of $\mathbf{f} = (x - y)\mathbf{i} + (y - z)\mathbf{j} + (z - x)\mathbf{k}$ and $\mathbf{g} = (x^2 + yz)\mathbf{i} + (y^2 + zx)\mathbf{j} + (z^2 + xy)\mathbf{k}$.

Answer: $\nabla \cdot \mathbf{f} = 3,\quad \nabla \times \mathbf{f} = [1, 1, 1];\quad \nabla \cdot \mathbf{g} = 2(x + y + z),\quad \nabla \times \mathbf{g} = \mathbf{0}$

PROBLEM 5-28 If $\phi = 3x^2 - yz$ and $\mathbf{f} = 3xyz^2\mathbf{i} + 2xy^3\mathbf{j} - x^2yz\mathbf{k}$, find, at $(1, -1, 1)$, (a) $\nabla\phi$, (b) $\nabla \cdot \mathbf{f}$, (c) $\nabla \times \mathbf{f}$, (d) $\mathbf{f} \cdot \nabla\phi$, (e) $\nabla \cdot (\phi\mathbf{f})$, (f) $\nabla \times (\phi\mathbf{f})$, and (g) $\nabla^2\phi$.

Answer: (a) $[6, -1, 1]$ (b) 4   (c) $[-1, -8, -5]$ (d) -15 (e) 1
(f) $[-3, -41, -35]$ (g) 6

PROBLEM 5-29 If $\mathbf{f} = 2z\mathbf{i} + x^2\mathbf{j} + x\mathbf{k}$ and $\phi = 2x^2y^2z^2$, find $(\mathbf{f} \times \nabla)\phi$ at the point $(1, -1, 1)$.

Answer: $[8, -4, -12]$

PROBLEM 5-30 For an arbitrary constant vector $\mathbf{a}$, and the position vector $\mathbf{r}$, show that

(a) $\nabla \cdot (\mathbf{a} \times \mathbf{r}) = 0$
(b) $\nabla \times (\mathbf{a} \times \mathbf{r}) = 2\mathbf{a}$

(c) $\nabla\left(\dfrac{\mathbf{a} \cdot \mathbf{r}}{r^3}\right) + \nabla \times \left(\dfrac{\mathbf{a} \times \mathbf{r}}{r^3}\right) = \mathbf{0}$

(d) $\nabla\left(\mathbf{a} \cdot \nabla\dfrac{1}{r}\right) + \nabla \times \left(\mathbf{a} \times \nabla\dfrac{1}{r}\right) = \mathbf{0}$

PROBLEM 5-31 If the second derivatives of the functions ϕ and ψ exist, then show that $\nabla^2(\phi\psi) = \phi\nabla^2\psi + 2\nabla\phi \cdot \nabla\psi + \psi\nabla^2\phi$.

PROBLEM 5-32 The vector field $\mathbf{f}$ is called **solenoidal** if $\mathbf{f}$ is differentiable and $\nabla \cdot \mathbf{f} = 0$. It's called **irrotational** if $\mathbf{f}$ is differentiable and $\nabla \times \mathbf{f} = \mathbf{0}$. Show that if $\mathbf{f}$ and $\mathbf{g}$ are irrotational, then $\mathbf{f} \times \mathbf{g}$ is solenoidal.

PROBLEM 5-33 If ϕ and ψ are scalar functions that have continuous second derivatives, show that $\nabla\phi \times \nabla\psi$ is solenoidal.

6 VECTOR INTEGRATION

THIS CHAPTER IS ABOUT

☑ **Line Integrals**
☑ **Surface Integrals**
☑ **Volume Integrals**
☑ **Divergence and Curl of a Vector**
☑ **Alternate Definitions of Gradient, Divergence, and Curl**

6-1. Line Integrals

In Section 4-4 a curve C, for $a \leqslant t \leqslant b$, is represented by

$$\mathbf{r}(t) = x(t)\mathbf{i} + y(t)\mathbf{j} + z(t)\mathbf{k} \tag{6.1}$$

where $\mathbf{r}$ is the position vector and t is any parameter. The **differential displacement vector** $d\mathbf{r}$ along curve C is defined by

DIFFERENTIAL DISPLACEMENT VECTOR
$$d\mathbf{r} = dx\,\mathbf{i} + dy\,\mathbf{j} + dz\,\mathbf{k} \tag{6.2}$$

Integrals that involve differential displacement vectors $d\mathbf{r}$ are called **line integrals**. Consider the following line integrals along a curve C that can be open or closed:

$$\int_C \phi\, d\mathbf{r} \tag{6.3}$$

LINE INTEGRALS
$$\int_C \mathbf{f} \cdot d\mathbf{r} \tag{6.4}$$

$$\int_C \mathbf{f} \times d\mathbf{r} \tag{6.5}$$

When the space curve C forms a closed path, line integral (6.3) is written as

$$\oint_C \phi\, d\mathbf{r} \tag{6.6}$$

In particular, if C is a simple closed plane curve and the direction of integration can be described, the symbols often used are

$$\oint_C \phi\, d\mathbf{r} \quad \text{or} \quad \oint_C \phi\, d\mathbf{r}$$

The first integral indicates movement along the closed curve C in the positive, or counterclockwise, direction; i.e., the movement along C is such that its enclosed region always lies to the left. The second integral indicates movement in the negative, or clockwise, direction, so the enclosed region lies to the right.

The line integral (6.4) is sometimes called the **scalar line integral** (or simply the *line integral*) of a vector field $\mathbf{f}$. The line integral of $\mathbf{f}$ around a closed curve $\mathbf{C}$ is called the **circulation** of $\mathbf{f}$ around C and is written as

CIRCULATION $\qquad\qquad\qquad\qquad\qquad \text{circ}\,\mathbf{f} \equiv \oint_C \mathbf{f} \cdot d\mathbf{r}$ $\qquad\qquad$ (6.7)

EXAMPLE 6-1: If $\phi = \phi(x, y, z)$, show that

$$\int_C \phi \, d\mathbf{r} = \mathbf{i} \int_C \phi \, dx + \mathbf{j} \int_C \phi \, dy + \mathbf{k} \int_C \phi \, dz \qquad (6.8)$$

Solution: Using the differential displacement vector $d\mathbf{r} = dx\,\mathbf{i} + dy\,\mathbf{j} + dz\,\mathbf{k}$ (6.2), we have

$$\int_C \phi \, d\mathbf{r} = \int_C \phi(dx\,\mathbf{i} + dy\,\mathbf{j} + dz\,\mathbf{k}) = \mathbf{i} \int_C \phi \, dx + \mathbf{j} \int_C \phi \, dy + \mathbf{k} \int_C \phi \, dz$$

Note: We can use the relation

$$\int_C \mathbf{i}\,\phi\,dx = \mathbf{i} \int_C \phi\,dx, \quad \text{etc.}$$

because $\mathbf{i}, \mathbf{j}$, and $\mathbf{k}$ of the rectangular coordinate system have constant magnitude and direction.

EXAMPLE 6-2: If $\mathbf{f} = f_1(x, y, z)\mathbf{i} + f_2(x, y, z)\mathbf{j} + f_3(x, y, z)\mathbf{k}$, show that

$$\int_C \mathbf{f} \cdot d\mathbf{r} = \int_C (f_1\,dx + f_2\,dy + f_3\,dz) \qquad (6.9)$$

Solution: Since $d\mathbf{r} = dx\,\mathbf{i} + dy\,\mathbf{j} + dz\,\mathbf{k}$,

$$\mathbf{f} \cdot d\mathbf{r} = f_1\,dx + f_2\,dy + f_3\,dz$$

and hence

$$\int_C \mathbf{f} \cdot d\mathbf{r} = \int_C (f_1\,dx + f_2\,dy + f_3\,dz)$$

EXAMPLE 6-3: If $\mathbf{f} = f_1\mathbf{i} + f_2\mathbf{j} + f_3\mathbf{k}$, show that

$$\int_C \mathbf{f} \times d\mathbf{r} = \mathbf{i} \int_C (f_2\,dz - f_3\,dy) + \mathbf{j} \int_C (f_3\,dx - f_1\,dz) + \mathbf{k} \int_C (f_1\,dy - f_2\,dx)$$
$$\qquad (6.10)$$

Solution: From (2.28), the vector product of $\mathbf{f}$ and $d\mathbf{r}$ can be written in determinant form:

$$\mathbf{f} \times d\mathbf{r} = \begin{vmatrix} \mathbf{i} & \mathbf{j} & \mathbf{k} \\ f_1 & f_2 & f_3 \\ dx & dy & dz \end{vmatrix}$$
$$= \mathbf{i}(f_2\,dz - f_3\,dy) + \mathbf{j}(f_3\,dx - f_1\,dz) + \mathbf{k}(f_1\,dy - f_2\,dx)$$

So the integral is

$$\int_C \mathbf{f} \times d\mathbf{r} = \mathbf{i} \int_C (f_2\,dz - f_3\,dy) + \mathbf{j} \int_C (f_3\,dx - f_1\,dz) + \mathbf{k} \int_C (f_1\,dy - f_2\,dx)$$

EXAMPLE 6-4: If $\mathbf{T}$ is the unit tangent vector along a curve C, show that

$$\int_C \mathbf{f} \cdot d\mathbf{r} = \int_C \mathbf{f} \cdot \mathbf{T}\,ds \qquad (6.11)$$

where s is the arc length of C measured from some fixed point on C. The curve C can be represented as in formula (4.45):

$$\mathbf{r}(s) = x(s)\mathbf{i} + y(s)\mathbf{j} + z(s)\mathbf{k}, \qquad a \leqslant s \leqslant b$$

Solution: From definition (4.46), the unit tangent vector $\mathbf{T}$ along curve C is given by

$$\mathbf{T} = \frac{d\mathbf{r}}{ds}$$

Then we have

$$d\mathbf{r} = \frac{d\mathbf{r}}{ds} ds = \mathbf{T}\, ds$$

Hence

$$\int_C \mathbf{f} \cdot d\mathbf{r} = \int_C \mathbf{f} \cdot \frac{d\mathbf{r}}{ds}\, ds = \int_C \mathbf{f} \cdot \mathbf{T}\, ds$$

Since the dot product $\mathbf{f} \cdot \mathbf{T}$ is equal to the component of $\mathbf{f}$ in the direction of $\mathbf{T}$, that is, the direction of C, eq. (6.11) shows that the line integral of $\mathbf{f}$ is equivalent to the integration of the tangential component of $\mathbf{f}$ along C with respect to the arc length.

6-2. Surface Integrals

In Section 4-5 a surface S is represented by a vector function

$$\mathbf{r}(u, v) = x(u, v)\mathbf{i} + y(u, v)\mathbf{j} + z(u, v)\mathbf{k} \tag{6.12}$$

where $\mathbf{r}$ is the position vector and u and v are parameters. The **differential element** $d\mathbf{S}$ of the surface area is defined by

**DIFFERENTIAL
SURFACE ELEMENT**
$$d\mathbf{S} = \mathbf{r}_u \times \mathbf{r}_v\, du\, dv \tag{6.13}$$

where $\mathbf{r}_u$ is the notation for $\partial\mathbf{r}/\partial u$ and $\mathbf{r}_v$ denotes $\partial\mathbf{r}/\partial v$. As shown in Chapter 4, the differential surface element (6.13) can be written as

$$d\mathbf{S} = \mathbf{n}\, dS \tag{6.14}$$

where $\mathbf{n}$ is the unit vector normal to the surface at the point that corresponds to the coordinates $\mathbf{u}$ and $\mathbf{v}$, and

$$\mathbf{n} = \frac{\mathbf{r}_u \times \mathbf{r}_v}{|\mathbf{r}_u \times \mathbf{r}_v|} \tag{6.15}$$

$$dS = |\mathbf{r}_u \times \mathbf{r}_v|\, du\, dv \tag{6.16}$$

Integrals that involve the differential element $d\mathbf{S}$ of surface area are called **surface integrals**. Consider the surface integrals

$$\iint_S \phi\, d\mathbf{S} \tag{6.17}$$

SURFACE INTEGRALS
$$\iint_S \mathbf{f} \cdot d\mathbf{S} \tag{6.18}$$

$$\iint_S \mathbf{f} \times d\mathbf{S} \tag{6.19}$$

each of which is over a surface S that can be open or closed. (A closed surface is one that has no boundary and completely encloses a bounded region in space.)

If S is a closed surface, the surface integrals are written as

$$\oiint_S \phi \, d\mathbf{S} \quad \text{or} \quad \oiint_S \mathbf{f} \cdot d\mathbf{S} \quad \text{or} \quad \oiint_S \mathbf{f} \times d\mathbf{S}$$

For closed surfaces, it is usual to assume that the positive direction of the normal is directed outward from the surface.

The surface integral (6.18) of a vector field $\mathbf{f}$ is called the **flux** of $\mathbf{f}$ through S. For example, if S is represented by $\mathbf{r}(u, v)$, then the flux of $\mathbf{f}$ through S is given by

FLUX OF f
$$\iint_S \mathbf{f} \cdot d\mathbf{S} = \iint_{R_{uv}} [\mathbf{f r}_u \mathbf{r}_v] \, du \, dv \tag{6.20}$$

where $[\mathbf{f r}_u \mathbf{r}_v] = \mathbf{f} \cdot \mathbf{r}_u \times \mathbf{r}_v$ is the scalar triple product and R_{uv} is the region where parameters u and v are defined. Since the differential element of the surface area (6.13) is

$$d\mathbf{S} = \mathbf{r}_u \times \mathbf{r}_v \, du \, dv$$

The flux of $\mathbf{f}$ through S is

$$\iint_S \mathbf{f} \cdot d\mathbf{S} = \iint_{R_{uv}} \mathbf{f} \cdot \mathbf{r}_u \times \mathbf{r}_v \, du \, dv = \iint_{R_{uv}} [\mathbf{f r}_u \mathbf{r}_v] \, du \, dv$$

EXAMPLE 6-5: If $\mathbf{f} = f_1(x, y, z)\mathbf{i} + f_2(x, y, z)\mathbf{j} + f_3(x, y, z)\mathbf{k}$, show that the surface integral of $\mathbf{f}$ can be expressed as

$$\iint_S \mathbf{f} \cdot d\mathbf{S} = \iint_S \mathbf{f} \cdot \mathbf{n} \, dS = \pm \iint_{R_{yz}} f_1 \, dy \, dz \pm \iint_{R_{zx}} f_2 \, dz \, dx \pm \iint_{R_{xy}} f_3 \, dx \, dy \tag{6.21}$$

where R_{yz}, R_{zx}, and R_{xy} are the projections of S onto the yz-, zx-, and xy-planes, respectively, and the signs of the integrals on the right-hand side of (6.21) are determined by the signs of $(\mathbf{n} \cdot \mathbf{i})$, $(\mathbf{n} \cdot \mathbf{j})$, and $(\mathbf{n} \cdot \mathbf{k})$, respectively.

Solution: If α, β, and γ are the angles between the rectangular coordinate axes and the unit vector $\mathbf{n}$, then from the direction cosines (see eq. (2.43)) we have

$$\mathbf{n} \cdot \mathbf{i} = \cos \alpha, \qquad \mathbf{n} \cdot \mathbf{j} = \cos \beta, \qquad \mathbf{n} \cdot \mathbf{k} = \cos \gamma$$

Hence from (2.56), $\mathbf{n}$ can be written as

$$\mathbf{n} = (\mathbf{n} \cdot \mathbf{i})\mathbf{i} + (\mathbf{n} \cdot \mathbf{j})\mathbf{j} + (\mathbf{n} \cdot \mathbf{k})\mathbf{k}$$
$$= (\cos \alpha)\mathbf{i} + (\cos \beta)\mathbf{j} + (\cos \gamma)\mathbf{k}$$

and consequently

$$d\mathbf{S} = \mathbf{n} \, dS = (\cos \alpha) \, dS \, \mathbf{i} + (\cos \beta) \, dS \, \mathbf{j} + (\cos \gamma) \, dS \, \mathbf{k} \tag{6.22}$$

Now we can write (see Figure 6-1)

$$dS(\mathbf{n} \cdot \mathbf{i}) = dS(\cos \alpha) = \pm \, dy \, dz$$
$$dS(\mathbf{n} \cdot \mathbf{j}) = dS(\cos \beta) = \pm \, dz \, dx$$
$$dS(\mathbf{n} \cdot \mathbf{k}) = dS(\cos \gamma) = \pm \, dx \, dy$$

where the signs are determined by those of $(\mathbf{n} \cdot \mathbf{i}) = \cos \alpha$, $(\mathbf{n} \cdot \mathbf{j}) = \cos \beta$, $(\mathbf{n} \cdot \mathbf{k}) = \cos \gamma$, respectively. Then

$$d\mathbf{S} = \mathbf{n} \, dS = (\pm \, dy \, dz)\mathbf{i} + (\pm \, dz \, dx)\mathbf{j} + (\pm \, dx \, dy)\mathbf{k} \tag{6.23}$$

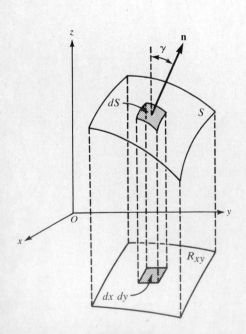

Figure 6-1

Thus if $\mathbf{f} = f_1\mathbf{i} + f_2\mathbf{j} + f_3\mathbf{k}$,

$$\iint_S \mathbf{f} \cdot d\mathbf{S} = \pm \iint_{R_{yz}} f_1 \, dy \, dz \pm \iint_{R_{zx}} f_2 \, dz \, dx + \iint_{R_{xy}} f_3 \, dx \, dy$$

where R_{yz}, R_{zx}, and R_{xy} are the projections of S onto the yz-, zx-, and xy-planes, respectively.

6-3. Volume Integrals

Since the element of volume dV is a scalar, consider two volume integrals over a region R:

VOLUME INTEGRALS

$$\iiint_R \phi \, dV \tag{6.24}$$

$$\iiint_R \mathbf{f} \, dV \tag{6.25}$$

In the rectangular coordinate system,

$$dV = dx \, dy \, dx \tag{6.26}$$

so the volume integral (6.24) can be rewritten as

$$\iiint_R \phi(x, y, z) \, dx \, dy \, dz \tag{6.27}$$

which is the ordinary triple integral of $\phi(x, y, z)$ over the region R. Setting $\phi(x, y, z) = 1$, the volume V of the region R is given by

$$V = \iiint_R dV = \iiint_R dx \, dy \, dz \tag{6.28}$$

If $\mathbf{f} = f_1\mathbf{i} + f_2\mathbf{j} + f_3\mathbf{k}$, then volume integral (6.25) can be resolved into its components:

$$\iiint_R \mathbf{f} \, dV = \mathbf{i} \iiint_R f_1 \, dV + \mathbf{j} \iiint_R f_2 \, dV + \mathbf{k} \iiint_R f_3 \, dV \tag{6.29}$$

6-4. Divergence and Curl of a Vector

A. Integral definitions of divergence and curl

The divergence of $\mathbf{f}$, written div $\mathbf{f}$ or $\mathbf{\nabla} \cdot \mathbf{f}$, can be defined by

DIVERGENCE OF f
(integral definition)

$$\mathbf{\nabla} \cdot \mathbf{f} = \lim_{\Delta V \to 0} \frac{1}{\Delta V} \oiint_S \mathbf{f} \cdot d\mathbf{S} \tag{6.30}$$

where ΔV is the volume of the region R bounded by a closed surface S. The volume ΔV always contains the point at which the divergence $\mathbf{\nabla} \cdot \mathbf{f}$ is to be evaluated as ΔV approaches zero.

The curl of $\mathbf{f}$, written curl $\mathbf{f}$ or $\mathbf{\nabla} \times \mathbf{f}$, can be defined by

CURL OF f
(integral definition)

$$\mathbf{\nabla} \times \mathbf{f} = \mathbf{n}_{max} \lim_{\Delta S \to 0} \frac{1}{\Delta S} \oint_C \mathbf{f} \cdot d\mathbf{r} \tag{6.31}$$

where ΔS is the surface bounded by a simple closed curve C and $\mathbf{n}_{max}$ is the unit normal vector associated with ΔS such that the orientation of the plane of ΔS will give a maximum value of

$$\frac{1}{\Delta S} \oint_C \mathbf{f} \cdot d\mathbf{r} \tag{6.32}$$

The curl of $\mathbf{f}$, $\mathbf{\nabla} \times \mathbf{f}$, is also defined by its component in a particular direction; that is, the component of $\mathbf{\nabla} \times \mathbf{f}$ in the $\mathbf{n}$ direction is given by

$$\mathbf{n} \cdot (\mathbf{\nabla} \times \mathbf{f}) = \lim_{\Delta S \to 0} \frac{1}{\Delta S} \oint_C \mathbf{f} \cdot d\mathbf{r} \tag{6.33}$$

where ΔS is a surface bounded by a simple closed curve C and $\mathbf{n}$ is the unit normal vector associated with ΔS, written $\Delta \mathbf{S} = \mathbf{n}\,\Delta S$.

As in the definition (6.30) of $\mathbf{\nabla} \cdot \mathbf{f}$, the surface ΔS contains the point at which $\mathbf{\nabla} \times \mathbf{f}$ is to be evaluated as ΔS approaches zero.

B. Physical interpretation of $\mathbf{\nabla} \cdot \mathbf{f}$

EXAMPLE 6-6: Give the physical interpretation of $\mathbf{\nabla} \cdot \mathbf{f}$ as defined by (6.30).

Solution: In Section 6-2, the surface integral $\iint_S \mathbf{f} \cdot d\mathbf{S}$ is defined as the flux of $\mathbf{f}$ passing through the surface S. Then $\oiint_S \mathbf{f} \cdot d\mathbf{S}$ is the total net outflow flux of $\mathbf{f}$ through the closed surface S. Hence (6.30) shows that the divergence of $\mathbf{f}$ at a point P is the limit of the net outflow flux per unit volume as S shrinks to the point P.

If the small volume that surrounds P contains a *source* or *sink* of a vector field $\mathbf{f}$, the flux of $\mathbf{f}$ will diverge from or converge to P, depending on whether $\mathbf{\nabla} \cdot \mathbf{f}$ is positive or negative. Hence $\mathbf{\nabla} \cdot \mathbf{f}$ can be considered a measure of vector source or sink strength at P.

EXAMPLE 6-7: Using the integral definition of the divergence of $\mathbf{f}$ (6.30), derive the formula for $\mathbf{\nabla} \cdot \mathbf{f}$ as given by the definition in (5.16):

$$\operatorname{div}\mathbf{f} = \mathbf{\nabla} \cdot \mathbf{f} = \frac{\partial f_1}{\partial x} + \frac{\partial f_2}{\partial y} + \frac{\partial f_3}{\partial z}$$

Solution: If the point (x, y, z) at the center of a small rectangular parallelepiped has edges Δx, Δy, and Δz, as shown in Figure 6-2, then $\Delta V = \Delta x\,\Delta y\,\Delta z$. If $\mathbf{f} = f_1\mathbf{i} + f_2\mathbf{j} + f_3\mathbf{k}$, then

on S_1 defined by face $ABCD$, $\quad\mathbf{n} = \mathbf{i}$, $\qquad d\mathbf{S} = dy\,dz\,\mathbf{i}$

on S_2 defined by face $HGFE$, $\quad\mathbf{n} = -\mathbf{i}$, $\qquad d\mathbf{S} = -dy\,dz\,\mathbf{i}$

on S_3 defined by face $BFGC$, $\quad\mathbf{n} = \mathbf{j}$, $\qquad d\mathbf{S} = dz\,dx\,\mathbf{j}$

on S_4 defined by face $DHEA$, $\quad\mathbf{n} = -\mathbf{j}$, $\qquad d\mathbf{S} = -dz\,dx\,\mathbf{j}$

on S_5 defined by face $CGHD$, $\quad\mathbf{n} = \mathbf{k}$, $\qquad d\mathbf{S} = dx\,dy\,\mathbf{k}$

on S_6 defined by face $EFBA$, $\quad\mathbf{n} = -\mathbf{k}$, $\qquad d\mathbf{S} = -dx\,dy\,\mathbf{k}$

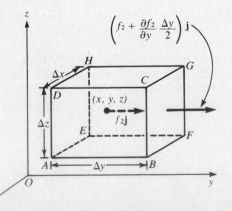

Figure 6-2

Hence if we ignore the differential contributions of higher order on each of the six faces, we obtain

$$\iint_{S_1} \mathbf{f} \cdot d\mathbf{S} = \left(f_1 + \frac{\partial f_1}{\partial x}\frac{\Delta x}{2}\right)\Delta y\,\Delta z$$

$$\iint_{S_2} \mathbf{f} \cdot d\mathbf{S} = -\left(f_1 - \frac{\partial f_1}{\partial x}\frac{\Delta x}{2}\right)\Delta y\,\Delta z$$

$$\iint_{S_3} \mathbf{f} \cdot d\mathbf{S} = \left(f_2 + \frac{\partial f_2}{\partial y}\frac{\Delta y}{2}\right)\Delta z\,\Delta x$$

$$\iint_{S_4} \mathbf{f} \cdot d\mathbf{S} = -\left(f_2 - \frac{\partial f_2}{\partial y}\frac{\Delta y}{2}\right)\Delta z\,\Delta x$$

$$\iint_{S_5} \mathbf{f} \cdot d\mathbf{S} = \left(f_3 + \frac{\partial f_3}{\partial z}\frac{\Delta z}{2}\right)\Delta x\,\Delta y$$

$$\iint_{S_6} \mathbf{f} \cdot d\mathbf{S} = -\left(f_3 - \frac{\partial f_3}{\partial z}\frac{\Delta z}{2}\right)\Delta x\,\Delta y$$

Adding all the above results, we get the total outward flux

$$\oiint_{S} \mathbf{f} \cdot d\mathbf{S} = \left(\frac{\partial f_1}{\partial x} + \frac{\partial f_2}{\partial y} + \frac{\partial f_3}{\partial z}\right)\Delta x\,\Delta y\,\Delta z = \left(\frac{\partial f_1}{\partial x} + \frac{\partial f_2}{\partial y} + \frac{\partial f_3}{\partial z}\right)\Delta V$$

So, by definition (6.30), we obtain

$$\mathbf{\nabla} \cdot \mathbf{f} = \lim_{\Delta V \to 0}\frac{1}{\Delta V}\oiint_{S}\mathbf{f} \cdot d\mathbf{S} = \lim_{\Delta V \to 0}\frac{1}{\Delta V}\left(\frac{\partial f_1}{\partial x} + \frac{\partial f_2}{\partial y} + \frac{\partial f_3}{\partial z}\right)\Delta V$$

$$= \left(\frac{\partial f_1}{\partial x} + \frac{\partial f_2}{\partial y} + \frac{\partial f_3}{\partial z}\right)$$

C. Physical interpretation of $\mathbf{\nabla} \times \mathbf{f}$

EXAMPLE 6-8: Give the physical interpretation of the curl of $\mathbf{f}$, $\mathbf{\nabla} \times \mathbf{f}$, as defined by the integral definition (6.31).

Solution: The integral on the right-hand side of (6.31), or $\oint_C \mathbf{f} \cdot d\mathbf{r}$ (6.32), is the circulation of $\mathbf{f}$ around curve C defined in (6.7). Hence (6.31) shows the magnitude of $\mathbf{\nabla} \times \mathbf{f}$ at a point P, that is, the intensity of circulation at P. In general, the circulation of $\mathbf{f}$ around C depends on the orientation of the plane in which C lies. The direction of $\mathbf{\nabla} \times \mathbf{f}$ is the direction in which maximum circulation occurs.

EXAMPLE 6-9: Using (6.33), derive the definition of the curl of $\mathbf{f}$ as the cross product of $\mathbf{\nabla}$ and $\mathbf{f}$ as given in determinant and component form in (5.28). Ignore the higher-order terms.

Solution: Construct a rectangle $EFGH$ centered at the point (x, y, z) so that rectangle $EFGH$ is parallel to the yz-plane and its sides are Δy and Δz, as shown in Figure 6-3. Then we have $\Delta\mathbf{S} = \mathbf{i}\,\Delta S = \mathbf{i}\,\Delta y\,\Delta z$. Now $\mathbf{f} = f_1\mathbf{i} + f_2\mathbf{j} + f_3\mathbf{k}$, so

$$\text{for side } EF, \qquad d\mathbf{r} = dy\,\mathbf{j}, \qquad \int_{EF}\mathbf{f} \cdot d\mathbf{r} = \left(f_2 - \frac{\partial f_2}{\partial z}\frac{\Delta z}{2}\right)\Delta y$$

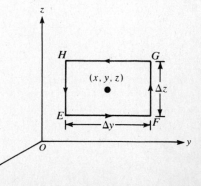

Figure 6-3

for side FG, $d\mathbf{r} = dz\,\mathbf{k}$, $\displaystyle\int_{FG} \mathbf{f}\cdot d\mathbf{r} = \left(f_3 + \frac{\partial f_3}{\partial y}\frac{\Delta y}{2}\right)\Delta z$

for side GH, $d\mathbf{r} = -dy\,\mathbf{j}$, $\displaystyle\int_{GH} \mathbf{f}\cdot d\mathbf{r} = -\left(f_2 + \frac{\partial f_2}{\partial z}\frac{\Delta z}{2}\right)\Delta y$

for side HE, $d\mathbf{r} = -dz\,\mathbf{k}$, $\displaystyle\int_{HE} \mathbf{f}\cdot d\mathbf{r} = -\left(f_3 - \frac{\partial f_3}{\partial y}\frac{\Delta y}{2}\right)\Delta z$

Adding all the above results, we get

$$\oint_C \mathbf{f}\cdot d\mathbf{r} = \left(\frac{\partial f_3}{\partial y} - \frac{\partial f_2}{\partial z}\right)\Delta y\,\Delta z = \left(\frac{\partial f_3}{\partial y} - \frac{\partial f_2}{\partial z}\right)\Delta S$$

Hence from the definition of curl $\mathbf{f}$ in (6.33), we obtain

$$\mathbf{i}\cdot(\nabla \times \mathbf{f}) = \lim_{\Delta S \to 0}\frac{1}{\Delta S}\oint_C \mathbf{f}\cdot d\mathbf{r} = \left(\frac{\partial f_3}{\partial y} - \frac{\partial f_2}{\partial z}\right)$$

Similarly, by performing integration around rectangles parallel to the zx- and xy-planes and adding results we obtain

$$\mathbf{j}\cdot(\nabla \times \mathbf{f}) = \left(\frac{\partial f_1}{\partial z} - \frac{\partial f_3}{\partial x}\right)$$

$$\mathbf{k}\cdot(\nabla \times \mathbf{f}) = \left(\frac{\partial f_2}{\partial x} - \frac{\partial f_1}{\partial y}\right)$$

Since for an arbitrary vector $\mathbf{A}$, $\mathbf{A} = (\mathbf{A}\cdot\mathbf{i})\mathbf{i} + (\mathbf{A}\cdot\mathbf{j})\mathbf{j} + (\mathbf{A}\cdot\mathbf{k})\mathbf{k}$ by (2.56), we obtain

$$\nabla \times \mathbf{f} = \left(\frac{\partial f_3}{\partial y} - \frac{\partial f_2}{\partial z}\right)\mathbf{i} + \left(\frac{\partial f_1}{\partial z} - \frac{\partial f_3}{\partial x}\right)\mathbf{j} + \left(\frac{\partial f_2}{\partial x} - \frac{\partial f_1}{\partial y}\right)\mathbf{k}$$

$$= \begin{vmatrix} \mathbf{i} & \mathbf{j} & \mathbf{k} \\ \dfrac{\partial}{\partial x} & \dfrac{\partial}{\partial y} & \dfrac{\partial}{\partial z} \\ f_1 & f_2 & f_3 \end{vmatrix}$$

6-5. Alternate Definitions of Gradient, Divergence, and Curl

A. Integral representation of the ∇ operator

Observation of the integral definition of div $\mathbf{f}$ (6.30) suggests that ∇ can be symbolically represented by

INTEGRAL REPRESENTATION OF ∇ $$\nabla = \lim_{\Delta V \to 0}\frac{1}{\Delta V}\oiint_S d\mathbf{S} \tag{6.34}$$

B. Alternate definitions of gradient, divergence, and curl

Using the integral representation (6.34), $\nabla\phi$, $\nabla\cdot\mathbf{f}$, and $\nabla \times \mathbf{f}$ can be defined as

$$\text{grad }\phi = \nabla\phi = \lim_{\Delta V \to 0}\frac{1}{\Delta V}\oiint_S d\mathbf{S}\,\phi \tag{6.35}$$

$$\text{div }\mathbf{f} = \nabla\cdot\mathbf{f} = \lim_{\Delta V \to 0}\frac{1}{\Delta V}\oiint_S d\mathbf{S}\cdot\mathbf{f} \tag{6.36}$$

$$\text{curl }\mathbf{f} = \nabla \times \mathbf{f} = \lim_{\Delta V \to 0}\frac{1}{\Delta V}\oiint_S d\mathbf{S} \times \mathbf{f} \tag{6.37}$$

where ΔV is the volume of the region R bounded by a closed surface S.
Note: Definition (6.36) is equal to definition (6.30). For verifications of (6.35) and (6.37), see Problems 6-20 and 6-21.

SUMMARY

1. Three forms of line integrals along a curve C are

$$\int_C \phi \, d\mathbf{r}, \qquad \int_C \mathbf{f} \cdot d\mathbf{r}, \qquad \int_C \mathbf{f} \times d\mathbf{r}$$

where $d\mathbf{r} = dx\,\mathbf{i} + dy\,\mathbf{j} + dz\,\mathbf{k}$ is the differential displacement vector.

2. Three forms of surface integrals over a surface S are

$$\iint_S \phi \, d\mathbf{S}, \qquad \iint_S \mathbf{f} \cdot d\mathbf{S}, \qquad \iint_S \mathbf{f} \times d\mathbf{S}$$

where $d\mathbf{S} = \mathbf{n}\,dS$ is the differential surface element and $\mathbf{n}$ is the unit vector normal to the surface.

3. Two forms of volume integrals over a region R are

$$\iiint_R \phi \, dV, \qquad \iiint_R \mathbf{f} \, dV$$

where $dV = dx\,dy\,dz$ is the element of volume.

4. The integral representation of $\mathbf{V}$ is given by

$$\mathbf{V} = \lim_{\Delta V \to 0} \frac{1}{\Delta V} \oiint_S d\mathbf{S}$$

where ΔV is the volume of the region R bounded by a closed surface S.

5. Using the integral representation of $\mathbf{V}$, the quantities $\mathbf{V}\phi$, $\mathbf{V} \cdot \mathbf{f}$, and $\mathbf{V} \times \mathbf{f}$ can be alternately defined as

$$\mathbf{V}\phi = \lim_{\Delta V \to 0} \frac{1}{\Delta V} \oiint_S d\mathbf{S}\,\phi$$

$$\mathbf{V} \cdot \mathbf{f} = \lim_{\Delta V \to 0} \frac{1}{\Delta V} \oiint_S d\mathbf{S} \cdot \mathbf{f}$$

$$\mathbf{V} \times \mathbf{f} = \lim_{\Delta V \to 0} \frac{1}{\Delta V} \oiint_S d\mathbf{S} \times \mathbf{f}$$

RAISE YOUR GRADES

Can you explain . . . ?

☑ how the line integral is defined and evaluated
☑ how the surface integral is defined and evaluated
☑ how to find the differential element of the surface area
☑ what the flux of $\mathbf{f}$ is
☑ how the divergence of $\mathbf{f}$, denoted by $\mathbf{V} \cdot \mathbf{f}$, is defined in integral form
☑ how the curl of $\mathbf{f}$, denoted by $\mathbf{V} \times \mathbf{f}$, is defined in integral form

SOLVED PROBLEMS

Line Integrals

PROBLEM 6-1   If $\phi = xy$, evaluate $\int_c \phi \, d\mathbf{r}$ from $(0,0,0)$ to $(1,1,0)$ along (a) the curve C specified by $y = x^2, z = 0$, and (b) the straight line C that joins $(0,0,0)$ and $(1,1,0)$.

Solution:
(a) Curve C can be represented parametrically by

$$x = t, \qquad y = t^2, \qquad z = 0, \qquad 0 \le t \le 1$$

Hence, since $dx = dt$ and $dy = 2t \, dt$, along this path

$$d\mathbf{r} = dx \, \mathbf{i} + dy \, \mathbf{j} + dz \, \mathbf{k} = dt \, \mathbf{i} + 2t \, dt \, \mathbf{j}$$

and function $\phi = (t)(t^2) = t^3$. Then

$$\int_{(0,0,0)}^{(1,1,0)} \phi \, d\mathbf{r} = \int_0^1 t^3 (\mathbf{i} + 2t\mathbf{j}) \, dt$$

$$= \mathbf{i} \int_0^1 t^3 \, dt + \mathbf{j} \int_0^1 2t^4 \, dt$$

$$= \tfrac{1}{4}\mathbf{i} + \tfrac{2}{5}\mathbf{j}$$

(b) Using the result of Example 3-4, the straight line C that connects $(0,0,0)$ and $(1,1,0)$ can be represented parametrically by $x = t, y = t, z = 0, 0 \le t \le 1$. So along this path

$$d\mathbf{r} = dx \, \mathbf{i} + dy \, \mathbf{j} + dz \, \mathbf{k} = dt \, \mathbf{i} + dt \, \mathbf{j}$$

and $\phi = (t)(t) = t^2$. Then

$$\int_{(0,0,0)}^{(1,1,0)} \phi \, d\mathbf{r} = \int_0^1 t^2 (\mathbf{i} + \mathbf{j}) \, dt$$

$$= \mathbf{i} \int_0^1 t^2 \, dt + \mathbf{j} \int_0^1 t^2 \, dt$$

$$= \tfrac{1}{3}\mathbf{i} + \tfrac{1}{3}\mathbf{j}$$

Note: The results of the same integral over different paths are different. The values of this type of line integral generally depend on the path of integration.

PROBLEM 6-2   Evaluate $\oint_C d\mathbf{r}$ around the circle represented by $x^2 + y^2 = a^2$, $z = 0$, where a is constant.

Solution: From the result of Example 4-9, the circle $x^2 + y^2 = a^2, z = 0$, can be represented parametrically by

$$x = a \cos t, \qquad y = a \sin t, \qquad z = 0, \qquad 0 \le t \le 2\pi$$

Then if we use the differentials $dx = -a \sin t \, dt$ and $dy = a \cos t \, dt$ and $dz = 0$, we get

$$d\mathbf{r} = dx \, \mathbf{i} + dy \, \mathbf{j} + dz \, \mathbf{k} = -a \sin t \, dt \, \mathbf{i} + a \cos t \, dt \, \mathbf{j}$$

Then

$$\oint_C d\mathbf{r} = \int_0^{2\pi} (-a\sin t\mathbf{i} + a\cos t\mathbf{j})\,dt$$

$$= \mathbf{i}\int_0^{2\pi} -a\sin t\,dt + \mathbf{j}\int_0^{2\pi} a\cos t\,dt$$

$$= \mathbf{i}a\cos t\,|_0^{2\pi} + \mathbf{j}a\sin t\,|_0^{2\pi}$$

$$= \mathbf{i}(a - a) + \mathbf{j}(0 - 0)$$

$$= \mathbf{0}$$

PROBLEM 6-3 If $\mathbf{f} = x^2\mathbf{i} + y\mathbf{j} + xyz\mathbf{k}$, evaluate $\int_C \mathbf{f} \cdot d\mathbf{r}$ from $(0,0,0)$ to $(1,1,1)$ along (a) a straight line that connects these two points and (b) a path C, as shown in Figure 6-4, which consists of three line segments $C_1, C_2,$ and C_3 that link these two points via $(1,0,0)$ and $(1,1,0)$.

Solution:
(a) Using the result of Example 3-4, we represent the straight line C that connects $(0,0,0)$ and $(1,1,1)$ parametrically:

$$x = t, \qquad y = t, \qquad z = t$$

Hence the curve C is represented by

$$\mathbf{r} = t\mathbf{i} + t\mathbf{j} + t\mathbf{k}, \qquad 0 \le t \le 1$$

Then along this path

$$d\mathbf{r} = dt\,\mathbf{i} + dt\,\mathbf{j} + dt\,\mathbf{k} \quad \text{and} \quad \mathbf{f} = t^2\mathbf{i} + t\mathbf{j} + t^3\mathbf{k}$$

and thus

$$\int_C \mathbf{f} \cdot d\mathbf{r} = \int_0^1 (t^2 + t + t^3)\,dt = \frac{1}{3} + \frac{1}{2} + \frac{1}{4} = \frac{13}{12}$$

(b) When path C consists of three line segments $C_1, C_2,$ and C_3 connecting $(0,0,0)$ and $(1,1,1)$ via $(1,0,0)$ and $(1,1,0)$,

$$\int_C \mathbf{f} \cdot d\mathbf{r} = \int_{C_1} \mathbf{f} \cdot d\mathbf{r} + \int_{C_2} \mathbf{f} \cdot d\mathbf{r} + \int_{C_3} \mathbf{f} \cdot d\mathbf{r}$$

On segment C_1, $d\mathbf{r} = dx\,\mathbf{i}$ and $\mathbf{f} = x^2\mathbf{i}$; hence

$$\int_{C_1} \mathbf{f} \cdot d\mathbf{r} = \int_0^1 x^2\,dx = \frac{1}{3}$$

On segment C_2, $d\mathbf{r} = dy\,\mathbf{j}$ and $\mathbf{f} = (1)^2\mathbf{i} + y\mathbf{j} = \mathbf{i} + y\mathbf{j}$; hence

$$\int_{C_2} \mathbf{f} \cdot d\mathbf{r} = \int_0^1 y\,dy = \frac{1}{2}$$

On segment C_3, $d\mathbf{r} = dz\,\mathbf{k}$ and $\mathbf{f} = (1)^2\mathbf{i} + (1)\mathbf{j} + (1)(1)z\mathbf{k} = \mathbf{i} + \mathbf{j} + z\mathbf{k}$; hence

$$\int_{C_3} \mathbf{f} \cdot d\mathbf{r} = \int_0^1 z\,dz = \frac{1}{2}$$

Adding these results, we obtain

$$\int_C \mathbf{f} \cdot d\mathbf{r} = \frac{1}{3} + \frac{1}{2} + \frac{1}{2} = \frac{4}{3}$$

[Note that the value of a line integral of a vector field also generally depends on the path of integration.]

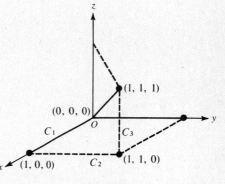

Figure 6-4

PROBLEM 6-4   If $\mathbf{f} = x\mathbf{i} + 2y\mathbf{j} + z\mathbf{k}$, evaluate $\int_C \mathbf{f} \cdot d\mathbf{r}$ from $(0,0,0)$ to $(1,1,1)$ along (a) a straight line that connects these two points and (b) a path C (shown in Figure 6-4) consisting of three line segments C_1, C_2, and C_3 that link these two points via $(1,0,0)$ and $(1,1,0)$.

Solution:

(a) From the results in the previous problem on the straight-line path,

$$d\mathbf{r} = dt\,\mathbf{i} + dt\,\mathbf{j} + dt\,\mathbf{k}$$

Since the parametric representation of $\mathbf{f}$ is $\mathbf{f} = t\mathbf{i} + 2t\mathbf{j} + t\mathbf{k}$, the integral is

$$\int_C \mathbf{f} \cdot d\mathbf{r} = \int_0^1 (t + 2t + t)\,dt = \int_0^1 4t\,dt = 2$$

(b) For a path C consisting of line segments C_1, C_2, and C_3,

$$\int_C \mathbf{f} \cdot d\mathbf{r} = \int_{C_1} \mathbf{f} \cdot d\mathbf{r} + \int_{C_2} \mathbf{f} \cdot d\mathbf{r} + \int_{C_3} \mathbf{f} \cdot d\mathbf{r}$$

On C_1, $d\mathbf{r} = dx\,\mathbf{i}$ and $\mathbf{f} = x\mathbf{i}$; hence

$$\int_{C_1} \mathbf{f} \cdot d\mathbf{r} = \int_0^1 x\,dx = \frac{1}{2}$$

On C_2, $d\mathbf{r} = dy\,\mathbf{j}$ and $\mathbf{f} = \mathbf{i} + 2y\mathbf{j}$; hence

$$\int_{C_2} \mathbf{f} \cdot d\mathbf{r} = \int_0^1 2y\,dy = 1$$

On C_3, $d\mathbf{r} = dz\,\mathbf{k}$ and $\mathbf{f} = \mathbf{i} + 2\mathbf{j} + z\mathbf{k}$; hence

$$\int_{C_3} \mathbf{f} \cdot d\mathbf{r} = \int_0^1 z\,dz = \frac{1}{2}$$

Adding these results, we obtain

$$\int_C \mathbf{f} \cdot d\mathbf{r} = \frac{1}{2} + 1 + \frac{1}{2} = 2$$

[Note that the value of the line integral is 2 on both paths.]

PROBLEM 6-5   Show that the value of the line integral given in Problem 6-4 is 2 for any path joining $(0,0,0)$ to $(1,1,1)$.

Solution: Since $\mathbf{f} = x\mathbf{i} + 2y\mathbf{j} + z\mathbf{k}$, by (6.9) we have

$$\int_C \mathbf{f} \cdot d\mathbf{r} = \int_C x\,dx + 2y\,dy + z\,dz$$

But we observe that $x\,dx + 2y\,dy + z\,dz$ can be expressed as a total (or exact) differential, namely

$$x\,dx + 2y\,dy + z\,dz = d\left(\frac{x^2}{2} + y^2 + \frac{z^2}{2}\right)$$

Thus we have

$$\int_C \mathbf{f} \cdot d\mathbf{r} = \int_{(0,0,0)}^{(1,1,1)} d\left(\frac{x^2}{2} + y^2 + \frac{z^2}{2}\right)$$

$$= \frac{x^2}{2} + y^2 + \frac{z^2}{2}\bigg|_{(0,0,0)}^{(1,1,1)}$$

$$= \tfrac{1}{2} + 1 + \tfrac{1}{2}$$

$$= 2$$

which shows that the result depends only on the end points $(0,0,0)$ and $(1,1,1)$ and is independent of the path of integration.

PROBLEM 6-6 Find $\oint_C \mathbf{r} \cdot d\mathbf{r}$ along the circle C represented by $x^2 + y^2 = a^2, z = 0$.

Solution: Referring to Problem 6-2, we write parametric equation for the curve as

$$\mathbf{r} = a\cos t\mathbf{i} + a\sin t\mathbf{j}, \qquad 0 \le t \le 2\pi$$

Therefore

$$d\mathbf{r} = -a\sin t\, dt\mathbf{i} + a\cos t\, dt\,\mathbf{j}$$

Then

$$\mathbf{r} \cdot d\mathbf{r} = -a^2\cos t\sin t\, dt + a^2\sin t\cos t\, dt = 0\, dt = 0$$

Hence we obtain

$$\oint_C \mathbf{r} \cdot d\mathbf{r} = 0$$

PROBLEM 6-7 Evaluate $\oint_C \mathbf{r} \times d\mathbf{r}$ along the circle C represented by $x^2 + y^2 = a^2, z = 0$.

Solution: Referring to Problem 6-2, we represent C parametrically by

$$\mathbf{r} = a\cos t\mathbf{i} + a\sin t\mathbf{j}, \qquad 0 \le t \le 2\pi$$

and again, $d\mathbf{r} = -a\sin t\, dt\,\mathbf{i} + a\cos t\, dt\,\mathbf{j}$. Thus

$$\mathbf{r} \times d\mathbf{r} = \begin{vmatrix} \mathbf{i} & \mathbf{j} & \mathbf{k} \\ a\cos t & a\sin t & 0 \\ -a\sin t\, dt & a\cos t\, dt & 0 \end{vmatrix}$$
$$- \mathbf{k}(a^2\cos^2 t + a^2\sin^2 t)\, dt$$
$$= a^2\, dt\,\mathbf{k}$$

Integrating this result over curve C, we obtain

$$\oint_C \mathbf{r} \times d\mathbf{r} = \int_0^{2\pi} \mathbf{k}a^2\, dt = \mathbf{k}a^2\int_0^{2\pi} dt = 2\pi a^2\mathbf{k}$$

[Note that the result is a vector in the z-direction with magnitude $2\pi a^2$, which is twice the area of a circle.]

Surface Integrals

PROBLEM 6-8 Evaluate $\iint_S d\mathbf{S}$ over the upper half $(z > 0)$ of the sphere $x^2 + y^2 + z^2 = a^2$.

Solution: From the result of Example 4-17 and using eq. (4.57), we can represent the sphere $x^2 + y^2 + z^2 = a^2$ parametrically in u and v by changing θ to u and ϕ to v. So we can represent the surface vectorially as

$$\mathbf{r}(u,v) = a\sin u\cos v\mathbf{i} + a\sin u\sin v\mathbf{j} + a\cos u\mathbf{k} \tag{a}$$

where $0 \le u \le \pi, 0 \le v \le 2\pi$. By (6.13) the differential element $d\mathbf{S}$ of the surface area is

$$d\mathbf{S} = \mathbf{r}_u \times \mathbf{r}_v\, du\, dv$$
$$= a^2\sin^2 u\cos v\, du\, dv\,\mathbf{i} + a^2\sin^2 u\sin v\, du\, dv\,\mathbf{j}$$
$$+ a^2\sin u\cos u\, du\, dv\,\mathbf{k}$$

Since S is the part of the sphere where $z > 0$, the range of u must be changed to $0 \leq u \leq \pi/2$ because $\cos u < 0$ for $\pi/2 < u \leq \pi$. Hence

$$\iint_S d\mathbf{S} = \int_0^{2\pi} \int_0^{\pi/2} \mathbf{r}_u \times \mathbf{r}_v \, du \, dv$$

$$= \mathbf{i} \int_0^{2\pi} \int_0^{\pi/2} a^2 \sin^2 u \cos v \, du \, dv + \mathbf{j} \int_0^{2\pi} \int_0^{\pi/2} a^2 \sin^2 u \sin v \, du \, dv$$

$$+ \mathbf{k} \int_0^{2\pi} \int_0^{\pi/2} \sin u \cos u \, du \, dv \qquad (b)$$

Now from elementary calculus we have

$$\int_0^{2\pi} \cos v \, dv = \int_0^{2\pi} \sin v \, dv = 0$$

$$\int_0^{2\pi} \int_0^{\pi/2} a^2 \sin^2 u \cos v \, du \, dv = a^2 \int_0^{2\pi} \cos v \, dv \int_0^{\pi/2} \sin^2 u \, du = 0$$

$$\int_0^{2\pi} \int_0^{\pi/2} a^2 \sin^2 u \sin v \, du \, dv = a^2 \int_0^{2\pi} \sin v \, dv \int_0^{\pi/2} \sin^2 u \, du = 0$$

$$\int_0^{2\pi} \int_0^{\pi/2} a^2 \sin u \cos u \, du \, dv = a^2 \int_0^{2\pi} dv \int_0^{\pi/2} \sin u \cos u \, du$$

$$= 2\pi a^2 \int_0^{\pi/2} \sin u \cos u \, du$$

$$= 2\pi a^2 \int_0^{\pi/2} \sin u \, d(\sin u)$$

$$= 2\pi a^2 \frac{1}{2} \sin^2 u \Big|_0^{\pi/2}$$

$$= \frac{2\pi a^2}{2}$$

$$= \pi a^2$$

By combining all these results into eq. (b), we get

$$\iint_S d\mathbf{S} = \pi a^2 \mathbf{k}$$

which shows that the resultant is a vector directed in the z-direction with magnitude πa^2, i.e., the area enclosed by the curve intersecting the sphere and the xy-plane (Figure 6-5).

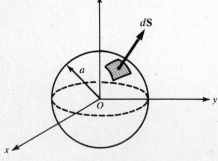

Figure 6-5

PROBLEM 6-9   Show that, over the entire sphere $x^2 + y^2 + z^2 = a^2$,

$$\oiint_S d\mathbf{S} = \mathbf{0} \qquad (a)$$

Solution: Let S_1 and S_2 be the parts of the sphere $x^2 + y^2 + z^2 = a^2$, where $z > 0$ and $z < 0$, respectively. Then

$$\oiint_S d\mathbf{S} = \iint_{S_1} d\mathbf{S} + \iint_{S_2} d\mathbf{S}$$

From the result of Problem 6-8, on the upper hemisphere

$$\iint_{S_1} d\mathbf{S} = \pi a^2 \mathbf{k} \tag{b}$$

For S_2, since $z < 0$, the range of u in eq. (a) of Problem 6-8 is $\pi/2 \leqslant u \leqslant \pi$. Then

$$\iint_{S_2} d\mathbf{S} = \int_0^{2\pi} \int_{\pi/2}^{\pi} \mathbf{r}_u \times \mathbf{r}_v \, du \, dv$$

and by calculations similar to those in Problem 6-8 we have

$$\int_0^{2\pi} \int_{\pi/2}^{\pi} a^2 \sin^2 u \cos v \, du \, dv = \int_0^{2\pi} \int_{\pi/2}^{\pi} a^2 \sin^2 u \sin v \, du \, dv = 0$$

and

$$\int_0^{2\pi} \int_{\pi/2}^{\pi} a^2 \sin u \cos u \, du \, dv = 2\pi a^2 \frac{1}{2} \sin^2 u \Big|_{\pi/2}^{\pi} = 2\pi a^2 \left(-\frac{1}{2} \right) = -\pi a^2$$

So over the lower hemisphere ($z < 0$) we obtain

$$\iint_{S_2} d\mathbf{S} = -\pi a^2 \mathbf{k} \tag{c}$$

Adding eq. (b) and eq. (c) in this problem, we obtain

$$\oiint_S d\mathbf{S} = \pi a^2 \mathbf{k} - \pi a^2 \mathbf{k} = \mathbf{0}$$

[Note that relation (a) is true for any closed surface. (See Problem 7-10 for an alternate proof.)]

PROBLEM 6-10: If $\mathbf{f} = (\cos u \cos v)\mathbf{i} + (\cos u \sin v)\mathbf{j} - (\sin u)\mathbf{k}$, where $0 \leqslant u \leqslant \pi$ and $0 \leqslant v \leqslant 2\pi$, evaluate $\iint_S \mathbf{f} \cdot d\mathbf{S}$ over that part of the sphere $x^2 + y^2 + z^2 = a^2$ for which $z > 0$.

Solution: From the result of Example 4-17 we have

$$\mathbf{r}_u = a(\cos u \cos v)\mathbf{i} + a(\cos u \sin v)\mathbf{j} - a(\sin u)\mathbf{k}$$
$$\mathbf{r}_v = -a(\sin u \sin v)\mathbf{i} + a(\sin u \cos v)\mathbf{j} + (0)\mathbf{k}$$

Using the definition of a scalar triple product (2.45), we have

$$[\mathbf{f}\mathbf{r}_u\mathbf{r}_v] = \begin{vmatrix} \cos u \cos v & \cos u \sin v & -\sin u \\ a \cos u \cos v & a \cos u \sin v & -a \sin u \\ -a \sin u \sin v & a \sin u \cos v & 0 \end{vmatrix} = 0$$

The value of this scalar triple product is zero because the first and second rows of the determinant are proportional. Consequently from (6.20), we obtain

$$\iint_S \mathbf{f} \cdot d\mathbf{S} = 0$$

PROBLEM 6-11 Find $\int_S \mathbf{r} \cdot d\mathbf{S}$, where S is the surface of the sphere $x^2 + y^2 + z^2 = a^2$.

Solution: At a point (x, y, z) on the surface of the sphere S, both the position vector $\mathbf{r} = x\mathbf{i} + y\mathbf{j} + z\mathbf{k}$ and the outer unit normal vector $\mathbf{n}$ to the surface S point directly away from the origin. Thus

$$\mathbf{n} = \mathbf{e}_r = \frac{\mathbf{r}}{|\mathbf{r}|}$$

Hence for points on the surface

$$\mathbf{r} \cdot \mathbf{n} = \mathbf{r} \cdot \frac{\mathbf{r}}{|\mathbf{r}|} = \frac{|\mathbf{r}|^2}{|\mathbf{r}|} = |\mathbf{r}| = a$$

and since the surface area of a sphere is $4\pi a^2$, we have from Example 4-17

$$\iint\limits_{S} \mathbf{r} \cdot d\mathbf{S} = \iint\limits_{S} \mathbf{r} \cdot \mathbf{n}\, dS = a \iint\limits_{S} dS = a(4\pi a^2) = 4\pi a^3$$

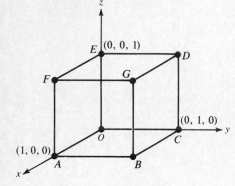

$(0, 0, 1)$

$(0, 1, 0)$

$(1, 0, 0)$

Figure 6-6
The unit cube.

PROBLEM 6-12 Evaluate $\iint_S \mathbf{r} \cdot d\mathbf{S}$, where S is the surface of the unit cube bounded by the planes $x = 0$, $x = 1$, $y = 0$, $y = 1$, $z = 0$, $z = 1$ (as shown in Figure 6-6), $\mathbf{r}$ is the position vector, and $\mathbf{n}$ is the unit normal vector having outward direction.

Solution: Divide the cube into the face areas S_1 through S_6:

on face S_1 defined by $AOCB$, $z = 0$, $\mathbf{n} = -\mathbf{k}$; hence

$$\mathbf{r} \cdot \mathbf{n} = (x\mathbf{i} + y\mathbf{j} + z\mathbf{k}) \cdot (-\mathbf{k}) = -z = 0$$

on face S_2 defined by $AFEO$, $y = 0$, $\mathbf{n} = -\mathbf{j}$; therefore

$$\mathbf{r} \cdot \mathbf{n} = (x\mathbf{i} + y\mathbf{j} + z\mathbf{k}) \cdot (-\mathbf{j}) = -y = 0$$

on face S_3 defined by $OEDC$, $z = 0$, $\mathbf{n} = -\mathbf{i}$; thus

$$\mathbf{r} \cdot \mathbf{n} = (x\mathbf{i} + y\mathbf{j} + z\mathbf{k}) \cdot (-\mathbf{i}) = -x = 0$$

on face S_4 defined by $AFGB$, $x = 1$, $\mathbf{n} = \mathbf{i}$; hence

$$\mathbf{r} \cdot \mathbf{n} = (x\mathbf{i} + y\mathbf{j} + z\mathbf{k}) \cdot (\mathbf{i}) = x = 1$$

on face S_5 defined by $BGDC$, $y = 1$, $\mathbf{n} = \mathbf{j}$; therefore

$$\mathbf{r} \cdot \mathbf{n} = (x\mathbf{i} + y\mathbf{j} + z\mathbf{k}) \cdot (\mathbf{j}) = y = 1$$

on face S_6 defined by $FEDG$, $z = 1$, $\mathbf{n} = \mathbf{k}$; thus

$$\mathbf{r} \cdot \mathbf{n} = (x\mathbf{i} + y\mathbf{j} + z\mathbf{k}) \cdot (\mathbf{k}) = z = 1$$

Then, because the integrals on S_1 through S_3 are equal to 0 and those on S_4 through S_6 are equal to 1, we obtain

$$\iint\limits_{S} \mathbf{r} \cdot d\mathbf{S} = \iint\limits_{S_1} \mathbf{r} \cdot \mathbf{n}\, dS + \iint\limits_{S_2} \mathbf{r} \cdot \mathbf{n}\, dS + \iint\limits_{S_3} \mathbf{r} \cdot \mathbf{n}\, dS$$

$$+ \iint\limits_{S_4} \mathbf{r} \cdot \mathbf{n}\, dS + \iint\limits_{S_5} \mathbf{r} \cdot \mathbf{n}\, dS + \iint\limits_{S_6} \mathbf{r} \cdot \mathbf{n}\, dS$$

$$= \iint\limits_{S_4} dS + \iint\limits_{S_5} dS + \iint\limits_{S_6} dS$$

$$= 1 + 1 + 1$$

$$= 3$$

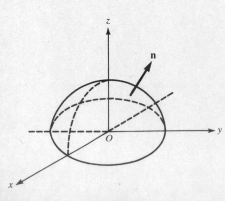

Figure 6-7

PROBLEM 6-13 If $\mathbf{f} = x\mathbf{i} + y\mathbf{j} + 2z\mathbf{k}$, evaluate $\iint_S \mathbf{f} \cdot d\mathbf{S}$, where S is that part of the surface of the paraboloid $x^2 + y^2 = 1 - z$ for which $z > 0$ (see Figure 6-7).

Solution: From the result of Example 5-3, we know that an outer unit normal vector $\mathbf{n}$ to the surface $\phi(x, y, z) = 0$ is given by

$$\mathbf{n} = \frac{\nabla \phi}{|\nabla \phi|}$$

where $\nabla\phi$ is the gradient of ϕ. So if $\phi(x, y, z) = x^2 + y^2 + z - 1 = 0$, then from definition (5.4) of the gradient, we have

$$\nabla\phi = \frac{\partial\phi}{\partial x}\mathbf{i} + \frac{\partial\phi}{\partial y}\mathbf{j} + \frac{\partial\phi}{\partial z}\mathbf{k} = 2x\mathbf{i} + 2y\mathbf{j} + \mathbf{k}$$

Hence

$$\mathbf{n} = \frac{\nabla\phi}{|\nabla\phi|} = \frac{2x\mathbf{i} + 2y\mathbf{j} + \mathbf{k}}{(4x^2 + 4y^2 + 1)^{1/2}}$$

The dot products of $\mathbf{n}$ with $\mathbf{i}$, $\mathbf{j}$, and $\mathbf{k}$ are

$$\mathbf{n}\cdot\mathbf{i} = \frac{2x}{(4x^2 + 4y^2 + 1)^{1/2}}$$

$$\mathbf{n}\cdot\mathbf{j} = \frac{2y}{(4x^2 + 4y^2 + 1)^{1/2}}$$

$$\mathbf{n}\cdot\mathbf{k} = \frac{1}{(4x^2 + 4y^2 + 1)^{1/2}}$$

Thus, $\mathbf{n}\cdot\mathbf{i}$ and $\mathbf{n}\cdot\mathbf{j}$ are positive or negative, depending upon whether x and y are positive or negative, while $\mathbf{n}\cdot\mathbf{k}$ is positive for all values of x and y. So if we use formula (6.21) given by

$$\iint_S \mathbf{f}\cdot d\mathbf{s} = \iint_S \mathbf{f}\cdot\mathbf{n}\, ds$$

$$= \pm\iint_{R_{yz}} f_1\, dy\, dz \pm \iint_{R_{zx}} f_2\, dz\, dx \pm \iint_{R_{xy}} f_3\, dx\, dy$$

to compute the integral, we have to subdivide S into two surfaces. To evaluate the first term, we must have a surface for $x > 0$ and another for $x < 0$. Similarly, to evaluate the second term, we must have a surface for $y > 0$ and another for $y < 0$.

We evaluate the first term $\iint_{R_{yz}} f_1\, dy\, dz$ as follows: For $x > 0$ we have $x = (1 - y^2 - z)^{1/2}$ on S. Hence

$$\iint_{R_{yz}} f_1\, dy\, dz = \iint_{R_{yz}} x\, dy\, dz$$

$$= \int_{y=-1}^{1}\int_{z=0}^{1-y^2} (1 - y^2 - z)^{1/2}\, dz\, dy$$

$$= \frac{2}{3}\int_{-1}^{1} (1 - y^2)^{3/2}\, dy$$

$$= \tfrac{2}{3}\left(\tfrac{3}{8}\pi\right)$$

$$= \tfrac{1}{4}\pi$$

For $x < 0$ we have $x = -(1 - y^2 - z)^{1/2}$ on S. Choosing the negative sign,

$$-\iint_{R_{yz}} f_1\, dy\, dz = -\iint_{R_{yz}} x\, dy\, dz$$

$$= \int_{y=-1}^{1}\int_{z=0}^{1-y^2} (1 - y^2 - z)^{1/2}\, dz\, dy$$

$$= \tfrac{1}{4}\pi$$

We evaluate the second term $\iint_{R_{zx}} f_2 \, dz \, dx$ as follows: For $y > 0$, $y = (1 - x^2 - z)^{1/2}$ on S; hence

$$
\iint_{R_{zx}} f_2 \, dz \, dx = \iint_{R_{zx}} y \, dz \, dx
$$

$$
= \int_{x=-1}^{1} \int_{z=0}^{1-x^2} (1 - x^2 - z)^{1/2} \, dz \, dx
$$

$$
= \tfrac{1}{4}\pi
$$

since the integral is the same as that for the R_{yz} case. For $y < 0$, $y = -(1 - x^2 - z)^{1/2}$ on S. Again choosing the negative sign,

$$
-\iint_{R_{zx}} f_2 \, dz \, dx = -\iint_{R_{zx}} y \, dz \, dx
$$

$$
= \int_{x=-1}^{1} \int_{z=0}^{1-x^2} (1 - x^2 - z)^{1/2} \, dz \, dx
$$

$$
= \tfrac{1}{4}\pi
$$

For the third term, since $\mathbf{n} \cdot \mathbf{k} > 0$,

$$
\iint_{R_{xy}} f_3 \, dx \, dy = \iint_{R_{xy}} 2z \, dx \, dy = 2 \int_{x=-1}^{1} \int_{y=-\sqrt{1-x^2}}^{\sqrt{1-x^2}} (1 - x^2 - y^2) \, dy \, dx
$$

$$
= 2\frac{4}{3} \int_{-1}^{1} (1 - x^2)^{3/2} \, dx
$$

$$
= \tfrac{8}{3}\left(\tfrac{3}{8}\pi\right)
$$

$$
= \pi
$$

Finally, we add the results for all three terms:

$$
\iint_{S} \mathbf{f} \cdot d\mathbf{S} = \left(\frac{1}{4}\pi + \frac{1}{4}\pi\right) + \left(\frac{1}{4}\pi + \frac{1}{4}\pi\right) + \pi = 2\pi
$$

PROBLEM 6-14 If S is represented by $z = z(x, y)$, show that the flux of $\mathbf{f}$ through S is

$$
\iint_{S} \mathbf{f} \cdot d\mathbf{S} = \iint_{S} \mathbf{f} \cdot \mathbf{n} \, dS = \iint_{R_{xy}} \mathbf{f} \cdot \mathbf{n} \sec\gamma \, dx \, dy = \iint_{R_{xy}} \mathbf{f} \cdot \frac{\mathbf{n}}{\mathbf{n} \cdot \mathbf{k}} \, dx \, dy \qquad \text{(a)}
$$

where

$$
\sec\gamma = \frac{1}{\mathbf{n} \cdot \mathbf{k}} = \left[\left(\frac{\partial z}{\partial x}\right)^2 + \left(\frac{\partial z}{\partial y}\right)^2 + 1\right]^{1/2}
$$

Solution: If S is represented by $z = z(x, y)$, the unit normal vector (see Problem 4-10) is given by

$$
\mathbf{n} = \frac{-\dfrac{\partial z}{\partial x}\mathbf{i} - \dfrac{\partial z}{\partial y}\mathbf{j} + \mathbf{k}}{\left[\left(\dfrac{\partial z}{\partial x}\right)^2 + \left(\dfrac{\partial z}{\partial y}\right)^2 + 1\right]^{1/2}}
$$

So if γ is the angle between vectors $\mathbf{n}$ and $\mathbf{k}$,

$$
\mathbf{n} \cdot \mathbf{k} = \cos\gamma = \frac{1}{\sec\gamma} = \frac{1}{\left[\left(\dfrac{\partial z}{\partial x}\right)^2 + \left(\dfrac{\partial z}{\partial y}\right)^2 + 1\right]^{1/2}}
$$

Next we find the differential element dS of the surface area S (from the result of Problem 4-13):

$$dS = \left[\left(\frac{\partial z}{\partial x}\right)^2 + \left(\frac{\partial z}{\partial y}\right)^2 + 1\right]^{1/2} dx\, dy$$

$$= \sec \gamma\, dx\, dy$$

$$= \frac{1}{(\mathbf{n} \cdot \mathbf{k})}\, dx\, dy$$

Hence the flux of $\mathbf{f}$ through S is

$$\iint_S \mathbf{f} \cdot d\mathbf{S} = \iint_S \mathbf{f} \cdot \mathbf{n}\, dS = \iint_{R_{xy}} \mathbf{f} \cdot \mathbf{n} \sec \gamma\, dx\, dy = \iint_{R_{xy}} \mathbf{f} \cdot \frac{\mathbf{n}}{\mathbf{n} \cdot \mathbf{k}}\, dx\, dy$$

PROBLEM 6-15 If $\mathbf{f} = x\mathbf{i} + y\mathbf{j} + 2z\mathbf{k}$, use eq. (a) of Problem 6-14 to evaluate $\iint_S \mathbf{f} \cdot d\mathbf{S}$, where S is that part of the surface of the paraboloid $x^2 + y^2 = 1 - z$ for which $z > 0$. (See Problem 6-13 and Figure 6-7.)

Solution: Since S is represented by $z = 1 - x^2 - y^2$ with $z > 0$, we know from the result of Problem 6-14 that the unit normal vector is

$$\mathbf{n} = \frac{-\dfrac{\partial z}{\partial x}\mathbf{i} - \dfrac{\partial z}{\partial y}\mathbf{j} + \mathbf{k}}{\left[\left(\dfrac{\partial z}{\partial x}\right)^2 + \left(\dfrac{\partial z}{\partial y}\right)^2 + 1\right]^{1/2}} = \frac{2x\mathbf{i} + 2y\mathbf{j} + \mathbf{k}}{(4x^2 + 4y^2 + 1)^{1/2}}$$

Now since $(\mathbf{k} \cdot \mathbf{i}) = (\mathbf{k} \cdot \mathbf{j}) = 0$ and $(\mathbf{k} \cdot \mathbf{k}) = 1$, we obtain

$$\mathbf{n} \cdot \mathbf{k} = \frac{1}{(4x^2 + 4y^2 + 1)^{1/2}}$$

and

$$\mathbf{f} \cdot \mathbf{n} = \frac{1}{(4x^2 + 4y^2 + 1)^{1/2}}\left[(x)(2x) + (y)(2y) + (2z)1\right] = \frac{2x^2 + 2y^2 + 2z}{(4x^2 + 4y^2 + 1)^{1/2}}$$

Since the projection of the surface of the given paraboloid onto the xy-plane is produced by setting $z = 0$ in its equation, we have $0 = 1 - x^2 - y^2$ or $x^2 + y^2 = 1$, which is the unit circle. We use this circle for R_{xy}.

Finally, with the use of eq. (a) of Problem 6-14 we obtain

$$\iint_S \mathbf{f} \cdot d\mathbf{S} = \iint_{R_{xy}} \frac{2x^2 + 2y^2 + 2z}{(2x^2 + 4y^2 + 1)^{1/2}}(4x^2 + 4y^2 + 1)^{1/2}\, dx\, dy$$

$$= \iint_{R_{xy}} [2x^2 + 2y^2 + 2(1 - x^2 - y^2)]\, dx\, dy$$

$$= 2\int_{x=-1}^{1}\int_{y=-\sqrt{1-x^2}}^{\sqrt{1-x^2}} dx\, dy$$

$$= 4\int_{-1}^{1}(1 - x^2)^{1/2}\, dx$$

$$= 4\frac{\pi}{2}$$

$$= 2\pi$$

PROBLEM 6-16 Evaluate $\iint_S \mathbf{r} \times d\mathbf{S}$, where S is the closed complete sphere represented by $x^2 + y^2 + z^2 = a^2$.

Solution: This sphere is centered at the origin and has radius a. From Problem 6-11, we see that the outer unit normal vector $\mathbf{n}$ at any point on S is given by

$$\mathbf{n} = \mathbf{e}_r = \frac{\mathbf{r}}{|\mathbf{r}|} = \frac{\mathbf{r}}{a}$$

So from relation (1.42), $\mathbf{r} \times \mathbf{n} = (1/a)\mathbf{r} \times \mathbf{r} = \mathbf{0}$. Hence

$$\iint_S \mathbf{r} \times d\mathbf{S} = \iint_S \mathbf{r} \times \mathbf{n}\, dS = \mathbf{0}$$

PROBLEM 6-17 Evaluate $\iint_S \mathbf{r} \times d\mathbf{S}$, where S is the surface of the unit cube bounded by the planes $x = 0$, $x = 1$, $y = 0$, $y = 1$, $z = 0$, $z = 1$; $\mathbf{r}$ is the position vector; and $\mathbf{n}$ is the unit normal vector having outward direction. (See Problem 6-12 and Figure 6-6.)

Solution: Referring to Figure 6-6, we see that on face S_1 defined by $AOCB$, $\mathbf{n} = -\mathbf{k}$, $d\mathbf{S} = \mathbf{n}\,dx\,dy = -dx\,dy\,\mathbf{k}$; hence

$$\mathbf{r} \times \mathbf{n} = (x\mathbf{i} + y\mathbf{j} + z\mathbf{k}) \times (-\mathbf{k}) = -y\mathbf{i} + x\mathbf{j}$$

So on face S_1

$$\iint_{S_1} \mathbf{r} \times d\mathbf{S} = \int_0^1 \int_0^1 (-y\mathbf{i} + x\mathbf{j})\,dx\,dy$$

$$= -\mathbf{i}\int_0^1 \int_0^1 y\,dy\,dx + \mathbf{j}\int_0^1 \int_0^1 x\,dx\,dy$$

$$= -\tfrac{1}{2}\mathbf{i} + \tfrac{1}{2}\mathbf{j}$$

On face S_2 defined by $AFEO$, $\mathbf{n} = -\mathbf{j}$, $d\mathbf{S} = \mathbf{n}\,dz\,dx = -dz\,dx\,\mathbf{j}$; therefore

$$\mathbf{r} \times \mathbf{n} = (x\mathbf{i} + y\mathbf{j} + z\mathbf{k}) \times (-\mathbf{j}) = z\mathbf{i} - x\mathbf{k}$$

Hence on S_2

$$\iint_{S_2} \mathbf{r} \times d\mathbf{S} = \int_0^1 \int_0^1 (z\mathbf{i} - x\mathbf{k})\,dz\,dx$$

$$= \mathbf{i}\int_0^1 \int_0^1 z\,dz\,dx - \mathbf{k}\int_0^1 \int_0^1 x\,dx\,dz$$

$$= \tfrac{1}{2}\mathbf{i} - \tfrac{1}{2}\mathbf{k}$$

On face S_3 defined by $OEDC$, $\mathbf{n} = -\mathbf{i}$, $d\mathbf{S} = \mathbf{n}\,dy\,dz = -dy\,dz\,\mathbf{i}$; thus

$$\mathbf{r} \times \mathbf{n} = (x\mathbf{i} + y\mathbf{j} + z\mathbf{k}) \times (-\mathbf{i}) = -z\mathbf{j} + y\mathbf{k}$$

Hence

$$\iint_{S_3} \mathbf{r} \times d\mathbf{S} = \int_0^1 \int_0^1 (-z\mathbf{j} + y\mathbf{k})\,dy\,dz$$

$$= -\mathbf{j}\int_0^1 \int_0^1 z\,dz\,dy + \mathbf{k}\int_0^1 \int_0^1 y\,dy\,dz$$

$$= -\tfrac{1}{2}\mathbf{j} + \tfrac{1}{2}\mathbf{k}$$

On face S_4 defined by $AFGB$, $\mathbf{n} = \mathbf{i}$, $d\mathbf{S} = \mathbf{n}\,dy\,dz = dy\,dz\,\mathbf{i}$; hence

$$\mathbf{r} \times \mathbf{n} = (x\mathbf{i} + y\mathbf{j} + z\mathbf{k}) \times \mathbf{i} = z\mathbf{j} - y\mathbf{k}$$

Thus

$$\iint\limits_{S_4} \mathbf{r} \times d\mathbf{S} = \int_0^1 \int_0^1 (z\mathbf{j} - y\mathbf{k})\, dy\, dz = \frac{1}{2}\mathbf{j} - \frac{1}{2}\mathbf{k}$$

On face S_5 defined by $BGDC$, $\mathbf{n} = \mathbf{j}$, $d\mathbf{S} = \mathbf{n}\, dz\, dx = dz\, dz\, \mathbf{j}$; thus

$$\mathbf{r} \times \mathbf{n} = (x\mathbf{i} + y\mathbf{j} + z\mathbf{k}) \times \mathbf{j} = -z\mathbf{i} + x\mathbf{k}$$

Hence

$$\iint\limits_{S_5} \mathbf{r} \times d\mathbf{S} = \int_0^1 \int_0^1 (-z\mathbf{i} + x\mathbf{k})\, dz\, dx = -\frac{1}{2}\mathbf{i} + \frac{1}{2}\mathbf{k}$$

On face S_6 defined by $FEDG$, $\mathbf{n} = \mathbf{k}$, $d\mathbf{S} = \mathbf{n}\, dx\, dy = dx\, dy\, \mathbf{k}$; therefore

$$\mathbf{r} \times \mathbf{n} = (x\mathbf{i} + y\mathbf{j} + z\mathbf{k}) \times \mathbf{k} = y\mathbf{i} - x\mathbf{j}$$

Hence

$$\iint\limits_{S_6} \mathbf{r} \times d\mathbf{S} = \int_0^1 \int_0^1 (y\mathbf{i} - x\mathbf{j})\, dx\, dy = \frac{1}{2}\mathbf{i} - \frac{1}{2}\mathbf{j}$$

Adding all these results, we get

$$\iint\limits_{S} \mathbf{r} \times d\mathbf{S} = 0$$

Volume Integrals

PROBLEM 6-18   Evaluate $\iiint_R \mathbf{V} \cdot \mathbf{r}\, dV$, where R is any region with volume V and $\mathbf{r}$ is the position vector.

Solution: From (5.17), we have $\mathbf{V} \cdot \mathbf{r} = 3$. So using the definition of volume in (6.28), we have

$$\iiint\limits_R \mathbf{V} \cdot \mathbf{r}\, dV = \iiint\limits_R 3\, dV = 3\iiint\limits_R dV = 3V$$

PROBLEM 6-19   Evaluate $\iiint_R \mathbf{V} \times \mathbf{f}\, dV$, where $\mathbf{f} = y\mathbf{i} - x\mathbf{j}$ and R is any three-dimensional region with volume V.

Solution: By definition (5.28), the curl of $\mathbf{f}$ is

$$\mathbf{V} \times \mathbf{f} = \begin{vmatrix} \mathbf{i} & \mathbf{j} & \mathbf{k} \\ \dfrac{\partial}{\partial x} & \dfrac{\partial}{\partial y} & \dfrac{\partial}{\partial z} \\ y & -x & 0 \end{vmatrix} = -2\mathbf{k}$$

So by formula (6.29)

$$\iiint\limits_R \mathbf{V} \times \mathbf{f}\, dV = -2\mathbf{k}\iiint\limits_R dV = -2V\mathbf{k}$$

Alternate Definitions of Gradient, Divergence, and Curl

PROBLEM 6-20   Show that the definition of gradient given in (6.35) is consistent with the definition of $\mathbf{V}\phi$ given by (5.4). Ignore the higher-order terms.

Solution: If we follow the procedure of Example 6-7 and use the same notations (see Figure 6-2), we have

$$\iint_{S_1} d\mathbf{S}\,\phi = \mathbf{i}\left(\phi + \frac{\partial\phi}{\partial x}\frac{\Delta x}{2}\right)\Delta y\,\Delta z$$

$$\iint_{S_2} d\mathbf{S}\,\phi = -\mathbf{i}\left(\phi - \frac{\partial\phi}{\partial x}\frac{\Delta x}{2}\right)\Delta y\,\Delta z$$

$$\iint_{S_3} d\mathbf{S}\,\phi = \mathbf{j}\left(\phi + \frac{\partial\phi}{\partial y}\frac{\Delta y}{2}\right)\Delta z\,\Delta x$$

$$\iint_{S_4} d\mathbf{S}\,\phi = -\mathbf{j}\left(\phi - \frac{\partial\phi}{\partial y}\frac{\Delta y}{2}\right)\Delta z\,\Delta x$$

$$\iint_{S_5} d\mathbf{S}\,\phi = \mathbf{k}\left(\phi + \frac{\partial\phi}{\partial z}\frac{\Delta z}{2}\right)\Delta x\,\Delta y$$

$$\iint_{S_6} d\mathbf{S}\,\phi = -\mathbf{k}\left(\phi - \frac{\partial\phi}{\partial z}\frac{\Delta z}{2}\right)\Delta x\,\Delta y$$

Adding all these results, we get

$$\oiint_{S} d\mathbf{S}\,\phi = \left[\mathbf{i}\left(\frac{\partial\phi}{\partial x}\right) + \mathbf{j}\left(\frac{\partial\phi}{\partial y}\right) + \mathbf{k}\left(\frac{\partial\phi}{\partial z}\right)\right]\Delta x\,\Delta y\,\Delta z$$

$$= \left[\mathbf{i}\left(\frac{\partial\phi}{\partial x}\right) + \mathbf{j}\left(\frac{\partial\phi}{\partial y}\right) + \mathbf{k}\left(\frac{\partial\phi}{\partial z}\right)\right]\Delta V$$

Finally, by formula (6.35), we obtain

$$\nabla\phi = \lim_{\Delta V \to 0}\frac{1}{\Delta V}\oiint d\mathbf{S}\,\phi = \mathbf{i}\frac{\partial\phi}{\partial x} + \mathbf{j}\frac{\partial\phi}{\partial y} + \mathbf{k}\frac{\partial\phi}{\partial z}$$

PROBLEM 6-21 Show that the definition of curl given in (6.37) is consistent with the definition of $\nabla \times \mathbf{f}$ given by (5.28). Ignore the higher-order terms.

Solution: As in Problem 6-20, we follow the procedure of Example 6-7 and use the same notations (see Figure 6-2). For the surface S_1 we have

$$\iint_{S_1} d\mathbf{S} \times \mathbf{f} = \iint_{S_1} \mathbf{i} \times (f_1\mathbf{i} + f_2\mathbf{j} + f_3\mathbf{k})\,dy\,dz$$

$$= \iint_{S_1} (\mathbf{k}f_2 - \mathbf{j}f_3)\,dy\,dz$$

$$= \mathbf{k}\left(f_2 + \frac{\partial f_2}{\partial x}\frac{\Delta x}{2}\right)\Delta y\,\Delta z - \mathbf{j}\left(f_3 + \frac{\partial f_3}{\partial x}\frac{\Delta x}{2}\right)\Delta y\,\Delta z$$

Similarly, for the other five surfaces

$$\iint_{S_2} d\mathbf{S} \times \mathbf{f} = -\mathbf{k}\left(f_2 - \frac{\partial f_2}{\partial x}\frac{\Delta x}{2}\right)\Delta y\,\Delta z + \mathbf{j}\left(f_3 - \frac{\partial f_3}{\partial x}\frac{\Delta x}{2}\right)\Delta y\,\Delta z$$

$$\iint_{S_3} d\mathbf{S} \times \mathbf{f} = -\mathbf{k}\left(f_1 + \frac{\partial f_1}{\partial y}\frac{\Delta y}{2}\right)\Delta z\,\Delta x + \mathbf{i}\left(f_3 + \frac{\partial f_3}{\partial y}\frac{\Delta y}{2}\right)\Delta z\,\Delta x$$

$$\iint\limits_{S_4} d\mathbf{S} \times \mathbf{f} = \mathbf{k}\left(f_1 - \frac{\partial f_1}{\partial y}\frac{\Delta y}{2}\right)\Delta z\,\Delta x - \mathbf{i}\left(f_3 - \frac{\partial f_3}{\partial y}\frac{\Delta y}{2}\right)\Delta z\,\Delta x$$

$$\iint\limits_{S_5} d\mathbf{S} \times \mathbf{f} = \mathbf{j}\left(f_1 + \frac{\partial f_1}{\partial z}\frac{\Delta z}{2}\right)\Delta x\,\Delta y - \mathbf{i}\left(f_2 + \frac{\partial f_2}{\partial z}\frac{\Delta z}{2}\right)\Delta x\,\Delta y$$

$$\iint\limits_{S_6} d\mathbf{S} \times \mathbf{f} = -\mathbf{j}\left(f_1 - \frac{\partial f_1}{\partial z}\frac{\Delta z}{2}\right)\Delta x\,\Delta y + \mathbf{i}\left(f_2 - \frac{\partial f_2}{\partial z}\frac{\Delta z}{2}\right)\Delta x\,\Delta y$$

Adding all these results yields

$$\oiint\limits_{S} d\mathbf{S} \times \mathbf{f} = \left[\mathbf{i}\left(\frac{\partial f_3}{\partial y} - \frac{\partial f_2}{\partial z}\right) + \mathbf{j}\left(\frac{\partial f_1}{\partial z} - \frac{\partial f_3}{\partial x}\right) + \mathbf{k}\left(\frac{\partial f_2}{\partial x} - \frac{\partial f_1}{\partial y}\right)\right]\Delta x\,\Delta y\,\Delta z$$

$$= \left[\mathbf{i}\left(\frac{\partial f_3}{\partial y} - \frac{\partial f_2}{\partial z}\right) + \mathbf{j}\left(\frac{\partial f_1}{\partial z} - \frac{\partial f_3}{\partial x}\right) + \mathbf{k}\left(\frac{\partial f_2}{\partial x} - \frac{\partial f_1}{\partial y}\right)\right]\Delta V$$

And finally, by definition (6.37), we obtain

$$\nabla \times \mathbf{f} = \lim_{\Delta V \to 0}\frac{1}{\Delta V}\oiint\limits_{S} d\mathbf{S} \times \mathbf{f} = \mathbf{i}\left(\frac{\partial f_3}{\partial y} - \frac{\partial f_2}{\partial z}\right) + \mathbf{j}\left(\frac{\partial f_1}{\partial z} - \frac{\partial f_3}{\partial x}\right) + \mathbf{k}\left(\frac{\partial f_2}{\partial x} - \frac{\partial f_1}{\partial y}\right)$$

Supplementary Exercises

PROBLEM 6-22 Evaluate $\int_C \phi\,d\mathbf{r}$ for $\phi = x^3 y + 2y$ from $(1,1,0)$ to $(2,4,0)$ along (a) the parabola $y = x^2$, $z = 0$, and (b) the straight line joining $(1,1,0)$ and $(2,4,0)$.

Answer: (a) $[\frac{91}{6}, \frac{359}{7}, 0]$ (b) $[\frac{161}{10}, \frac{483}{10}, 0]$

PROBLEM 6-23 Evaluate $\int_C \mathbf{f} \cdot d\mathbf{r}$ for $\mathbf{f} = y\mathbf{i} + (x + z)^2\mathbf{j} + (x - z)^2\mathbf{k}$ from $(0,0,0)$ to $(2,4,0)$ along (a) the parabola $y = x^2$, $z = 0$, and (b) the straight line $y = 2x$.

Answer: (a) $\frac{32}{3}$ (b) $\frac{28}{3}$

PROBLEM 6-24 Evaluate $\int_C \mathbf{f} \cdot d\mathbf{r}$ for $\mathbf{f} = 2xy^2\mathbf{i} + 2(x^2 y + y)\mathbf{j}$ from $(0,0,0)$ to $(2,4,0)$ along (a) the parabola $y = x^2$, $z = 0$; and (b) the straight line $y = 2x$.

Answer: (a) 80   (b) 80

PROBLEM 6-25 Evaluate $\int_C \mathbf{f} \cdot d\mathbf{r}$ for $\mathbf{f} = 3x\mathbf{i} + (2xz - y)\mathbf{j} + z\mathbf{k}$ from $(0,0,0)$ to $(2,1,3)$ along (a) the curve of $x = 2t^2$, $y = t$, $z = 4t^2 - t$, where $0 \leqslant t \leqslant 1$, and (b) the straight line from $(0,0,0)$ to $(2,1,3)$.

Answer: (a) $\frac{61}{5}$ (b) 14

PROBLEM 6-26 Evaluate $\int_C \mathbf{f} \times d\mathbf{r}$ for $\mathbf{f} = y\mathbf{i} + x\mathbf{j}$ from $(0,0,0)$ to $(3,9,0)$ along the curve given by $y = x^3/3$, $z = 0$.

Answer: $[0, 0, 36]$

PROBLEM 6-27 If $\mathbf{a}$ is a constant vector, show that $\oint \mathbf{a} \cdot d\mathbf{r} = 0$ and $\oint \mathbf{a} \times d\mathbf{r} = \mathbf{0}$.

PROBLEM 6-28 If $\mathbf{f} = x\mathbf{i} + y\mathbf{j} + z\mathbf{k}$, evaluate $\iint_S \mathbf{f} \cdot d\mathbf{S}$, where S is the cylindrical surface represented by $\mathbf{r} = \cos u\mathbf{i} + \sin u\mathbf{j} + v\mathbf{k}, 0 \le u \le 2\pi, 0 \le v \le 1$, $d\mathbf{S} = \mathbf{n}\, dS$, and $\mathbf{n}$ is the outer unit normal vector.

Answer: 2π

PROBLEM 6-29 If $\mathbf{f} = 4xz\mathbf{i} + xyz^2\mathbf{j} + 3z\mathbf{k}$, evaluate $\iint_S \mathbf{f} \cdot d\mathbf{S}$, where S is the surface bounded by $z^2 = x^2 + y^2, z = 0, z = 4, d\mathbf{S} = \mathbf{n}\, dS$, and $\mathbf{n}$ is the outer unit normal vector.

Answer: 320

PROBLEM 6-30 If $\mathbf{f} = (y + z)\mathbf{i} + (z + x)\mathbf{j} + (x + y)\mathbf{k}$ and S is the surface of the cube bounded by $x = 0, y = 0, z = 0, x = 1, y = 1, z = 1$, evaluate (a) $\iint_S \mathbf{f}\, dS$, (b) $\iint_S \mathbf{f} \cdot d\mathbf{S}$, and (c) $\iint_S \mathbf{f} \times d\mathbf{S}$.

Answer: (a) $[6,6,6]$ (b) 0   (c) **0**

PROBLEM 6-31 Evaluate $\iiint_R \mathbf{r}\, dV$, where $\mathbf{r}$ is the position vector and R is the region bounded by the surface $x = 0, y = 0, y = 6, z = 4$, and $z = x^2$.

Answer: $[24, 96, \frac{384}{5}]$

PROBLEM 6-32 Using the equivalent integral representation (6.34) of $\mathbf{V}$, verify that $\mathbf{V} \times (\phi\mathbf{f}) = \phi\mathbf{V} \times \mathbf{f} + (\mathbf{V}\phi) \times \mathbf{f}$.

7 INTEGRAL THEOREMS

THIS CHAPTER IS ABOUT

- ☑ **Divergence Theorem (or Gauss' Theorem)**
- ☑ **Green's Theorems**
- ☑ **Volume-to-Surface Integral Transformations**
- ☑ **Stokes' Theorem**
- ☑ **Surface-to-Line Integral Transformations**
- ☑ **Irrotational and Solenoidal Fields**

7-1. Divergence Theorem (or Gauss' Theorem)

The definition of divergence

$$\text{div}\,\mathbf{f} = \lim_{\Delta V \to 0} \frac{1}{\Delta V} \oiint_S \mathbf{f} \cdot d\mathbf{S} \tag{7.1}$$

gives the value of $\nabla \cdot \mathbf{f}$ at a point (see Section 6-4). The **divergence theorem**, also called **Gauss' theorem**, is obtained by extending this definition of divergence to a finite region. It states that if $\mathbf{f}$ is a continuous vector function in a region R with volume V and bounded by a closed surface S, then

DIVERGENCE THEOREM
$$\iiint_R \nabla \cdot \mathbf{f}\, dV = \oiint_S \mathbf{f} \cdot d\mathbf{S} \tag{7.2}$$

EXAMPLE 7-1: Give the physical interpretation of the divergence theorem (7.2).

Solution: As shown in Example 6-6 (the physical interpretation of $\nabla \cdot \mathbf{f}$), the divergence of a vector field $\mathbf{f}$ at a given point is the density of outward flux flow from that point. The divergence theorem (7.2) states that the total outward flux flow from a closed surface S equals the integral of the divergence throughout the region R bounded by S.

EXAMPLE 7-2: Verify the divergence theorem (7.2).

Solution: Consider a finite closed surface S that encloses a region R with volume V. Divide R into N subregions of volumes $\Delta V_1, \Delta V_2, \ldots, \Delta V_N$. At an arbitrarily chosen point (x_i, y_i, z_i) within ΔV_i, where $i = 1, 2, \ldots, N$, the divergence definition (7.1) gives

$$\nabla \cdot \mathbf{f} = \frac{1}{\Delta V_i} \oiint_{\Delta S_i} \mathbf{f} \cdot d\mathbf{S} + \varepsilon_i$$

where $\varepsilon_i \to 0$ as $\Delta V_i \to 0$. Thus

$$\nabla \cdot \mathbf{f}\, \Delta V_i = \oiint_{\Delta S_i} \mathbf{f} \cdot d\mathbf{S} + \varepsilon_i \Delta V_i \tag{7.3}$$

The sum of these terms over the entire volume V is

$$\sum_{i=1}^{N} \nabla \cdot \mathbf{f} \, \Delta V_i = \sum_{i=1}^{N} \oiint_{\Delta S_i} \mathbf{f} \cdot d\mathbf{S} + \sum_{i=1}^{N} \varepsilon_i \, \Delta V_i$$

Now consider the limit of this expression as $N \to \infty$. The surface boundary ΔS_i of each ΔV_i consists of numerous segments that are either part of the boundary S or of two adjacent subregions. The surface integrals for the adjacent boundary surfaces cancel out because the outer normals have opposite directions over the common boundary surface. Thus only the surface integral over S is left (see Figure 7-1). Also, by the definition of a multiple integral,

$$\lim_{N \to \infty} \sum_{i=1}^{N} \nabla \cdot \mathbf{f} \, \Delta V_i = \iiint_R \nabla \cdot \mathbf{f} \, dV$$

Because

$$\lim_{N \to \infty} \sum_{i=1}^{N} \oiint_{\Delta S_i} \mathbf{f} \cdot d\mathbf{S} = \oiint_S \mathbf{f} \cdot d\mathbf{S}$$

we have

$$\iiint_R \nabla \cdot \mathbf{f} \, dV = \oiint_S \mathbf{f} \cdot d\mathbf{S} + \lim_{N \to \infty} \sum_{i=1}^{N} \varepsilon_i \, \Delta V_i \tag{7.4}$$

For the second term on the right-hand side of (7.4),

$$\left| \sum_{i=1}^{N} \varepsilon_i \, \Delta V_i \right| \leqslant \sum_{i=1}^{N} |\varepsilon_i| \, \Delta V_i \leqslant |\varepsilon_m| \sum_{i=1}^{N} \Delta V_i = |\varepsilon_m| V$$

where $\varepsilon_m = \max \varepsilon_i$. But $\varepsilon_m \to 0$ as $N \to \infty$ and $\Delta V_i \to 0$. Hence

$$\lim_{N \to \infty} \sum_{i=1}^{N} \varepsilon_i \, \Delta V_i \to 0$$

Thus, upon taking the appropriate limits, we obtain the divergence theorem (7.2):

$$\iiint_R \nabla \cdot \mathbf{f} \, dV = \oiint_S \mathbf{f} \cdot d\mathbf{S}$$

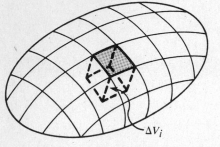

Figure 7-1
Verification of the divergence theorem.

EXAMPLE 7-3: If $\mathbf{f} = P(x, y, z)\mathbf{i} + Q(x, y, z)\mathbf{j} + R(x, y, z)\mathbf{k}$ and R is the region bounded by a closed surface S, show that the divergence theorem (7.2) expressed in rectangular coordinates is

DIVERGENCE THEOREM (rectangular coordinates)

$$\iiint_R \left(\frac{\partial P}{\partial x} + \frac{\partial Q}{\partial y} + \frac{\partial R}{\partial z} \right) dx \, dy \, dz$$

$$= \iint_S (P \, dy \, dz + Q \, dz \, dx + R \, dx \, dy) \tag{7.5}$$

Solution: If we write

$$\mathbf{f} = P\mathbf{i} + Q\mathbf{j} + R\mathbf{k}$$
$$\mathbf{n} = \cos \alpha \, \mathbf{i} + \cos \beta \, \mathbf{j} + \cos \gamma \, \mathbf{k}$$

then, in general, for any surface S (see Example 6-5)

$$\mathbf{n} \cdot \mathbf{i} \, dS = \cos \alpha \, dS = dy \, dz$$
$$\mathbf{n} \cdot \mathbf{j} \, dS = \cos \beta \, ds = dz \, dx$$
$$\mathbf{n} \cdot \mathbf{k} \, dS = \cos \gamma \, dS = dx \, dy$$

Since the divergence of **f** is

$$\nabla \cdot \mathbf{f} = \frac{\partial P}{\partial x} + \frac{\partial Q}{\partial y} + \frac{\partial R}{\partial z}$$

Then

$$\iiint_R \nabla \cdot \mathbf{f}\, dV = \iiint_R \left(\frac{\partial P}{\partial x} + \frac{\partial Q}{\partial y} + \frac{\partial R}{\partial z}\right) dx\, dy\, dz$$

We also write

$$\oiint_S \mathbf{f} \cdot d\mathbf{S} = \oiint_S (P\mathbf{i} + Q\mathbf{j} + R\mathbf{k}) \cdot \mathbf{n}\, d\mathbf{S}$$

$$= \oiint_S (P\cos\alpha + Q\cos\beta + R\cos\gamma)\, dS$$

$$= \oiint_S (P\, dy\, dz + Q\, dz\, dx + R\, dx\, dy)$$

Hence expression (7.2) reduces to expression (7.5):

$$\iiint_R \left(\frac{\partial P}{\partial x} + \frac{\partial Q}{\partial y} + \frac{\partial R}{\partial z}\right) dx\, dy\, dz = \oiint_S (P\, dy\, dz + Q\, dz\, dx + R\, dx\, dy)$$

EXAMPLE 7-4: Show that if **r** is the position vector, then

$$\oiint_S \mathbf{r} \cdot d\mathbf{S} = 3V \qquad (7.6)$$

where V is the volume of the region R bounded by closed surface S.

Solution: The divergence $\nabla \cdot \mathbf{r}$ is

$$\nabla \cdot \mathbf{r} = \frac{\partial x}{\partial x} + \frac{\partial y}{\partial y} + \frac{\partial z}{\partial z} = 3$$

Thus applying Gauss' theorem (7.2),

$$\oiint_S \mathbf{r} \cdot d\mathbf{S} = \iiint_R \nabla \cdot \mathbf{r}\, dV = 3\iiint_R dV = 3V$$

7-2. Green's Theorems

Green's first theorem (or *identity*) states that if ϕ and ψ are scalar functions having continuous second derivatives in a region R bounded by a closed surface S, then

GREEN'S FIRST THEOREM
$$\iiint_R (\phi \nabla^2 \psi + \nabla\phi \cdot \nabla\psi)\, dV = \oiint_S \phi \nabla\psi \cdot d\mathbf{S} \qquad (7.7)$$

Green's second theorem (or *identity*) states that if ϕ are ψ are scalar functions having continuous second derivatives in a region R bounded by a closed surface S, then

GREEN'S SECOND THEOREM
$$\iiint_R (\phi \nabla^2 \psi - \psi \nabla^2 \phi)\, dV = \oiint_S (\phi \nabla\psi - \psi \nabla\phi) \cdot d\mathbf{S} \qquad (7.8)$$

Green's third theorem (or *identity*) states that if $\mathbf{f}(\nabla \cdot \mathbf{g})$ and $\mathbf{f} \times (\nabla \times \mathbf{g})$ are vector functions having continuous second derivatives in a region R bounded by a closed surface S, then

GREEN'S
THIRD
THEOREM

$$\iiint_R [\mathbf{f} \cdot \nabla^2 \mathbf{g} - \mathbf{g} \cdot \nabla^2 \mathbf{f}]\, dV$$

$$= \oiint_S [\mathbf{f} \times (\nabla \times \mathbf{g}) + \mathbf{f}(\nabla \cdot \mathbf{g}) - \mathbf{g} \times (\nabla \times \mathbf{f}) - \mathbf{g}(\nabla \cdot \mathbf{f})] \cdot d\mathbf{S} \quad (7.9)$$

where $\nabla^2 \mathbf{f} = \nabla(\nabla \cdot \mathbf{f}) - \nabla \times (\nabla \times \mathbf{f})$ from (5.45).

Note: Green's third theorem is the vector equivalent to Green's second theorem (7.8), which relates two scalars.

EXAMPLE 7-5: Show that Green's first theorem (7.7) and second theorem (7.8) can be rewritten, respectively, as

$$\iiint_R (\phi \nabla^2 \psi + \nabla\phi \cdot \nabla\psi)\, dV = \oiint_S \phi \frac{\partial \psi}{\partial n}\, dS \quad (7.10)$$

$$\iiint_R (\phi \nabla^2 \psi - \psi \nabla^2 \phi)\, dV = \oiint_S \left(\phi \frac{\partial \psi}{\partial n} - \psi \frac{\partial \phi}{\partial n} \right) dS \quad (7.11)$$

where $\partial\psi/\partial n$ and $\partial\phi/\partial n$ are the normal derivatives.

Solution: From the result of Example 5-1, we know that $\partial\phi/\partial s = \nabla\phi \cdot \mathbf{T}$ (5.5). So we have

$$\nabla\psi \cdot \mathbf{n} = \frac{\partial \psi}{\partial n} \quad \text{and} \quad \nabla\phi \cdot \mathbf{n} = \frac{\partial \phi}{\partial n} \quad (7.12)$$

Thus

$$\nabla\psi \cdot d\mathbf{S} = \nabla\psi \cdot \mathbf{n}\, dS = \frac{\partial \psi}{\partial n}\, dS$$

and

$$\nabla\phi \cdot d\mathbf{S} = \nabla\phi \cdot \mathbf{n}\, dS = \frac{\partial \phi}{\partial n}\, dS$$

and (7.7) and (7.8) can be rewritten as (7.10) and (7.11), respectively.

EXAMPLE 7-6: Verify Green's first theorem (7.7).

Solution: Using the vector identity (5.41), $\nabla \cdot (\phi\mathbf{f}) = \phi\nabla \cdot \mathbf{f} + \mathbf{f} \cdot \nabla\phi$, where $\mathbf{f} = \nabla\psi$, we have

$$\nabla \cdot (\phi \nabla\psi) = \phi \nabla \cdot \nabla\psi + \nabla\psi \cdot \nabla\phi = \phi \nabla^2 \psi + \nabla\phi \cdot \nabla\psi \quad (7.13)$$

Integrating over the region R, we obtain

$$\iiint_R \nabla \cdot (\phi \nabla\psi)\, dV = \iiint_R (\phi \nabla^2 \psi + \nabla\phi \cdot \nabla\psi)\, dV \quad (7.14)$$

Applying the divergence theorem (7.2), we have

$$\iiint_R \nabla \cdot (\phi \nabla\psi)\, dV = \oiint_S \phi \nabla\psi \cdot d\mathbf{S} \quad (7.15)$$

Combining these, we obtain Green's first theorem (7.7):

$$\iiint\limits_R (\phi \, \nabla^2 \psi + \nabla\phi \cdot \nabla\psi) \, dV = \oiint\limits_S \phi \, \nabla\psi \cdot d\mathbf{S}$$

EXAMPLE 7-7: Verify Green's second theorem (7.8).

Solution: Interchanging ϕ and ψ in (7.7), we have

$$\iiint\limits_R (\psi \, \nabla^2 \phi + \nabla\psi \cdot \nabla\phi) \, dV = \oiint\limits_S \psi \, \nabla\phi \cdot d\mathbf{S} \qquad (7.16)$$

Then by subtracting (7.16) from (7.7), we obtain Green's second theorem (7.8):

$$\iiint\limits_R (\phi \, \nabla^2 \psi - \psi \, \nabla^2 \phi) \, dV = \oiint\limits_S (\phi \, \nabla\psi - \psi \, \nabla\phi) \cdot d\mathbf{S}$$

EXAMPLE 7-8: If ψ is harmonic in a region R enclosed by S, prove that

$$\oiint\limits_S \nabla\psi \cdot d\mathbf{S} = \oiint\limits_S \frac{\partial\psi}{\partial n} \, dS = 0 \qquad (7.17)$$

Solution: If we set $\phi = 1$ in Green's first theorem (7.7), we obtain

$$\oiint\limits_S \nabla\psi \cdot d\mathbf{S} = \iiint\limits_R \nabla^2\psi \, dV \qquad (7.18)$$

since $\nabla(1) = 0$. Since ψ is harmonic, then by definition (5.22), $\nabla^2\psi = 0$. Hence, (7.18) reduces to

$$\oiint\limits_S \nabla\psi \cdot d\mathbf{S} = \oiint\limits_S \nabla\psi \cdot \mathbf{n} \, dS = \oiint\limits_S \frac{\partial\psi}{\partial n} \, dS = 0$$

7-3. Volume-to-Surface Integral Transformations

The divergence theorem (7.2) represents a volume-to-surface integral transformation involving the divergence of a vector; that is,

$$\iiint\limits_R \nabla \cdot \mathbf{f} \, dV = \oiint\limits_S d\mathbf{S} \cdot \mathbf{f}$$

By extending the integral definitions of the gradient and a curl of a vector (see (6.35) and (6.37)) to finite volumes, we obtain the gradient and curl theorems.

The **gradient theorem** states that if ϕ is a continuous scalar function in a region R bounded by a closed surface S, then

GRADIENT THEOREM
$$\iiint\limits_R \nabla\phi \, dV = \oiint\limits_S d\mathbf{S} \, \phi \qquad (7.19)$$

The **curl theorem** states that if $\mathbf{f}$ is a continuous vector function in a region R bounded by a closed surface S, then

CURL THEOREM
$$\iiint\limits_R \nabla \times \mathbf{f} \, dV = \oiint\limits_S d\mathbf{S} \times \mathbf{f} \qquad (7.20)$$

Now we have three theorems—the divergence theorem (7.2), the gradient theorem (7.19), and the curl theorem (7.20)–that can be stated in one condensed notational form:

VOLUME-TO-SURFACE INTEGRAL TRANSFORMATION

$$\iiint_R \nabla * a \, dV = \oiint_S dS * a \tag{7.21}$$

where a is any scalar or vector quantity, and the asterisk ($*$) represents any acceptable form of multiplication, i.e., the dot, cross, or simple product.

EXAMPLE 7-9: Prove the gradient theorem (7.19).

Solution: Let $\mathbf{f} = \phi\mathbf{a}$, where $\mathbf{a}$ is any constant vector. Applying the divergence theorem (7.2), we obtain

$$\iiint_R \nabla \cdot (\phi\mathbf{a}) \, dV = \oiint_S \phi\mathbf{a} \cdot d\mathbf{S} \tag{7.22}$$

But by vector identity (5.41),

$$\nabla \cdot (\phi\mathbf{a}) = \phi \nabla \cdot \mathbf{a} + \mathbf{a} \cdot \nabla\phi$$

Since $\mathbf{a}$ is a constant vector, $\nabla \cdot \mathbf{a} = 0$ and hence $\nabla \cdot (\phi\mathbf{a}) = \mathbf{a} \cdot \nabla\phi$. Thus by transposing and factoring, (7.22) can be written as

$$\mathbf{a} \cdot \left(\iiint_R \nabla\phi \, dV - \oiint_S \phi \, d\mathbf{S} \right) = 0 \tag{7.23}$$

Since $\mathbf{a}$ is any constant vector, the expression in parentheses must be $\mathbf{0}$, and the gradient theorem (7.19) is proved.

EXAMPLE 7-10: Prove the curl theorem (7.20).

Solution: If $\mathbf{a}$ is any constant vector and we replace $\mathbf{f}$ in the divergence theorem (7.2) by $\mathbf{f} \times \mathbf{a}$, we obtain

$$\iiint_R \nabla \cdot (\mathbf{f} \times \mathbf{a}) \, dV = \oiint_S (\mathbf{f} \times \mathbf{a}) \cdot d\mathbf{S} \tag{7.24}$$

But by vector identity (5.43),

$$\nabla \cdot (\mathbf{f} \times \mathbf{a}) = \mathbf{a} \cdot (\nabla \times \mathbf{f}) - \mathbf{f} \cdot (\nabla \times \mathbf{a})$$

Since $\mathbf{a}$ is a constant vector, $\nabla \times \mathbf{a} = \mathbf{0}$. This now reduces to

$$\nabla \cdot (\mathbf{f} \times \mathbf{a}) = \mathbf{a} \cdot (\nabla \times \mathbf{f})$$

From the permutation rule of the triple scalar product (see Problem 1-16), we have

$$\mathbf{f} \times \mathbf{a} \cdot d\mathbf{S} = \mathbf{a} \cdot d\mathbf{S} \times \mathbf{f}$$

Thus (7.24) can be written as

$$\mathbf{a} \cdot \left(\iiint_R \nabla \times \mathbf{F} \, dV - \oiint_S d\mathbf{S} \times \mathbf{f} \right) = 0 \tag{7.25}$$

Again, because $\mathbf{a}$ is any constant vector, the expression in parentheses $\mathbf{0}$, and the curl theorem (7.20) is proved.

7-4. Stokes' Theorem

Stokes' Theorem states that if S is a surface bounded by a simple closed curve C and $\mathbf{f}$ is a vector function that has continuous first partial derivatives both on the surface S and its boundary C, then

STOKES' THEOREM
$$\iint_S \nabla \times \mathbf{f} \cdot d\mathbf{S} = \oint_C \mathbf{f} \cdot d\mathbf{r} \tag{7.26}$$

EXAMPLE 7-11: Give the physical interpretation of Stokes' theorem.

Solution: The curl of a vector field $\mathbf{f}$ is the intensity of circulation at a point for $\mathbf{f}$ (see Example 6-8). Stokes' theorem (7.26) states that the total circulation around a closed curve C is equal to the flux of curl $\mathbf{f}$ through a surface S enclosed by C.

EXAMPLE 7-12: Prove Stokes' theorem (7.26).

Solution: Consider a surface S bounded by a simple closed curve C. Divide S into N subregions so small that they can be considered to be planar with areas $\Delta S_1, \Delta S_2, \ldots, \Delta S_N$. At arbitrarily chosen points (x_i, y_i, z_i) within ΔS_i, definition (6.33) of the curl gives

$$\mathbf{n} \cdot \nabla \times \mathbf{f}\, \Delta S_i = \oint_{C_i} \mathbf{f} \cdot d\mathbf{r} + \varepsilon_i \Delta S_i$$

where $\varepsilon_i \to 0$ as $\Delta S_i \to 0$, and $\mathbf{n}$ is the unit normal vector associated with ΔS_i. (See Figure 7-2.) Summation over the entire surface S yields

$$\sum_{i=1}^N \mathbf{n} \cdot \nabla \times \mathbf{f}\, \Delta S_i = \sum_{i=1}^N \oint_{C_i} \mathbf{f} \cdot d\mathbf{r} + \sum_{i=1}^N \varepsilon_i \Delta S_i$$

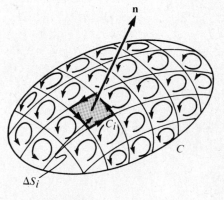

Figure 7-2
Proof of Stokes' theorem.

Now consider the limit of this expression as $N \to \infty$. The boundary C_i of each ΔS_i consists of a number of pieces that are either part of the boundary C or part of the boundaries of the adjacent subregions. The line integrals along the adjacent boundary curves cancel, since the $d\mathbf{r}$ vectors are directed in opposite directions. This leaves only the line integral along the boundary C. Hence we obtain

$$\lim_{N \to \infty} \sum_{i=1}^N \oint_{C_i} \mathbf{f} \cdot d\mathbf{r} = \oint_C \mathbf{f} \cdot d\mathbf{r}$$

as well as

$$\lim_{N \to \infty} \sum_{i=1}^N \mathbf{n} \cdot \nabla \times \mathbf{f}\, \Delta S_i = \iint_S \mathbf{n} \cdot \nabla \times \mathbf{f}\, dS = \iint_S \nabla \times \mathbf{f} \cdot d\mathbf{S}$$

Thus as $N \to \infty$,

$$\iint_S \nabla \times \mathbf{f} \cdot d\mathbf{S} = \oint_C \mathbf{f} \cdot d\mathbf{r} + \lim_{N \to \infty} \sum_{i=1}^N \varepsilon_i \Delta S_i$$

For the remaining term,

$$\left| \sum_{i=1}^N \varepsilon_i \Delta S_i \right| \leqslant \sum_{i=1}^N |\varepsilon_i|\, \Delta S_i \leqslant |\varepsilon_m| \sum_{i=1}^N \Delta S_i = |\varepsilon_m| S$$

where $\varepsilon_m = \max \varepsilon_i$. But $\varepsilon_m \to 0$ as $N \to \infty$ and $\Delta S_i \to 0$. Hence

$$\lim_{N \to \infty} \sum_{i=1}^N \varepsilon_i \Delta S_i \to 0$$

Thus in the limit,

$$\iint_S \mathbf{\nabla} \times \mathbf{f} \cdot d\mathbf{S} = \oint_C \mathbf{f} \cdot d\mathbf{r}$$

EXAMPLE 7-13: Show that Stokes' theorem (7.26) expressed in rectangular coordinates is

$$\oint_C P\,dx + Q\,dy + R\,dz$$

$$= \iint_S \left(\frac{\partial R}{\partial y} - \frac{\partial Q}{\partial z}\right) dy\,dz + \left(\frac{\partial P}{\partial z} - \frac{\partial R}{\partial x}\right) dz\,dx + \left(\frac{\partial Q}{\partial x} - \frac{\partial P}{\partial y}\right) dx\,dy \qquad (7.27)$$

where $\mathbf{f} = P(x, y, z)\mathbf{i} + Q(x, y, z)\mathbf{j} + R(x, y, z)\mathbf{k}$.

Solution: If $\mathbf{f} = P\mathbf{i} + Q\mathbf{j} + R\mathbf{k}$, then $\mathbf{f} \cdot d\mathbf{r} = P\,dx + Q\,dy + R\,dz$ and

$$\mathbf{\nabla} \times \mathbf{f} = \begin{vmatrix} \mathbf{i} & \mathbf{j} & \mathbf{k} \\ \dfrac{\partial}{\partial x} & \dfrac{\partial}{\partial y} & \dfrac{\partial}{\partial z} \\ P & Q & R \end{vmatrix}$$

$$= \left(\frac{\partial R}{\partial y} - \frac{\partial Q}{\partial z}\right)\mathbf{i} + \left(\frac{\partial P}{\partial z} - \frac{\partial R}{\partial x}\right)\mathbf{j} + \left(\frac{\partial Q}{\partial x} - \frac{\partial P}{\partial y}\right)\mathbf{k}$$

Now, as in Example 7-3,

$$\iint_S \mathbf{\nabla} \times \mathbf{f} \cdot d\mathbf{S} = \iint_S \mathbf{\nabla} \times \mathbf{f} \cdot \mathbf{n}\,dS$$

$$= \iint_S \left[\left(\frac{\partial R}{\partial y} - \frac{\partial Q}{\partial z}\right)\mathbf{i} + \left(\frac{\partial P}{\partial z} - \frac{\partial R}{\partial x}\right)\mathbf{j} + \left(\frac{\partial Q}{\partial x} - \frac{\partial P}{\partial y}\right)\mathbf{k}\right] \cdot \mathbf{n}\,dS$$

$$= \iint_S \left[\left(\frac{\partial R}{\partial y} - \frac{\partial Q}{\partial z}\right)\cos\alpha + \left(\frac{\partial P}{\partial z} - \frac{\partial R}{\partial x}\right)\cos\beta\right.$$

$$\left. + \left(\frac{\partial Q}{\partial x} - \frac{\partial P}{\partial y}\right)\cos\gamma\right] dS$$

$$= \iint_S \left(\frac{\partial R}{\partial y} - \frac{\partial Q}{\partial z}\right) dy\,dz + \left(\frac{\partial P}{\partial z} - \frac{\partial R}{\partial x}\right) dz\,dx + \left(\frac{\partial Q}{\partial x} - \frac{\partial P}{\partial y}\right) dx\,dy$$

So Stokes' theorem (7.26) reduces to

$$\oint_C P\,dx + Q\,dy + R\,dz$$

$$= \iint_S \left(\frac{\partial R}{\partial y} - \frac{\partial Q}{\partial z}\right) dy\,dz + \left(\frac{\partial P}{\partial z} - \frac{\partial R}{\partial x}\right) dz\,dx + \left(\frac{\partial Q}{\partial x} - \frac{\partial P}{\partial y}\right) dx\,dy$$

EXAMPLE 7-14: Prove Green's theorem for a plane: If P, Q, $\partial P/\partial y$, and $\partial Q/\partial x$ are continuous functions in a region S in the xy-plane bounded by a closed curve C, then

GREEN'S THEOREM FOR A PLANE
$$\oint_C P\,dx + Q\,dy = \iint_S \left(\frac{\partial Q}{\partial x} - \frac{\partial P}{\partial y}\right) dx\,dy \qquad (7.28)$$

This theorem can also be expressed in vector form: if $\mathbf{f} = P\mathbf{i} + Q\mathbf{j}$, then

$$\oint_C \mathbf{f} \cdot d\mathbf{r} = \iint_S \nabla \times \mathbf{f} \cdot \mathbf{k}\, dx\, dy \qquad (7.29)$$

Solution: If $\mathbf{f} = P\mathbf{i} + Q\mathbf{j}$ and $d\mathbf{S} = \mathbf{k}\, dx\, dy$, then Green's theorem for a plane (7.28) should follow directly from Stokes' theorem (7.26), as we show here.

Now if $\mathbf{f} = P\mathbf{i} + Q\mathbf{j}$, then

$$\nabla \times \mathbf{f} = \begin{vmatrix} \mathbf{i} & \mathbf{j} & \mathbf{k} \\ \dfrac{\partial}{\partial x} & \dfrac{\partial}{\partial y} & \dfrac{\partial}{\partial z} \\ P & Q & 0 \end{vmatrix} = -\frac{\partial Q}{\partial z}\mathbf{i} + \frac{\partial P}{\partial z}\mathbf{j} + \left(\frac{\partial Q}{\partial x} - \frac{\partial P}{\partial y}\right)\mathbf{k}$$

and

$$\mathbf{f} \cdot d\mathbf{r} = P\, dx + Q\, dy$$

Thus

$$\nabla \times \mathbf{f} \cdot d\mathbf{S} = \nabla \times \mathbf{f}\ \ \mathbf{k}\, dx\, dy = \left(\frac{\partial Q}{\partial x} - \frac{\partial P}{\partial y}\right) dx\, dy$$

Hence by Stokes' theorem (7.26), we have

$$\oint_C P\, dx + Q\, dy = \iint_S \left(\frac{\partial Q}{\partial x} - \frac{\partial P}{\partial y}\right) dx\, dy$$

The form given in equation (7.29) follows directly from these computations.

EXAMPLE 7-15: Show that $\nabla \times \mathbf{f} = \mathbf{0}$ is a necessary and sufficient condition for $\oint_C \mathbf{f} \cdot d\mathbf{r} = 0$ around any closed curve C.

Solution: For the sufficient condition we have $\nabla \times \mathbf{f} = \mathbf{0}$. Then by Stokes' theorem (7.26) we conclude that

$$\oint_C \mathbf{f} \cdot d\mathbf{r} = \iint_S (\nabla \times \mathbf{f}) \cdot d\mathbf{S} = 0$$

For the necessary condition assume that $\oint_C \mathbf{f} \cdot d\mathbf{r} = 0$ for any closed curve C and that $\nabla \times \mathbf{f} \neq \mathbf{0}$ at some point P. Then if $\nabla \times \mathbf{f}$ is continuous, there is some region about P where $\nabla \times \mathbf{f} \neq \mathbf{0}$. Choose a small plane surface S in this region and a unit normal vector to S parallel to $\nabla \times \mathbf{f}$; that is, $\nabla \times \mathbf{f} \cong a\mathbf{n}$, where $a > 0$. If C is the boundary of S, then by Stokes' theorem (7.26),

$$\oint_C \mathbf{f} \cdot d\mathbf{r} = \iint_S \nabla \times \mathbf{f} \cdot d\mathbf{S} \cong \iint_S a\mathbf{n} \cdot \mathbf{n}\, dS = a \iint_S dS = aS > 0$$

which contradicts the assumption that $\oint_C \mathbf{f} \cdot d\mathbf{r} = 0$.

7-5. Surface-to-Line Integral Transformations

Stokes' theorem (7.26) represents a surface-to-line integral transformation involving the curl of a vector; that is,

$$\iint_S \nabla \times \mathbf{f} \cdot d\mathbf{S} = \oint_C \mathbf{f} \cdot d\mathbf{r}$$

By the scalar triple product identity (1.53) we can write

$$\iint_S \mathbf{\nabla} \times \mathbf{f} \cdot d\mathbf{S} = \iint_S d\mathbf{S} \cdot \mathbf{\nabla} \times \mathbf{f} = \iint_S (d\mathbf{S} \times \mathbf{\nabla}) \cdot \mathbf{f}$$

Hence Stokes' theorem (7.26) can be stated as

$$\iint_S (d\mathbf{S} \times \mathbf{\nabla}) \cdot \mathbf{f} = \oint_C d\mathbf{r} \cdot \mathbf{f} \tag{7.30}$$

By extending the integral definitions of the curl of the gradient and the curl of the curl to finite surfaces, we obtain, respectively, the following theorems:

If S is a finite surface bounded by a simple closed curve C and ϕ is a scalar function with continuous derivatives, then

$$\iint_S d\mathbf{S} \times \mathbf{\nabla}\phi = \oint_C \phi\, d\mathbf{r} \tag{7.31}$$

If S is a finite surface bounded by a closed curve C and $\mathbf{f}$ is a vector function with continuous derivatives, then

$$\iint_S (d\mathbf{S} \times \mathbf{\nabla}) \times \mathbf{f} = \oint_C d\mathbf{r} \times \mathbf{f} \tag{7.32}$$

Now the three theorems (7.30), (7.31), and (7.32) can be stated in condensed form as

SURFACE-TO-LINE INTEGRAL TRANSFORMATION

$$\iint_S (d\mathbf{S} \times \mathbf{\nabla}) * \mathfrak{a} = \oint_C d\mathbf{r} * \mathfrak{a} \tag{7.33}$$

where $\mathfrak{a}$ is any scalar or vector quantity and the asterisk ($*$) represents any acceptable form of multiplication, i.e., dot, cross, or simple products.

EXAMPLE 7-16: Verify theorem (7.31).

Solution: Let $\mathbf{f} = \phi\mathbf{a}$, where $\mathbf{a}$ is any constant nonzero vector. Applying Stokes' theorem (7.26), we have

$$\iint_S \mathbf{\nabla} \times (\phi\mathbf{a}) \cdot d\mathbf{S} = \oint_C \phi\mathbf{a} \cdot d\mathbf{r} \tag{7.34}$$

From vector identity (5.42),

$$\mathbf{\nabla} \times (\phi\mathbf{a}) = \phi\mathbf{\nabla} \times \mathbf{a} - \mathbf{a} \times \mathbf{\nabla}\phi$$

Since $\mathbf{a}$ is a constant vector, $\mathbf{\nabla} \times \mathbf{a} = \mathbf{0}$; hence

$$\mathbf{\nabla} \times (\phi\mathbf{a}) = \mathbf{\nabla}\phi \times \mathbf{a}$$

From the permutation rule of the triple scalar product (see Problem 1-16), we have

$$\iint_S (\mathbf{\nabla}\phi \times \mathbf{a}) \cdot d\mathbf{S} = \iint_S \mathbf{a} \cdot d\mathbf{S} \times \mathbf{\nabla}\phi$$

So (7.34) reduces to

$$\mathbf{a} \cdot \iint_S d\mathbf{S} \times \mathbf{\nabla}\phi = \mathbf{a} \cdot \oint_C \phi\, d\mathbf{r}$$

or

$$\mathbf{a} \cdot \left(\iint_S d\mathbf{S} \times \nabla\phi - \oint_C \phi\, d\mathbf{r} \right) = 0 \qquad (7.35)$$

Since **a** is any constant nonzero vector, the expression in parentheses vanishes, and theorem (7.31) is proved.

EXAMPLE 7-17: Verify theorem (7.32).

Solution: If **a** is any constant nonzero vector and we replace **f** in (7.26) by **f** × **a**, we have

$$\iint_S \nabla \times (\mathbf{f} \times \mathbf{a}) \cdot d\mathbf{S} = \oint_C (\mathbf{f} \times \mathbf{a}) \cdot d\mathbf{r} \qquad (7.36)$$

From the result of Problem 5-23 we have

$$\nabla \times (\mathbf{f} \times \mathbf{a}) = \mathbf{f}(\nabla \cdot \mathbf{a}) - \mathbf{a}(\nabla \cdot \mathbf{f}) + (\mathbf{a} \cdot \nabla)\mathbf{f} - (\mathbf{f} \cdot \nabla)\mathbf{a}$$

Since **a** is a constant vector, $\nabla \cdot \mathbf{a} = 0$ and $(\mathbf{f} \cdot \nabla)\mathbf{a} = \mathbf{0}$. Hence

$$\nabla \times (\mathbf{f} \times \mathbf{a}) = (\mathbf{a} \cdot \nabla)\mathbf{f} - \mathbf{a}(\nabla \cdot \mathbf{f})$$

Thus

$$\iint_S \nabla \times (\mathbf{f} \times \mathbf{a}) \cdot d\mathbf{S} = \iint_S [(\mathbf{a} \cdot \nabla)\mathbf{f} \cdot d\mathbf{S} - \mathbf{a} \cdot d\mathbf{S}(\nabla \cdot \mathbf{f})]$$

$$= \mathbf{a} \cdot \iint_S [\nabla_\mathbf{f}(\mathbf{f} \cdot d\mathbf{S}) - d\mathbf{S}(\nabla \cdot \mathbf{f})]$$

where $\nabla_\mathbf{f}$ indicates that ∇ operates only on **f**.

Applying the middle-term rule for the triple vector product (1.60) to $(d\mathbf{S} \times \nabla) \times \mathbf{f}$, we have

$$(d\mathbf{S} \times \nabla) \times \mathbf{f} = \nabla_\mathbf{f}(\mathbf{f} \cdot d\mathbf{S}) - d\mathbf{S}(\nabla \cdot \mathbf{f})$$

and using this, we can rewrite the above expression as

$$\iint_S \nabla \times (\mathbf{f} \times \mathbf{a}) \cdot d\mathbf{S} = \mathbf{a} \cdot \iint_S (d\mathbf{S} \times \nabla) \times \mathbf{f}$$

From the permutation rule of the triple scalar product (see Problem 1-16), we have

$$\oint_C (\mathbf{f} \times \mathbf{a}) \cdot d\mathbf{r} = \oint_C \mathbf{a} \cdot (d\mathbf{r} \times \mathbf{f}) = \mathbf{a} \cdot \oint_C d\mathbf{r} \times \mathbf{f}$$

So (7.36) reduces to

$$\mathbf{a} \cdot \iint_S (d\mathbf{S} \times \nabla) \times \mathbf{f} = \mathbf{a} \cdot \oint_C d\mathbf{r} \times \mathbf{f}$$

or by transposing and factoring

$$\mathbf{a} \cdot \left(\iint_S (d\mathbf{S} \times \nabla) \times \mathbf{f} - \oint_C d\mathbf{r} \times \mathbf{f} \right) = 0 \qquad (7.37)$$

Since **a** is any constant nonzero vector, the expression in parentheses vanishes, and theorem (7.32) is proved.

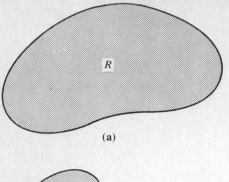

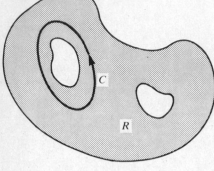

(a)

(b)

Figure 7-3
(a) A simply connected region; (b) a region that is not simply connected.

7-6. Irrotational and Solenoidal Fields

A. Simply connected region

A region R is said to be **connected** if any two points in R can be joined by an arc so that every point on the arc belongs to the region R. (Think of the region as being in "one piece.") A region R is said to be **simply connected** if every closed curve in R can be continuously shrunk to a point in R. The region R in Figure 7-3a is simply connected. However, the region R in Figure 7-3b is not simply connected because the closed curve C that surrounds one of the "holes" cannot be continuously shrunk to a point without leaving R.

B. Scalar potential

A **scalar potential function** ϕ is a single-valued function for which there exists a continuous vector field $\mathbf{f}$ in a simply connected region that satisfies the relation

SCALAR POTENTIAL $$\mathbf{f} = \nabla\phi = \operatorname{grad}\phi \tag{7.38}$$

EXAMPLE 7-18: Show that the necessary and sufficient condition for the line integral $\int_P^Q \mathbf{f}\cdot d\mathbf{r}$ to be independent of the path of integration from the point P to the point Q is that the continuous vector field $\mathbf{f}$ satisfy relation (7.38).

Solution: To prove sufficiency, assume that $\mathbf{f} = \nabla\phi$. Then by the result of Problem 5-4,

$$\mathbf{f}\cdot d\mathbf{r} = \nabla\phi\cdot d\mathbf{r} = d\phi$$

and hence

$$\int_P^Q \mathbf{f}\cdot d\mathbf{r} = \int_P^Q d\phi = \phi(Q) - \phi(P) \tag{7.39}$$

If ϕ is a single-valued function, the right-hand side of (7.39) has a definite value that depends only on the endpoints P and Q and not on the path. This shows that the line integral is simply the potential difference between the path endpoints.

To prove necessity, assume that $\int_P^Q \mathbf{f}\cdot d\mathbf{r}$ is independent of the path of integration. Let

$$\phi(Q) = \int_P^Q \mathbf{f}\cdot d\mathbf{r}$$

where P is a fixed point and Q is a variable point in R. Since the line integral is independent of the path, Q moves along a curve through P on which the unit tangent vector $\mathbf{T}$ is continuous (see Figure 7-4). Along this curve

$$\phi = \int_P^Q \mathbf{f}\cdot d\mathbf{r} = \int_P^Q \mathbf{f}\cdot\frac{d\mathbf{r}}{ds}\,ds = \int_P^Q \mathbf{f}\cdot\mathbf{T}\,ds$$

and ϕ is a function of the arc length s. Thus

$$\frac{\partial\phi}{\partial s} = \mathbf{f}\cdot\frac{d\mathbf{r}}{ds} = \mathbf{f}\cdot\mathbf{T} \tag{7.40}$$

The curve from P to Q could be chosen to have any given direction at Q. Hence (7.40) shows that ϕ has a continuous directional derivative in any direction. But by (5.5) we have

$$\frac{\partial\phi}{\partial s} = \nabla\phi\cdot\mathbf{T}$$

Now comparing (7.40) and (5.5), we obtain

$$(\mathbf{f} - \nabla\phi)\cdot\mathbf{T} = 0 \tag{7.41}$$

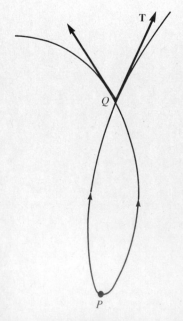

Figure 7-4

Since (7.41) holds for every direction of **T**, we conclude that

$$\mathbf{f} - \nabla\phi = 0 \quad \text{or} \quad \mathbf{f} = \nabla\phi$$

EXAMPLE 7-19: If $\mathbf{f} = \nabla\phi$ everywhere in a simply connected region R and C is any closed curve in R, show that

$$\oint_C \mathbf{f} \cdot d\mathbf{r} = 0 \tag{7.42}$$

Solution: If $\mathbf{f} = \nabla\phi$ everywhere in R, then from Example 7-18 the line integral $\int_C \mathbf{f} \cdot d\mathbf{r}$ is independent of the path of integration. If the path of integration is closed, then $P = Q$ in (7.39). Thus

$$\oint_C \mathbf{f} \cdot d\mathbf{r} = \oint_C \nabla\phi \cdot d\mathbf{r} = \int_P^P d\phi = \phi(P) - \phi(P) = 0$$

C. Irrotational field

A vector field **f** is said to be **irrotational** in a region R if

IRROTATIONAL FIELD
$$\nabla \times \mathbf{f} = 0 \tag{7.43}$$

everywhere in R. For a vector field **f** to be irrotational in a simply connected region R, any of the following three conditions is necessary and sufficient:

(1) $\nabla \times \mathbf{f} = 0$
(2) **f** is the gradient of a scalar field; i.e., $\mathbf{f} = \nabla\phi$
(3) for every simple closed curve C in R, $\oint_c \mathbf{f} \cdot d\mathbf{r} = 0$

Note: Condition (2) follows by identity (5.31) and condition (3) follows from the result of Example 7-19.

D. Vector potential

A **vector potential function A** in a simply connected region R in space is a vector field for which there exists a vector field **f** satisfying

VECTOR POTENTIAL
$$\mathbf{f} = \nabla \times \mathbf{A} \tag{7.44}$$

A simply connected region R in space is a region in which every closed surface S in R must bound a volume V also in R. (Briefly, R is a simply connected region if it has no holes.)

EXAMPLE 7-20: Show that there is no unique vector potential **A** for which condition (7.44) holds.

Solution: Let $\mathbf{A}' = \mathbf{A} + \nabla\psi$, where ψ is an arbitrary scalar function. Then by identity (5.31) we have

$$\nabla \times \mathbf{A}' = \nabla \times (\mathbf{A} + \nabla\psi) = \nabla \times \mathbf{A} + \nabla \times \nabla\psi = \nabla \times \mathbf{A} \tag{7.45}$$

So, there is no unique **A** for which (7.44) holds. (See Example 5-14.)

E. Solenoidal field

A vector field **f** is said to be **solenoidal** if everywhere in a simply connected region R

SOLENOIDAL FIELD
$$\nabla \cdot \mathbf{f} = 0 \tag{7.46}$$

If a solenoidal field **f** is defined in the region R, then **f** has the following properties:

(1) $\nabla \cdot \mathbf{f} = 0$

(2) $\mathbf{f}$ is the curl of some vector potential function, i.e., $\mathbf{f} = \nabla \times \mathbf{A}$

(3) for every closed surface S in R, $\oiint_S \mathbf{f} \cdot d\mathbf{S} = 0$

Note: Condition (2) follows from the identity $\nabla \cdot (\nabla \times \mathbf{A}) = 0$ (5.32), and condition (3) follows from the divergence theorem (7.2).

F. The uniqueness theorem

The **uniqueness theorem** states that a vector $\mathbf{f}$ is uniquely determined in a region R enclosed by a surface S if the normal component of $\mathbf{f}$ is specified on S and if $\nabla \cdot \mathbf{f}$ and $\nabla \times \mathbf{f}$ are specified throughout R.

EXAMPLE 7-21: Verify the uniqueness theorem.

Proof: We prove this theorem by contradiction. Assume that $\mathbf{f}$ and $\mathbf{g}$ are two distinct vectors that satisfy the given conditions; i.e, if $\mathbf{n}$ is the unit normal vector to S, we have

$$\nabla \cdot \mathbf{f} = \nabla \cdot \mathbf{g}, \quad \nabla \times \mathbf{f} = \nabla \times \mathbf{g} \quad \text{in } R$$

and

$$\mathbf{f} \cdot \mathbf{n} = \mathbf{g} \cdot \mathbf{n} \quad \text{on } S$$

Then setting $\mathbf{h} = \mathbf{f} - \mathbf{g}$, we get

$$\left. \begin{array}{l} \nabla \cdot \mathbf{h} = \nabla \cdot \mathbf{f} - \nabla \cdot \mathbf{g} = 0 \\ \nabla \times \mathbf{h} = \nabla \times \mathbf{f} - \nabla \times \mathbf{g} = \mathbf{0} \end{array} \right\} \quad \text{in } R \qquad (7.47)$$

and

$$\mathbf{h} \cdot \mathbf{n} = \mathbf{f} \cdot \mathbf{n} - \mathbf{g} \cdot \mathbf{n} = 0 \quad \text{on } S \qquad (7.48)$$

Since $\nabla \times \mathbf{h} = \mathbf{0}$, the vector $\mathbf{h}$ is irrotational i.e., there exists a scalar function ϕ such that

$$\mathbf{h} = \nabla\phi \qquad (7.49)$$

Now

$$\nabla \cdot \mathbf{h} = \nabla \cdot (\nabla\phi) = \nabla^2\phi = 0 \quad \text{in } R \qquad (7.50)$$

and

$$\mathbf{h} \cdot \mathbf{n} = \nabla\phi \cdot \mathbf{n} = \frac{\partial\phi}{\partial n} = 0 \quad \text{on } S \qquad (7.51)$$

Next, setting $\psi = \phi$ in (7.10), we have

$$\iiint_R (\phi \nabla^2\phi + \nabla\phi \cdot \nabla\phi)\, d\mathbf{V} = \oiint_S \phi \frac{\partial\phi}{\partial n}\, dS \qquad (7.52)$$

or

$$\iiint_R \phi \nabla^2\phi\, dV + \iiint_R |\nabla\phi|^2\, dV = \oiint_S \phi \frac{\partial\phi}{\partial n}\, dS \qquad (7.53)$$

Applying the conditions (7.50) and (7.51), eq. (7.53) reduces to

$$\iiint_R |\nabla\phi|^2\, dV = 0 \qquad (7.54)$$

But since the integrand $|\nabla \phi|^2$ is nonnegative, we must have $|\nabla \phi|^2 = 0$, or

$$\mathbf{h} = \nabla \phi = \mathbf{0} \tag{7.55}$$

Hence, $\mathbf{f} = \mathbf{g}$, which contradicts our original assumption that $\mathbf{f}$ and $\mathbf{g}$ are distinct. Therefore the theorem is valid.

SUMMARY

1. The divergence theorem (or Gauss' theorem) is

$$\iiint_R \nabla \cdot \mathbf{f} \, dV = \oiint_S \mathbf{f} \cdot d\mathbf{S}$$

2. Green's first theorem is

$$\iiint_R (\phi \, \nabla^2 \psi + \nabla \phi \cdot \nabla \psi) \, dV = \oiint_S \phi \, \nabla \psi \cdot d\mathbf{S}$$

3. Green's second theorem is

$$\iiint_R (\phi \, \nabla^2 \psi - \psi \, \nabla^2 \phi) \, dV = \oiint_S (\phi \, \nabla \psi - \psi \, \nabla \phi) \cdot d\mathbf{S}$$

4. Green's third theorem is

$$\iiint_R [\mathbf{f} \cdot \nabla^2 \mathbf{g} - \mathbf{g} \cdot \nabla^2 \mathbf{f}] \, dV$$

$$= \oiint_S [\mathbf{f} \times (\nabla \times \mathbf{g}) + \mathbf{f}(\nabla \cdot \mathbf{g}) - \mathbf{g} \times (\nabla \times \mathbf{f}) - \mathbf{g}(\nabla \cdot \mathbf{f})] \cdot d\mathbf{S}$$

5. The gradient theorem is

$$\iiint_R \nabla \phi \, dV = \oiint_S d\mathbf{S} \, \phi$$

6. The curl theorem is

$$\iiint_R \nabla \times \mathbf{f} \, dV = \oiint_S d\mathbf{S} \times \mathbf{f}$$

7. Volume-to-surface integral transformation is given by

$$\iiint_R \nabla * \mathfrak{a} \, dV = \oiint_S d\mathbf{S} * \mathfrak{a}$$

where $\mathfrak{a}$ is any scalar or vector quantity and the asterisk (*) represents any acceptable form of multiplication (dot, cross, or simple product).

8. Stokes' theorem is

$$\iint_S \nabla \times \mathbf{f} \cdot d\mathbf{S} = \oint_C \mathbf{f} \cdot d\mathbf{r}$$

9. Surface-to-line integral transformation is given by

$$\iint_S (d\mathbf{S} \times \nabla) * \mathfrak{a} = \oint_C d\mathbf{r} * \mathfrak{a}$$

where a is any scalar or vector quantity and the asterisk ($*$) represents any acceptable form of multiplication, (dot, cross, or simple product).

10. For a given vector $\mathbf{f}$, if $\mathbf{f} = \nabla\phi$, then ϕ is called the scalar potential, and if $\mathbf{f} = \nabla \times \mathbf{A}$, then $\mathbf{A}$ is called the vector potential.

11. A vector field $\mathbf{f}$ is irrotational in a region R if $\nabla \times \mathbf{f} = \mathbf{0}$.

12. A vector field $\mathbf{f}$ is solenoidal in a region R if $\nabla \cdot \mathbf{f} = 0$.

RAISE YOUR GRADES

Can you explain . . . ?

☑ how to evaluate a surface integral $\oiint_S \mathbf{f} \cdot d\mathbf{S}$ without actually integrating over the closed surface S

☑ how to evaluate a line integral $\oint_C \mathbf{f} \cdot d\mathbf{r}$ without actually integrating over the closed curve C

☑ what the necessary and sufficient condition is for the line integral $\int_P^Q \mathbf{f} \cdot d\mathbf{r}$ to be independent of the path of integration

☑ how to use the simple test to find whether a vector field is the gradient of a scalar field ϕ.

☑ what the necessary and sufficient condition is for a vector field to be irrotational

☑ what the necessary and sufficient condition is for a vector field to be solenoidal

---

SOLVED PROBLEMS

Divergence Theorem (or Gauss' Theorem)

PROBLEM 7-1   Show that

$$\iiint_R \frac{1}{r^2}\, dV = \oiint_S \frac{\mathbf{r} \cdot d\mathbf{S}}{r^2}$$

where S encloses the region R, $\mathbf{r}$ is the position vector, and $|\mathbf{r}| = r$.

Solution: From the result of Problem 5-18, we have

$$\nabla \cdot [r^{n-1}\mathbf{r}] = (n + 2)r^{n-1}$$

If $n = -1$,

$$\nabla \cdot \left(\frac{\mathbf{r}}{r^2}\right) = (-1 + 2)r^{-2} = \frac{1}{r^2}$$

Hence, applying the divergence theorem (7.2), we obtain

$$\iiint_R \nabla \cdot \left(\frac{\mathbf{r}}{r^2}\right) dV = \iiint_R \frac{1}{r^2}\, dV = \oiint_S \frac{\mathbf{r} \cdot d\mathbf{S}}{r^2}$$

PROBLEM 7-2   Find $\iint_S \mathbf{r} \cdot d\mathbf{S}$ over the surface S of the sphere $x^2 + y^2 + z^2 = a^2$. (See Problem 6-11.)

Solution: The volume V of a sphere with radius a is $V = \frac{4}{3}\pi a^3$. So by formula (7.6),

$$\oiint_S \mathbf{r} \cdot d\mathbf{S} = 3V = 3\frac{4}{3}\pi a^3 = 4\pi a^3$$

PROBLEM 7-3   Evaluate $\iint_s \mathbf{r} \cdot d\mathbf{S}$ where S is the surface of the unit cube bounded by the planes $x = 0$, $x = 1$, $y = 0$, $y = 1$, $z = 0$, $z = 1$ (as shown in Figure 6-6), and $\mathbf{r}$ is the position vector. (See Problem 6-12.)

Solution: Since the volume of the cube is 1, by formula (7.6), we obtain

$$\oiint_S \mathbf{r} \cdot d\mathbf{S} = 3V = 3 \cdot 1 = 3$$

PROBLEM 7-4   Show that for any closed surface S

$$\oiint_S \mathbf{V} \times \mathbf{f} \cdot d\mathbf{S} = 0$$

Solution: By the divergence theorem (7.2) we have

$$\oiint_S \mathbf{V} \times \mathbf{f} \cdot d\mathbf{S} = \iiint_R \mathbf{V} \cdot (\mathbf{V} \times \mathbf{f}) \, dV$$

But the divergence of the curl of $\mathbf{f}$ is $\mathbf{V} \cdot (\mathbf{V} \times \mathbf{f}) = 0$ by condition (5.32). So

$$\oiint_S \mathbf{V} \times \mathbf{f} \cdot d\mathbf{S} = 0$$

PROBLEM 7-5   Using the divergence theorem (7.5), evaluate

$$I = \iint_S x \, dy \, dz + y \, dz \, dx + 2z \, dx \, dy$$

where S is a closed surface that consists of the surface of the paraboloid $x^2 + y^2 = 1 - z$, $0 < z < 1$, and the disk $x^2 + y^2 \leqslant 1$, $z = 0$, as shown in Figure 7-5. (See Problem 6-13.)

Solution: From the divergence theorem expressed in rectangular coordinates (7.5) we have

$$I = \iiint_R \left[\frac{\partial x}{\partial x} + \frac{\partial y}{\partial y} + \frac{\partial(2z)}{\partial z} \right] dx \, dy \, dz = \iiint 4 \, dx \, dy \, dz$$

$$= 4 \iiint_R dV = 4V$$

$$= 4 \int_0^1 \pi (\sqrt{1 - z})^2 \, dz$$

$$= 4 \int_0^1 \pi (1 - z) \, dz$$

$$= 4 \left(\pi - \frac{\pi}{2} \right)$$

$$= 2\pi$$

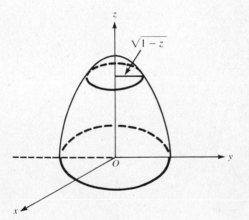

Figure 7-5

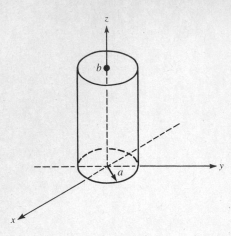

Figure 7-6

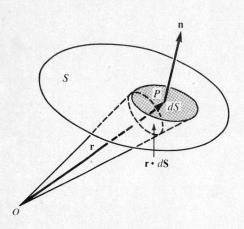

Figure 7-7
Solid angle.

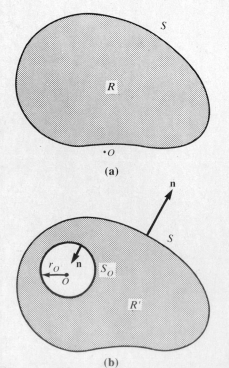

(a)

(b)

Figure 7-8

which is the same result obtained in Problem 6-13. [Note that the contribution from the disk $x^2 + y^2 \leqslant 1$, $z = 0$, is zero, since on this disk $\mathbf{n} = -\mathbf{k}$ and $\mathbf{f} \cdot \mathbf{n}|_{z=0} = -2z|_{z=0} = 0$.]

PROBLEM 7-6 Using the divergence theorem (7.5), evaluate

$$I = \iint_S (x^3 \, dy \, dz + x^2 y \, dz \, dx + x^2 z \, dx \, dy)$$

where S is the closed surface that consists of the cylinder $x^2 + y^2 = a^2$, $0 \leqslant z \leqslant b$, and the circular disks $x^2 + y^2 \leqslant a^2$, $z = 0$, and $x^2 + y^2 \leqslant a^2$, $z = b$, as shown in Figure 7-6.

Solution: From the divergence theorem (7.5) we have

$$I = \iiint_R \left(\frac{\partial x^3}{\partial x} + \frac{\partial x^2 y}{\partial y} + \frac{\partial x^2 z}{\partial z} \right) dx \, dy \, dz = 5 \iiint_R x^2 \, dx \, dy \, dz$$

$$= 5 \int_0^b \int_{-a}^a \int_{-\sqrt{a^2-y^2}}^{\sqrt{a^2-y^2}} x^2 \, dx \, dy \, dz = (5)(4) \int_0^b \int_0^a \int_0^{\sqrt{a^2-y^2}} x^2 \, dx \, dy \, dz$$

$$= 20 \int_0^b \int_0^a \frac{1}{3} (a^2 - y^2)^{3/2} \, dy \, dz = \frac{20b}{3} \int_0^a (a^2 - y^2)^{3/2} \, dy = \frac{20b}{3} \frac{3}{16} \pi a^4$$

$$= \frac{5}{4} \pi a^4 b$$

PROBLEM 7-7 In Figure 7-7 $\mathbf{r}$ is the position vector that represents a point P on the surface S. The *solid angle* $d\Omega$ subtended at the origin O by an element dS of the surface S is defined as

$$d\Omega = \frac{\mathbf{r} \cdot d\mathbf{S}}{r^3}$$

The total solid angle Ω subtended by the surface S is given by

$$\Omega = \iint_S \frac{\mathbf{r} \cdot d\mathbf{S}}{r^3}$$

Show that the total solid angle subtended by a closed surface S at the origin O is zero if O lies outside the region R bounded by S, and that it is 4π if O lies inside R.

Solution: From the divergence theorem (7.2) we have

$$\oiint_S \frac{\mathbf{r} \cdot d\mathbf{S}}{r^3} = \iiint_R \mathbf{\nabla} \cdot \left(\frac{\mathbf{r}}{r^3} \right) dV$$

However, by the result of Problem 5-18, $\mathbf{\nabla} \cdot (\mathbf{r}/r^3) = 0$ if $r \neq 0$. Now if O lies outside R, then $r \neq 0$, as shown in Figure 7-8a. Hence when O lies outside R, the total solid angle subtended by S is

$$\Omega = \oiint_S \frac{\mathbf{r} \cdot d\mathbf{S}}{r^3} = 0$$

If O lies inside R, we construct a small sphere S_O of radius r_O and consider the region R' bounded by S and S_O, as shown in Figure (7-8b). If we apply the divergence theorem (7.2), we have

$$\oiint_{S+S_O} \frac{\mathbf{r} \cdot d\mathbf{S}}{r^3} = \iint_S \frac{\mathbf{r} \cdot d\mathbf{S}}{r^3} + \iint_{S_O} \frac{\mathbf{r} \cdot d\mathbf{S}}{r^3} = \iiint_{R'} \mathbf{\nabla} \cdot \left(\frac{\mathbf{r}}{r^3} \right) dV = 0$$

since $r \neq 0$ within R'. Thus

$$\oiint_{S} \frac{\mathbf{r} \cdot d\mathbf{S}}{r^3} = - \oiint_{S_O} \frac{\mathbf{r} \cdot d\mathbf{S}}{r^3}$$

At points on S_O, $r = r_O$ and the positive or outward normal is directed toward O; that is, $\mathbf{n} = -\mathbf{r}/r_O$. Hence

$$\oiint_{S_O} \frac{\mathbf{r} \cdot d\mathbf{S}}{r^3} = \iint_{S_O} \frac{\mathbf{r} \cdot \mathbf{n} \, dS}{r^3} = - \oiint_{S_O} \frac{dS}{r_O^2} = -\frac{1}{r_O^2} \oiint_{S_O} dS = -\frac{1}{r_O^2} 4\pi r_O^2 = -4\pi$$

and when O is inside R, the total angle subtended by S is

$$\Omega = \oiint_{S} \frac{\mathbf{r} \cdot d\mathbf{S}}{r^3} = - \oiint_{S_O} \frac{\mathbf{r} \cdot d\mathbf{S}}{r^3} = -(-4\pi) = 4\pi$$

Green's Theorems

PROBLEM 7-8   If ϕ and ψ are harmonic in a region R enclosed by S, prove that

$$\oiint_{S} (\phi \, \nabla \psi - \psi \, \nabla \phi) \cdot d\mathbf{S} = \oiint_{S} \left(\phi \, \frac{\partial \psi}{\partial n} - \psi \, \frac{\partial \phi}{\partial n} \right) dS = 0$$

Solution: If ϕ and ψ are harmonic, then by definition (5.22) we have $\nabla^2 \phi = \nabla^2 \psi = 0$. Thus by Green's second theorem (7.8) we have

$$\oiint_{S} (\phi \, \nabla \psi - \psi \, \nabla \phi) \cdot d\mathbf{S} = \iiint_{R} (\phi \, \nabla^2 \psi - \psi \, \nabla^2 \phi) \, dV = 0$$

PROBLEM 7-9   Derive Green's third theorem (7.9).

Solution: Applying the divergence theorem (7.2) to the vectors $\mathbf{f}(\nabla \cdot \mathbf{g})$ and $\mathbf{f} \times (\nabla \times \mathbf{g})$, we have

$$\iiint_{R} \nabla \cdot [\mathbf{f}(\nabla \cdot \mathbf{g})] \, dV = \oiint_{S} \mathbf{f}(\nabla \cdot \mathbf{g}) \cdot d\mathbf{S} \qquad \text{(a)}$$

and

$$\iiint_{R} \nabla \cdot [\mathbf{f} \times (\nabla \times \mathbf{g})] \, dV = \oiint_{S} \mathbf{f} \times (\nabla \times \mathbf{g}) \cdot d\mathbf{S} \qquad \text{(b)}$$

From the vector identities (5.41) and (5.43) we have

$$\nabla \cdot [\mathbf{f}(\nabla \cdot \mathbf{g})] = (\nabla \cdot \mathbf{g})(\nabla \cdot \mathbf{f}) + \mathbf{f} \cdot \nabla(\nabla \cdot \mathbf{g}) \qquad \text{(c)}$$

$$\nabla \cdot [\mathbf{f} \times (\nabla \times \mathbf{g})] = (\nabla \times \mathbf{g}) \cdot (\nabla \times \mathbf{f}) - \mathbf{f} \cdot \nabla \times (\nabla \times \mathbf{g}) \qquad \text{(d)}$$

Hence

$$\iiint_{R} \nabla \cdot [\mathbf{f}(\nabla \cdot \mathbf{g})] \, dV = \iiint_{R} [(\nabla \cdot \mathbf{g})(\nabla \cdot \mathbf{f}) + \mathbf{f} \cdot \nabla(\nabla \cdot \mathbf{g})] \, dV \qquad \text{(e)}$$

$$\iiint_{R} \nabla \cdot [\mathbf{f} \times (\nabla \times \mathbf{g})] \, dV = \iiint_{R} [(\nabla \times \mathbf{g}) \cdot (\nabla \times \mathbf{f}) - \mathbf{f} \cdot \nabla \times (\nabla \times \mathbf{g})] \, dV \qquad \text{(f)}$$

Interchanging the roles of **f** and **g** and subtracting, we have

$$\iiint_R [\mathbf{f} \cdot \nabla(\nabla \cdot \mathbf{g}) - \mathbf{g} \cdot \nabla(\nabla \cdot \mathbf{f})]\, dV = \oiint_S [\mathbf{f}(\nabla \cdot \mathbf{g}) - \mathbf{g}(\nabla \cdot \mathbf{f})] \cdot d\mathbf{S} \qquad (g)$$

$$\iiint_R [\mathbf{f} \cdot \nabla \times (\nabla \times \mathbf{g}) - \mathbf{g} \cdot \nabla \times (\nabla \times \mathbf{f})]\, dV$$

$$= -\oiint_S [\mathbf{f} \times (\nabla \times \mathbf{g}) - \mathbf{g} \times (\nabla \times \mathbf{f})] \cdot d\mathbf{S} \qquad (h)$$

Adding eq. (g) and eq. (h) and using the identity $\nabla^2 \mathbf{f} = \nabla(\nabla \cdot \mathbf{f}) - \nabla \times (\nabla \times \mathbf{f})$ (5.45), we obtain

$$\iiint_R (\mathbf{f} \cdot \nabla^2 \mathbf{g} - \mathbf{g} \cdot \nabla^2 \mathbf{f})\, dV$$

$$= \oiint_S [\mathbf{f} \times (\nabla \times \mathbf{g}) + \mathbf{f}(\nabla \cdot \mathbf{g}) - \mathbf{g} \times (\nabla \times \mathbf{f}) - \mathbf{g}(\nabla \cdot \mathbf{f})] \cdot d\mathbf{S}$$

which is Green's third theorem (7.9).

Volume-to-Surface Integral Transformations

PROBLEM 7-10   Show that for a closed surface S

$$\oiint_S d\mathbf{S} = \mathbf{0}$$

Solution: By the gradient theorem (7.19) we have

$$\oiint_S \phi\, d\mathbf{S} = \iiint_R \nabla\phi\, dV$$

If $\phi = 1$, then $\nabla\phi = \mathbf{0}$; hence

$$\oiint_S d\mathbf{S} = \mathbf{0}$$

PROBLEM 7-11   If **r** is a position vector, show that for a closed surface S

$$\oiint_S \mathbf{r} \times d\mathbf{S} = \mathbf{0}$$

Solution: By the curl theorem (7.20) we have

$$\oiint_S d\mathbf{S} \times \mathbf{r} = \iiint_R \nabla \times \mathbf{r}\, dV$$

But by expression (5.29) $\nabla \times \mathbf{r} = \mathbf{0}$: hence

$$\oiint_S \mathbf{r} \times d\mathbf{S} = -\oiint_S d\mathbf{S} \times \mathbf{r} = \mathbf{0}$$

Stokes' Theorem

PROBLEM 7-12 Show that if **r** is the position vector, then

$$\oint_C \mathbf{r} \cdot d\mathbf{r} = 0$$

Solution: From expression (5.29) $\nabla \times \mathbf{r} = \mathbf{0}$. So by Stokes' theorem (7.26) we have

$$\oint_C \mathbf{r} \cdot d\mathbf{r} = \iint_S \nabla \times \mathbf{r} \cdot d\mathbf{S} = 0$$

PROBLEM 7-13 Using Stokes' theorem (7.26), show that for any closed surface S

$$\oiint_S \nabla \times \mathbf{f} \cdot d\mathbf{S} = 0$$

Solution: Let a closed curve C divide a closed surface S into two subsurfaces S_1 and S_2, as shown in Figure 7-9. Applying Stokes' theorem (7.26) to both S_1 and S_2, we obtain

$$\iint_{S_1} \nabla \times \mathbf{f} \cdot d\mathbf{S} = \oint_C \mathbf{f} \cdot d\mathbf{r}$$

$$\iint_{S_2} \nabla \times \mathbf{f} \cdot d\mathbf{S} = \oint_C \mathbf{f} \cdot d\mathbf{r} = -\oint_C \mathbf{f} \cdot d\mathbf{r}$$

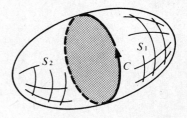

Figure 7-9

Hence for the closed surface S we have

$$\oiint_S \nabla \times \mathbf{f} \cdot d\mathbf{S} = \iint_{S_1} \nabla \times \mathbf{f} \cdot d\mathbf{S} + \iint_{S_2} \nabla \times \mathbf{f} \cdot d\mathbf{S}$$

$$= \oint_C \mathbf{f} \cdot d\mathbf{r} + \oint_C \mathbf{f} \cdot d\mathbf{r}$$

$$= 0$$

Alternate Solution: Consider an almost closed surface S with a small opening bounded by a simple closed curve C, as shown in Figure 7-10. Applying Stokes' theorem (7.26), we have

$$\iint_S \nabla \times \mathbf{f} \cdot d\mathbf{S} = \oint_C \mathbf{f} \cdot d\mathbf{r}$$

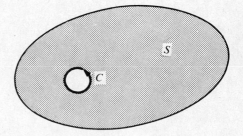

Figure 7-10

Now let the small opening become smaller and smaller so that in the limit it shrinks to a point as the bounding curve C goes to zero. Then the surface becomes a closed surface and the line integral goes to zero. Hence,

$$\oiint_S \nabla \times \mathbf{f} \cdot d\mathbf{S} = 0$$

PROBLEM 7-14 If C is a closed curve, show that

$$\oint_C \nabla \phi \cdot d\mathbf{r} = 0 \qquad \text{(a)}$$

Solution: By Stokes' theorem (7.26) we have

$$\oint_C \nabla\phi \cdot d\mathbf{r} = \iint_S \nabla \times (\nabla\phi) \cdot d\mathbf{S}$$

where S is the surface enclosed by C. But from expression (5.31), $\nabla \times \nabla\phi = \mathbf{0}$, so that eq. (a) is established.

PROBLEM 7-15 If $\mathbf{f} = 4y\mathbf{i} + x\mathbf{j} + 2z\mathbf{k}$, evaluate $I = \iint_S \nabla \times \mathbf{f} \cdot d\mathbf{S}$ over the hemisphere $x^2 + y^2 + z^2 = a^2, z \geqslant 0$.

Solution: Since I is in the form of the surface integral in Stokes' theorem (7.26),

$$I = \oint_C \mathbf{f} \cdot d\mathbf{r} = \oint_C 4y\,dx + x\,dy + 2z\,dz$$

where C is the circle $x^2 + y^2 = a^2, z = 0$, directed as shown in Figure 7-11.

The parametric representation of C (see Example 4-9) is $x = a\cos t$, $y = a\sin t, z = 0$, where $0 \leqslant t \leqslant 2\pi$. Taking differentials, we get $dx = -a\sin t\,dt$, $dy = a\cos t\,dt$, $\mathrm{dz} = 0$. Hence,

$$I = \int_0^{2\pi} 4(a\sin t)(-a\sin t\,dt) + (a\cos t)(a\cos t\,dt)$$

$$= a^2 \int_0^{2\pi} (-4\sin^2 t + \cos^2 t)\,dt$$

$$= a^2 \int_0^{2\pi} (1 - 5\sin^2 t)\,dt$$

$$= a^2 \int_0^{2\pi} \left[1 - 5\frac{1}{2}(1 - \cos 2t) \right] dt$$

$$= a^2 \int_0^{2\pi} \left(-\frac{3}{2} + \frac{5}{2}\cos 2t \right) dt$$

$$= -3a^2\pi$$

PROBLEM 7-16 Show that Green's theorem for a plane (7.28) can also be expressed as

$$\oint_C \mathbf{f} \cdot \mathbf{n}\,ds = \iint_R \nabla \cdot \mathbf{f}\,dx\,dy$$

where $\mathbf{f} = Q\mathbf{i} - P\mathbf{j}$, $\mathbf{n}$ is the outward unit normal vector to C (as shown in Figure 7-12) and s is the arc length.

Solution: From (4.46), the unit tangent vector $\mathbf{T}$ to C is given by

$$\mathbf{T} = \frac{d\mathbf{r}}{ds} = \frac{dx}{ds}\mathbf{i} + \frac{dy}{ds}\mathbf{j}$$

Now, from Figure 7-12,

$$\mathbf{n} = \mathbf{T} \times \mathbf{k} = \left(\frac{dx}{ds}\mathbf{i} + \frac{dy}{ds}\mathbf{j} \right) \times \mathbf{k} = \frac{dy}{ds}\mathbf{i} - \frac{dx}{ds}\mathbf{j}$$

Accordingly,

$$\mathbf{f} \cdot \mathbf{n} = (Q\mathbf{i} - P\mathbf{j}) \cdot \left(\frac{dy}{ds}\mathbf{i} - \frac{dx}{ds}\mathbf{j} \right) = Q\frac{dy}{ds} + P\frac{dx}{ds}$$

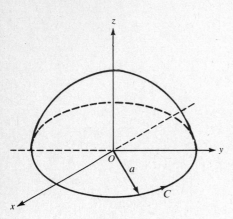

Figure 7-11

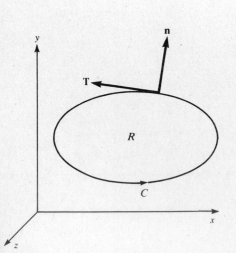

Figure 7-12

Thus,

$$\oint_C \mathbf{f} \cdot \mathbf{n} \, ds = \oint_C \left(Q \frac{dy}{ds} + P \frac{dx}{ds} \right) ds = \oint_C P \, dx + Q \, dy$$

Next,

$$\mathbf{\nabla} \cdot \mathbf{f} = \frac{\partial Q}{\partial x} + \frac{\partial (-P)}{\partial y} = \frac{\partial Q}{\partial x} - \frac{\partial P}{\partial y}$$

Hence Green's theorem for a plane (7.28) can be expressed as

$$\oint_C \mathbf{f} \cdot \mathbf{n} \, ds = \iint_R \mathbf{\nabla} \cdot \mathbf{f} \, dx \, dy$$

PROBLEM 7-17   Show that the area A of a region R in the xy-plane bounded by a simple closed curve C is given by

$$A = \oint_C x \, dy = \oint_C -y \, dx = \frac{1}{2} \oint_C x \, dy - y \, dx$$

Solution: If $P(x, y) = 0$ and $Q(x, y) = x$ in (7.28), then

$$\oint_C x \, dy = \iint_R dx \, dy = A$$

If we let $P = -y$ and $Q = 0$ in (7.28), then

$$\oint_C -y \, dx = \iint_R dx \, dy = A.$$

Adding the above results, we obtain $A = \frac{1}{2} \oint x \, dy - y \, dx$. If we let $P = -y$ and $Q = x$ in (7.28),

$$\oint_C (-y) \, dx + x \, dy = \oint_C x \, dy - y \, dx = \iint_R \left[\frac{\partial x}{\partial x} - \frac{\partial (-y)}{\partial y} \right] dx \, dy$$

$$= 2 \iint_R dx \, dy$$

$$= 2A$$

Hence,

$$A = \frac{1}{2} \oint x \, dy - y \, dx$$

PROBLEM 7-18   Determine the area A of a planar region R bounded by an ellipse C whose major and minor axes are $2a$ and $2b$, respectively.

Solution: The equation of an ellipse C centered at the origin having a major axis $2a$ and a minor axis $2b$ is

$$\frac{x^2}{a^2} + \frac{y^2}{b^2} = 1, \qquad z = 0$$

The curve C can be represented parametrically by

$$x = a \cos t, \qquad y = b \sin t, \qquad z = 0, \qquad 0 \leqslant t \leqslant 2\pi$$

Then, $dx = -a \sin t \, dt$, $dy = b \cos t \, dt$. So by the result of Problem 7-17, we have

$$A = \frac{1}{2} \oint x \, dy - y \, dx$$

$$= \frac{1}{2} \int_0^{2\pi} (a \cos t)(b \cos t \, dt) - (b \sin t)(-a \sin t \, dt)$$

$$= \frac{1}{2} \int_0^{2\pi} ab(\cos^2 t + \sin^2 t) \, dt$$

$$= \frac{1}{2} ab \int_0^{2\pi} dt$$

$$= \pi ab$$

Surface-to-Line Integral Transformations

PROBLEM 7-19   Prove that if $\mathbf{r}$ is a position vector, then

$$\oint_C d\mathbf{r} = \mathbf{0}$$

Solution: Using expression (7.31),

$$\oint_C \phi \, d\mathbf{r} = \iint_S d\mathbf{S} \times \nabla \phi$$

First set $\phi = 1$. Then we have $\nabla \phi = \mathbf{0}$, so, $d\mathbf{S} \times \nabla \phi = \mathbf{0}$ and so

$$\oint_C d\mathbf{r} = \mathbf{0}$$

PROBLEM 7-20   Show that by integrating around a closed curve C in the xy-plane we obtain

$$\left| \oint_C \mathbf{r} \times d\mathbf{r} \right| = 2A$$

where $\mathbf{r}$ is the position vector and A is the area enclosed by the curve C.

Solution: If we set $\mathbf{f} = \mathbf{r}$ and $d\mathbf{S} = \mathbf{k} \, dx \, dy$ in (7.32), we have

$$\oint_C d\mathbf{r} \times \mathbf{r} = \iint_S (\mathbf{k} \times \nabla) \times \mathbf{r} \, dx \, dy$$

Since for position vector $\mathbf{r}$, $\nabla \cdot \mathbf{r} = 3$, using expansion (1.60) gives us the expression

$$(\mathbf{k} \times \nabla) \times \mathbf{r} = \nabla(\mathbf{k} \cdot \mathbf{r}) - \mathbf{k}(\nabla \cdot \mathbf{r}) = \nabla z - 3\mathbf{k} = \mathbf{k} - 3\mathbf{k} = -2\mathbf{k}$$

So

$$\oint_C \mathbf{r} \times d\mathbf{r} = -\oint_C d\mathbf{r} \times \mathbf{r} = 2\mathbf{k} \iint_S dx \, dy = 2A\mathbf{k}$$

and

$$\left| \oint_C \mathbf{r} \times d\mathbf{r} \right| = 2A$$

PROBLEM 7-21   If S is an open surface bounded by a simple closed curve C, show that,

$$\oint_C \mathbf{r} \times d\mathbf{r} = 2 \iint_S d\mathbf{S}$$

Solution: If in Stokes' theorem (7.26) we set $\mathbf{f} = \mathbf{r} \times \mathbf{a}$, where $\mathbf{a}$ is a nonzero constant vector, then

$$\oint_C (\mathbf{r} \times \mathbf{a}) \cdot d\mathbf{r} = \iint_S \mathbf{\nabla} \times (\mathbf{r} \times \mathbf{a}) \cdot d\mathbf{S} \qquad (a)$$

Since $\mathbf{a}$ is a constant vector, from the result of Example 7-17 we have

$$\mathbf{\nabla} \times (\mathbf{r} \times \mathbf{a}) = (\mathbf{a} \cdot \mathbf{\nabla})\mathbf{r} - \mathbf{a}(\mathbf{\nabla} \cdot \mathbf{r})$$

Now from the result of Problem 5-14 and expression (5-17) we have

$$(\mathbf{a} \cdot \mathbf{\nabla})\mathbf{r} = \mathbf{a} \quad \text{and} \quad \mathbf{\nabla} \cdot \mathbf{r} = 3$$

So we can write

$$\mathbf{\nabla} \times (\mathbf{r} \times \mathbf{a}) = \mathbf{a} - 3\mathbf{a} = -2\mathbf{a}$$

Also, by the permutation rule of the triple scalar product (2.46), $(\mathbf{r} \times \mathbf{a}) \cdot d\mathbf{r} = -\mathbf{a} \cdot (\mathbf{r} \times d\mathbf{r})$. Hence, eq. (a) can be written as

$$\mathbf{a} \cdot \oint_C \mathbf{r} \times d\mathbf{r} = 2\mathbf{a} \cdot \iint_S d\mathbf{S}$$

Since $\mathbf{a}$ is any nonzero constant vector, we conclude that

$$\oint_C \mathbf{r} \times d\mathbf{r} = 2 \iint_S d\mathbf{S}$$

Irrotational and Solenoidal Fields

PROBLEM 7-22   If $\mathbf{f}$ is a solenoidal vector field, show that there exists a vector potential function $\mathbf{A}$ such that

$$\mathbf{f} = \mathbf{\nabla} \times \mathbf{A}$$

Solution: We can demonstrate the existence of $\mathbf{A}$ in this case by actually calculating it. Let $\mathbf{f} = f_1\mathbf{i} + f_2\mathbf{j} + f_3\mathbf{k}$ and $\mathbf{A} = A_1\mathbf{i} + A_2\mathbf{j} + A_3\mathbf{k}$. Now we need to show that there exist scalar functions A_1, A_2, and A_3 such that, for the given solenoidal vector field $\mathbf{f}$,

$$\mathbf{f} = \mathbf{\nabla} \times \mathbf{A}$$

Then match the components in the definition of the curl of $\mathbf{A}$ (5.28) with the corresponding components of $\mathbf{f}$:

$$f_1 = \frac{\partial A_3}{\partial y} - \frac{\partial A_2}{\partial z}$$

$$f_2 = \frac{\partial A_1}{\partial z} - \frac{\partial A_3}{\partial x}$$

$$f_3 = \frac{\partial A_2}{\partial x} - \frac{\partial A_1}{\partial y}$$

To find any $\mathbf{A}$, assume that $A_1 = 0$, to create conditions for the equations to be solved. This assumption gives us the set

$$f_1 = \frac{\partial A_3}{\partial y} - \frac{\partial A_2}{\partial z} \qquad (a)$$

$$f_2 = \frac{\partial A_3}{\partial x} \tag{b}$$

$$f_3 = \frac{\partial A_2}{\partial x} \tag{c}$$

Integration of eqs. (b) and (c) gives

$$A_2 = \int_{x_0}^{x} f_3 \, dx + g_2(y, z)$$

$$A_3 = -\int_{x_0}^{x} f_2 \, dx + g_3(y, z)$$

where g_2 and g_3 are arbitrary functions of y and z, but not of x. The difference of the specific partial derivatives of A_3 and A_2 delineated in eq. (a) is

$$\frac{\partial A_3}{\partial y} - \frac{\partial A_2}{\partial z} = -\int_{x_0}^{x} \left(\frac{\partial f_2}{\partial y} + \frac{\partial f_3}{\partial z} \right) dx + \frac{\partial g_3}{\partial y} - \frac{\partial g_2}{\partial z}$$

Since it is given that $\mathbf{f}$ is solenoidal, we have $\nabla \cdot \mathbf{f} = 0$, which by the definition of divergence (5.16) gives

$$\frac{\partial f_1}{\partial x} = -\left(\frac{\partial f_2}{\partial y} + \frac{\partial f_3}{\partial z} \right)$$

Hence,

$$\frac{\partial A_3}{\partial y} - \frac{\partial A_2}{\partial z} = \int_{x_0}^{x} \frac{\partial f_1}{\partial x} \, dx + \frac{\partial g_3}{\partial z} - \frac{\partial g_2}{\partial z}$$

$$= f_1(x, y, z) - f_1(x_0, y, z) + \frac{\partial g_3}{\partial z} - \frac{\partial g_2}{\partial z}$$

Using this computation, we can rewrite eq. (a) as

$$f_1 = f_1(x, y, z) - f_1(x_0, y, z) + \frac{\partial g_3}{\partial y} - \frac{\partial g_2}{\partial z} \tag{d}$$

Since by assumption g_2 and g_3 are arbitrary functions of y and z, we can satisfy eq. (d) if we choose

$$g_2 = 0 \tag{e}$$

and

$$g_3 = \int_{y_0}^{y} f_1(x_0, y, z) \, dy \tag{f}$$

where y_0 is a constant. With g_2 and g_3 given by eq. (e) and eq. (f) and with A_1 chosen earlier to be zero, we can construct $\mathbf{A} = A_1 \mathbf{i} + A_2 \mathbf{j} + A_3 \mathbf{k}$ as

$$\mathbf{A} = \mathbf{j} \int_{x_0}^{x} f_3(x, y, z) \, dx + \mathbf{k} \left[\int_{y_0}^{y} f_1(x_0, y, z) \, dy - \int_{x_0}^{x} f_2(x, y, z) \, dx \right]$$

[*Note:* In this proof, several arbitrary selections have been made and so the vector $\mathbf{A}$ produced here is not unique. Note that, for ease of choice in the integrals, the simply connected region R in which $\mathbf{A}$ was constructed has been assumed to be a rectangular parallelepiped.]

PROBLEM 7-23 Let

$$\mathbf{f} = -\frac{y}{x^2 + y^2} \mathbf{i} + \frac{x}{x^2 + y^2} \mathbf{j}$$

(a) Show that $\mathbf{f}$ can be expressed as $\nabla\phi$, where $\phi = \tan^{-1}(y/x)$. (b) Evaluate the integral $I = \oint_C \mathbf{f} \cdot d\mathbf{r}$, if C is a circle of radius a on the xy-plane (1) when its center is at the origin and (2) when its center is at point $(\alpha, \beta, 0)$ with $\alpha^2 + \beta^2 > a^2$.

Solution:

(a) Since $\phi = \tan^{-1}(y/x)$, we use elementary calculus to compute

$$\nabla\phi = \frac{\partial}{\partial x}\left[\tan^{-1}\left(\frac{y}{x}\right)\right]\mathbf{i} + \frac{\partial}{\partial y}\left[\tan^{-1}\left(\frac{y}{x}\right)\right]\mathbf{j}$$

$$= \frac{1}{1 + (y/x)^2}\left(\frac{-y}{x^2}\right)\mathbf{i} + \frac{1}{1 + (y/x)^2}\left(\frac{1}{x}\right)\mathbf{j}$$

$$= \frac{-y}{x^2 + y^2}\mathbf{i} + \frac{x}{x^2 + y^2}\mathbf{j}$$

$$= \mathbf{f}$$

(b) Evaluating the integral:

(1) When the center of C is at the origin,

$$I = \oint_C \mathbf{f} \cdot d\mathbf{r} = \oint_C \nabla\phi \cdot d\mathbf{r} = \oint_C d\phi = \oint_C d\left[\tan^{-1}\left(\frac{y}{x}\right)\right]$$

The parametric representation for C is

$$x = a\cos t, \qquad y = a\sin t, \qquad z = 0$$

Thus,

$$I = \int_0^{2\pi} d\left[\tan^{-1}\left(\frac{y}{x}\right)\right] = \int_0^{2\pi} d[\tan^{-1}(\tan t)] = \int_0^{2\pi} dt = 2\pi \neq 0$$

Note that the result is not zero because the region defined, for which $\mathbf{f} = \nabla\phi$, is not a simply connected region. Since the function $\phi = \tan^{-1}(y/x)$ is not defined at the origin, we know that the region has a hole at the origin.

(2) When the center of C is at $(\alpha, \beta, 0)$, with the condition that $\alpha^2 + \beta^2 > a^2$, its parametric representation is

$$x = \alpha + a\cos t, \qquad y = \beta + a\sin t, \qquad z = 0$$

The condition ensures that this represents a circle in the plane which does not enclose the origin. So the integral

$$I = \int_0^{2\pi} d\left[\tan^{-1}\left(\frac{y}{x}\right)\right] = \tan^{-1}\left(\frac{\beta + a\sin t}{\alpha + a\cos t}\right)\Big|_0^{2\pi}$$

$$= \tan^{-1}\left(\frac{\beta}{\alpha + a}\right) - \tan^{-1}\left(\frac{\beta}{\alpha + a}\right)$$

$$= 0$$

PROBLEM 7-24 If R is a simply connected region, show that the necessary and sufficient condition for $\nabla \times \mathbf{f} = \mathbf{0}$ is that $\mathbf{f} = \nabla\phi$ where ϕ is a scalar function with continuous partial derivatives.

Solution: If $\mathbf{f} = \nabla\phi$, then by condition (5.31)

$$\nabla \times \mathbf{f} = \nabla \times \nabla\phi = \mathbf{0}$$

On the other hand, if $\nabla \times \mathbf{f} = \mathbf{0}$, then from Stokes' theorem (7.26)

$$\oint_C \mathbf{f} \cdot d\mathbf{r} = 0$$

for every simple closed curve C in R. From the solution to Example 7-18, we can see that the line integral of **f** is independent of the path of integration and that **f** can be expressed as $\mathbf{f} = \nabla\phi$.

[*Note:* The result of Problem 7-24 provides a simple test for determining whether or not a given vector field **f** is the gradient of some scalar field ϕ.]

PROBLEM 7-25   Let two vector fields be given by $\mathbf{f} = 3y^2\mathbf{i} + z\mathbf{j} + 2y\mathbf{k}$ and $\mathbf{g} = yz\mathbf{i} + xz\mathbf{j} + xy\mathbf{k}$. Determine whether these vector fields are the gradients of some scalar fields.

Solution: Since $\mathbf{f} = 3y^2\mathbf{i} + z\mathbf{j} + 2y\mathbf{k}$,

$$\nabla \times \mathbf{f} = \begin{vmatrix} \mathbf{i} & \mathbf{j} & \mathbf{k} \\ \dfrac{\partial}{\partial x} & \dfrac{\partial}{\partial y} & \dfrac{\partial}{\partial z} \\ 3y^2 & z & 2y \end{vmatrix} = \mathbf{i} - 6y\mathbf{j} \neq \mathbf{0}$$

So **f** is not the gradient of a scalar field.

Since $\mathbf{g} = yz\mathbf{i} + xz\mathbf{j} + xy\mathbf{k}$,

$$\nabla \times \mathbf{g} = \begin{vmatrix} \mathbf{i} & \mathbf{j} & \mathbf{k} \\ \dfrac{\partial}{\partial x} & \dfrac{\partial}{\partial y} & \dfrac{\partial}{\partial z} \\ yz & xz & xy \end{vmatrix}$$
$$= (x - x)\mathbf{i} + (y - y)\mathbf{j} + (z - z)\mathbf{k}$$
$$= \mathbf{0}$$

So **g** is the gradient of a scalar field. [Note that $\mathbf{g} = \nabla(xyz + \text{constant})$.]

PROBLEM 7-26   Show that if there exists a scalar $\lambda \neq 0$ such that $\lambda\mathbf{f}$ is irrotational, then

$$\mathbf{f} \cdot \nabla \times \mathbf{f} = 0$$

Solution: If $\lambda\mathbf{f}$ is irrotational, there must be a scalar function ϕ such that

$$\lambda\mathbf{f} = \nabla\phi$$

Taking the curl of both sides, we have

$$\nabla \times (\lambda\mathbf{f}) = \nabla \times (\nabla\phi) = \mathbf{0}$$

But by expansion (5.42), we have

$$\nabla \times (\lambda\mathbf{f}) = \lambda\nabla \times \mathbf{f} + \nabla\lambda \times \mathbf{f} = \mathbf{0}$$

or

$$\lambda\nabla \times \mathbf{f} = -\nabla\lambda \times \mathbf{f} = \mathbf{f} \times \nabla\lambda$$

If we dot-product both sides with **f**, using (1.51), we obtain

$$\lambda\mathbf{f} \cdot \nabla \times \mathbf{f} = \mathbf{f} \cdot \mathbf{f} \times \nabla\lambda = 0$$

Since λ is not zero,

$$\mathbf{f} \cdot \nabla \times \mathbf{f} = 0$$

[Note that the converse is also true.]

Supplementary Exercises

PROBLEM 7-27 Show that if $\mathbf{f} = \nabla\phi$ and $\nabla \cdot \mathbf{f} = 0$ for a region R bounded by S,

$$\iiint_R |\mathbf{f}|^2 \, dV = \oiint_S \phi\mathbf{f} \cdot d\mathbf{S}$$

PROBLEM 7-28 Evaluate $\oiint_S \mathbf{r} \cdot d\mathbf{S}$ over the surface of the ellipsoid

$$\frac{x^2}{a^2} + \frac{y^2}{b^2} + \frac{z^2}{c^2} = 1$$

Answer: $4\pi \, abc$

PROBLEM 7-29 If $\mathbf{f} = ax\mathbf{i} + by\mathbf{j} + cz\mathbf{k}$, evaluate $\oiint_S \mathbf{f} \cdot d\mathbf{S}$ over any closed surface S that encloses a region of volume V.

Answer: $(a + b + c)V$

PROBLEM 7-30 If $\mathbf{f} = y\mathbf{i} + x\mathbf{j} + z^2\mathbf{k}$, evaluate $\iiint_R \nabla \cdot \mathbf{f} \, dV$, where R is the region bounded by $z = (1 - x^2 - y^2)^{1/2}$ and $z = 0$.

Answer: $\pi/2$

PROBLEM 7-31 If $\mathbf{f} = u(x, y)\mathbf{i} + v(x, y)\mathbf{j}$, prove that

$$\oint_C \mathbf{f} \times d\mathbf{r} = \mathbf{k}\iint_S \nabla \cdot \mathbf{f} \, dx \, dy$$

where C is a closed curve in the xy-plane that bounds a region S. [*Hint:* Apply the divergence theorem (7.2) to $\mathbf{f}$ over a cylinder of base S with $z = 1$.]

PROBLEM 7-32 Show that the volume enclosed by any closed surface S is

$$V = \frac{1}{6}\oiint_S \nabla(r^2) \cdot d\mathbf{S}$$

PROBLEM 7-33 Using the divergence theorem expressed in rectangular coordinates (7.5), evaluate

$$\iint_S x \, dy \, dz + y \, dz \, dx + z \, dx \, dy$$

where S is the surface of the sphere $x^2 + y^2 + z^2 = 1$, and $\mathbf{n}$ is the outer unit normal vector.

Answer: 4π

PROBLEM 7-34 Let S be the boundary surface of a region R whose volume is V in space, and let $\mathbf{n}$ be its outer unit normal vector. Prove that

$$V = \iint_S x \, dy \, dz = \iint_S y \, dz \, dx = \iint_S z \, dx \, dy$$

$$= \frac{1}{3}\iint_S x \, dy \, dz + y \, dz \, dx + z \, dx \, dy$$

$$= \frac{1}{3}\oiint_S \mathbf{r} \cdot d\mathbf{S}$$

PROBLEM 7-35   If there exists $h(x, y, z)$ such that $\nabla^2 \phi = h\phi$ and $\nabla^2 \psi = h\psi$ in a region R bounded by S, prove that

$$\oiint_S (\phi \nabla \psi - \psi \nabla \phi) \cdot d\mathbf{S} = 0$$

PROBLEM 7-36   If ϕ is harmonic in a region R bounded by S, show that

$$\oiint_S \phi \nabla \phi \cdot d\mathbf{S} = \iiint_R |\nabla \phi|^2 \, dV$$

PROBLEM 7-37   Show that

$$\iiint_R \mathbf{f} \cdot \nabla \phi \, dV = \oiint_S \phi \mathbf{f} \cdot d\mathbf{S} - \iiint_R \phi \nabla \cdot \mathbf{f} \, dV$$

PROBLEM 7-38   Prove that

$$5 \iiint_R r^4 \mathbf{r} \, dV = \oiint_S r^5 \, d\mathbf{S}$$

PROBLEM 7-39   If $\mathbf{a}$ is an arbitrary constant vector and V is the volume of a region R bounded by S, prove that

$$\oiint_S \mathbf{n} \times (\mathbf{a} \times \mathbf{r}) \, dS = 2V\mathbf{a}$$

PROBLEM 7-40   If $\mathbf{f} = (x^2 + y^2)y\mathbf{i} - (x^2 + y^2)x\mathbf{j} + (a^3 + z^3)\mathbf{k}$,   evaluate $\oint_C \mathbf{f} \cdot d\mathbf{r}$ where C is the circle $x^2 + y^2 = a^2, z = 0$,

Answers: $-4\pi a^4$

PROBLEM 7-41   If $\mathbf{f} = (x^2 + y^2 - 4)\mathbf{i} + 3xy\mathbf{j} + (2xz + z^2)\mathbf{k}$,   evaluate $\iint_S \nabla \times \mathbf{f} \cdot d\mathbf{S}$ where S is the surface defined by $z = 4 - (x^2 + y^2), (z \geq 0)$, and $\mathbf{n}$ is the outer unit normal vector.

Answer: 0

PROBLEM 7-42   The sphere $x^2 + y^2 + z^2 = a^2$ intersects the positive x-, y-, and z-axes at points A, B, and C, respectively. The simple closed curve K consists of the three circular arcs AB, BC, and CA. If $\mathbf{f} = (y + z)\mathbf{i} + (z + x)\mathbf{j} + (x + z)\mathbf{k}$, evaluate $\oint_K \mathbf{f} \cdot d\mathbf{r}$ directly. [*Hint:* Use Stokes' theorem (7.26) to verify this line integral by evaluating a surface integral over (a) the surface ABC of the octant of the sphere in the positive quadrants, and (b) the three quadrants of the circular arcs in the coordinate planes.]

Answer: $-\frac{1}{4}\pi a^2$

PROBLEM 7-43   By Stokes' theorem (7.26) prove that $\nabla \times (\nabla \phi) = \mathbf{0}$.

PROBLEM 7-44   Prove that

$$\oint_C \phi \nabla \psi \cdot d\mathbf{r} = \iint_S \nabla \phi \times \nabla \psi \cdot d\mathbf{S}$$

PROBLEM 7-45 Using Green's theorem for a plane (7.28), evaluate

$$I = \oint_C (3x + 4y)\,dx + (2x - 3y)\,dy$$

where C is the circle $x^2 + y^2 = 4$.

Answer: -8π

PROBLEM 7-46 Using Green's theorem for a plane (7.28), evaluate

$$I = \oint_C ay\,dx + bx\,dy$$

where a and b are arbitrary constants and C is an arbitrary closed regular curve.

Answer: $(b - a)A$, where A is the area of the region bounded by C

PROBLEM 7-47 By changing variables from (x, y) to (u, v) according to the transformation $x = x(u, v)$, $y = y(u, v)$, show that the area A of a simply connected region R bounded by a closed regular curve C is

$$A = \iint_R \frac{\partial(x, y)}{\partial(u, v)}\,du\,dv = \iint_R J\left[\frac{(x, y)}{(u, v)}\right]du\,dv$$

where the Jacobian of the transformation is

$$J\left[\frac{(x, y)}{(u, v)}\right] = \frac{\partial(x, y)}{\partial(u,v)} = \begin{vmatrix} \dfrac{\partial x}{\partial u} & \dfrac{\partial y}{\partial u} \\ \dfrac{\partial x}{\partial v} & \dfrac{\partial y}{\partial v} \end{vmatrix}$$

PROBLEM 7-48 Show that

$$\iint_S d\mathbf{S} \times \mathbf{r} = \frac{1}{2}\oint_C r^2\,d\mathbf{r}$$

PROBLEM 7-49 Show that

$$\iint_S \frac{\mathbf{r}}{r^3} \times d\mathbf{S} = \oint_C \frac{d\mathbf{r}}{r}$$

PROBLEM 7-50 Establish the independence of the path of integration and evaluate

(a) $\displaystyle\int_{(1,-1)}^{(4,2)} \frac{y\,dx - x\,dy}{x^2}$

(b) $\displaystyle\int_{(0,1,0)}^{(\pi,0,1)} \sin x\,dx + y^2\,dy + e^z\,dz$

Answer: (a) $-\frac{3}{2}$ (b) $e + \frac{2}{3}$

PROBLEM 7-51 If C is the curve $x^2 + y^2 = a^2$, prove that

$$\oint_C (\cos x \sinh y + xy^2)\,dx + (\sin x \cosh y + x^2 y)\,dy = 0$$

PROBLEM 7-52 Find a scalar potential function ϕ for the vector field

$$\mathbf{f} = (y + z\cos xz)\mathbf{i} + x\mathbf{j} + (x\cos xz)\mathbf{k}$$

Answer: $\phi = yx + \sin xz + k$, where k is an arbitrary constant

PROBLEM 7-53   Show that a constant vector **a** has a scalar potential $\phi = \mathbf{a} \cdot \mathbf{r}$ and a vector potential $\mathbf{A} = (\mathbf{a} \times \mathbf{r})/2$

PROBLEM 7-54   Consider any vector field **f** such that $\nabla \cdot \mathbf{f} = \rho$ and $\nabla \times \mathbf{f} = \mathbf{c}$. The scalar ρ and the vector **c** are given functions of x, y, and z, and ρ can be interpreted as a source density, and **c** as a circulation density. Show that if both source and circulation densities vanish at infinity, the vector field **f** can be expressed as the sum of two parts, one irrotational and the other solenoidal. (This is called **Helmboltz' theorem**.)

PROBLEM 7-55   Show that if ϕ is harmonic in a simply connected region whose boundary is the closed curve C, then

$$\oint_C \nabla \phi \cdot \mathbf{n} \, ds = 0$$

where **n** is the outer unit normal vector to C. [*Hint*: Use the result of Problem 7-16.]

PROBLEM 7-56   If $\mathbf{h} = \frac{1}{2} \nabla \times \mathbf{g}$ and $\mathbf{g} = \nabla \times \mathbf{f}$, show that

$$\frac{1}{2} \iiint_R g^2 \, dV = \frac{1}{2} \oiint_S (\mathbf{f} \times \mathbf{g}) \cdot d\mathbf{S} + \iiint_R \mathbf{f} \cdot \mathbf{h} \, dV$$

where R is the region bounded by a closed surface S.

8 CURVILINEAR ORTHOGONAL COORDINATES

THIS CHAPTER IS ABOUT

☑ **Curvilinear Coordinates**
☑ **Orthogonal Curvilinear Coordinates**
☑ **Gradient, Divergence, and Curl in Orthogonal Curvilinear Coordinates**
☑ **Special Coordinate Systems**

8-1. Curvilinear Coordinates

So far in our vector analysis we've restricted ourselves almost completely to a rectangular (or Cartesian) coordinate system, which has the unique advantage that all three base vectors $\mathbf{i}, \mathbf{j}, \mathbf{k}$ are constant unit vectors. But it is often useful to employ other coordinate systems. In this chapter, we'll develop expressions for the gradient, divergence, and curl in these other coordinate systems.

A. Transformations of coordinate systems

The **transformation** between the coordinates of a point (x, y, z) in the Cartesian coordinate system and those of a point (u, v, w) in a three-dimensional space is defined by

$$x = x(u, v, w), \qquad y = y(u, v, w), \qquad z = z(u, v, w) \tag{8.1}$$

The transformation (8.1) is *single-valued* and its **Jacobian** is

$$J = \frac{\partial(x, y, z)}{\partial(u, v, w)} = \begin{vmatrix} \dfrac{\partial x}{\partial u} & \dfrac{\partial x}{\partial v} & \dfrac{\partial x}{\partial w} \\[2mm] \dfrac{\partial y}{\partial v} & \dfrac{\partial y}{\partial v} & \dfrac{\partial y}{\partial w} \\[2mm] \dfrac{\partial z}{\partial w} & \dfrac{\partial z}{\partial v} & \dfrac{\partial z}{\partial w} \end{vmatrix} \neq 0 \tag{8.2}$$

A theorem from elementary calculus indicates that transformation (8.1) can be locally solved uniquely for u, v, w, in terms of x, y, z; and written as

$$u = u(x, y, z), \qquad v = v(x, y, z), \qquad w = w(x, y, z) \tag{8.3}$$

Hence any point (x, y, z) in space has unique corresponding coordinates (u, v, w).

B. Curvilinear coordinates

Coordinate surfaces are families of surfaces obtained by setting the coordinate equations equal to a constant; e.g., if c_1, c_2, c_3 are constants, then three families of surfaces are obtained:

$$u(x, y, z) = c_1, \qquad v(x, y, z) = c_2, \qquad w(x, y, z) = c_3$$

Thus if v and w are constant, the transformation (8.1) represents the u-curve; if u and w are constant, (8-1) represents the v-curve; and if u and v are constant, (8-1) represents the w-curve. These three curves are referred to as

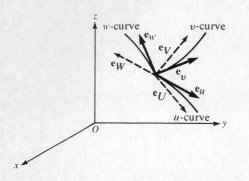

Figure 8-1
Base vectors for general curvilinear coordinates.

coordinate curves. Since the three coordinate curves are generally not straight lines, as in the rectangular coordinate system, such coordinate systems are said to have **curvilinear coordinates**.

C. Base vectors for a curvilinear coordinate system

In the rectangular coordinate system a set of mutually orthogonal base vectors is $\mathbf{i}, \mathbf{j}, \mathbf{k}$, as was shown in Chapter 2. Similarly, we can introduce a set of base vectors appropriate for curvilinear coordinate systems. In the most general case at each point P in space there are two sets of base vectors (see Figure 8-1). The first set is the set of unit vectors $\mathbf{e}_u, \mathbf{e}_v, \mathbf{e}_w$ tangent at point P to the coordinate curves through P. The second is the set of unit vectors $\mathbf{e}_U, \mathbf{e}_V, \mathbf{e}_W$ normal at P to the coordinate surfaces through P.

The unit vectors $\mathbf{e}_u, \mathbf{e}_v, \mathbf{e}_w$ (or $\mathbf{e}_U, \mathbf{e}_V, \mathbf{e}_W$) generally vary in orientation from point to point.

EXAMPLE 8-1: Show that if $\mathbf{r} = x\mathbf{i} + y\mathbf{j} + z\mathbf{k}$, the three unit vectors $\mathbf{e}_u, \mathbf{e}_v, \mathbf{e}_w$ at P tangent to the coordinate u-, v-, w-curves can be expressed as

$$\mathbf{e}_u = \frac{1}{h_u}\frac{\partial \mathbf{r}}{\partial u}, \qquad \mathbf{e}_v = \frac{1}{h_v}\frac{\partial \mathbf{r}}{\partial v}, \qquad \mathbf{e}_w = \frac{1}{h_w}\frac{\partial \mathbf{r}}{\partial w} \qquad (8.4)$$

where

$$h_u = \left|\frac{\partial \mathbf{r}}{\partial u}\right|, \qquad h_v = \left|\frac{\partial \mathbf{r}}{\partial v}\right|, \qquad h_w = \left|\frac{\partial \mathbf{r}}{\partial w}\right| \qquad (8.5)$$

Solution: If $\mathbf{r}$ is the position vector from the origin of a rectangular coordinate system to the point $P(x, y, z)$, then $\mathbf{r} = x\mathbf{i} + y\mathbf{j} + z\mathbf{k} = \mathbf{r}(x, y, z)$. Using (8.1), vector $\mathbf{r}$ can also be expressed as a function of $u, v,$ and w:

$$\mathbf{r} = \mathbf{r}(u, v, w)$$

where (u, v, w) are the curvilinear coordinates of point P. If v and w are constant, then transformation (8.1) is the parametric equation of the u-curve, with parameter u. Hence according to (4.34) a tangent vector at P to the u-curve is given by $\partial \mathbf{r}/\partial u$. Similarly, $\partial \mathbf{r}/\partial v$ and $\partial \mathbf{r}/\partial w$ are tangent at P to the v- and w-curves, respectively.

In general the lengths $\left|\dfrac{\partial \mathbf{r}}{\partial u}\right|, \left|\dfrac{\partial \mathbf{r}}{\partial v}\right|,$ and $\left|\dfrac{\partial \mathbf{r}}{\partial w}\right|$ are not equal to one. Hence if we denote these by

$$h_u = \left|\frac{\partial \mathbf{r}}{\partial u}\right|, \qquad h_v = \left|\frac{\partial \mathbf{r}}{\partial v}\right|, \qquad h_w = \left|\frac{\partial \mathbf{r}}{\partial w}\right|$$

we can express unit vectors in the tangential directions by

$$\mathbf{e}_u = \frac{1}{h_u}\frac{\partial \mathbf{r}}{\partial u}, \qquad \mathbf{e}_v = \frac{1}{h_v}\frac{\partial \mathbf{r}}{\partial v}, \qquad \mathbf{e}_w = \frac{1}{h_w}\frac{\partial \mathbf{r}}{\partial w}$$

Note: In general h_u, h_v, h_w are functions of u, v, w; and $h_u \neq 0, h_v \neq 0, h_w \neq 0$. Hence $\mathbf{e}_u, \mathbf{e}_v, \mathbf{e}_w$ are also functions of u, v, w.

EXAMPLE 8-2: Show that the three unit vectors $\mathbf{e}_U, \mathbf{e}_V, \mathbf{e}_W$ normal at P to the coordinate u-, v-, w-surfaces, respectively, can be expressed as

$$\mathbf{e}_U = \frac{1}{H_u}\nabla u, \qquad \mathbf{e}_V = \frac{1}{H_v}\nabla v, \qquad \mathbf{e}_W = \frac{1}{H_w}\nabla w \qquad (8.6)$$

where

$$H_u = |\nabla u|, \qquad H_v = |\nabla v|, \qquad H_w = |\nabla w| \qquad (8.7)$$

Solution: From the result of Example 5-3, we see that the vector ∇u is perpendicular to the surface $u(x, y, z) = c_1$ at point P. Similarly, ∇v and ∇w are normal to the v- and w-surfaces, respectively, at this point. So if we define

$$H_u = |\nabla u|, \qquad H_v = |\nabla v|, \qquad H_w = |\nabla w|$$

we obtain the unit vectors

$$\mathbf{e}_U = \frac{1}{H_u} \nabla u, \qquad \mathbf{e}_V = \frac{1}{H_v} \nabla v, \qquad \mathbf{e}_W = \frac{1}{H_w} \nabla w$$

8-2. Orthogonal Curvilinear Coordinates

A curvilinear coordinate system is called *orthogonal* if the coordinate curves are everywhere orthogonal. In this case the three vectors $\mathbf{e}_u, \mathbf{e}_v, \mathbf{e}_w$ are mutually orthogonal at every point. This is expressed by the conditions

$$\mathbf{e}_u \cdot \mathbf{e}_v = \mathbf{e}_v \cdot \mathbf{e}_w = \mathbf{e}_w \cdot \mathbf{e}_u = 0 \tag{8.8}$$

We'll also assume that $\mathbf{e}_u, \mathbf{e}_v, \mathbf{e}_w$ form a right-handed system (see Example 1-13), for which

$$[\mathbf{e}_u \mathbf{e}_v \mathbf{e}_w] = 1 \tag{8.9}$$

Note: It will be shown in Problem 8-3 that in the orthogonal system $\mathbf{e}_u = \mathbf{e}_U$, $\mathbf{e}_v = \mathbf{e}_V$, and $\mathbf{e}_w = \mathbf{e}_W$.

A. Scale factors

We call the quantities h_u, h_v, and h_w the **scale factors** or **metric coefficients** because we must multiply them by the differential coefficients du, dv, and dw to obtain arc lengths. The scale factors depend on the coordinates and have no dimension. But their *products* with the differential coordinates have the dimension of length.

If in an orthogonal curvilinear coordinate system s_u, s_v, s_w represent arc lengths along the u-, v-, w-curves, then

$$ds_u = h_u \, du, \qquad ds_v = h_v \, dv, \qquad ds_w = h_w \, dw \tag{8.10}$$
$$(ds)^2 = (h_u \, du)^2 + (h_v \, dv)^2 + (h_w \, dw)^2 \tag{8.11}$$
$$dV = h_u h_v h_w \, du \, dv \, dw \tag{8.12}$$

where ds represents the infinitesimal arc length between the points (u, v, w) and $(u + du, v + dv, w + dw)$, and dV is the volume element.

EXAMPLE 8-3: Verify (8.10), (8.11), and (8.12)

Solution:

(*8.10*): Using the arc lengths s_u, s_v, s_w to obtain

$$\frac{\partial \mathbf{r}}{\partial u} = \frac{\partial \mathbf{r}}{\partial s_u} \frac{ds_u}{du}, \qquad \frac{\partial \mathbf{r}}{\partial v} = \frac{\partial \mathbf{r}}{\partial s_v} \frac{ds_v}{dv}, \qquad \frac{\partial \mathbf{r}}{\partial w} = \frac{\partial \mathbf{r}}{\partial s_w} \frac{ds_w}{dw} \tag{8.13}$$

The unit tangent vector to curve C of arc length s at any point P is the derivative of the position vector $\mathbf{r}$ to P with respect to s. Thus

$$\frac{\partial \mathbf{r}}{\partial s_u} = \mathbf{e}_u, \qquad \frac{\partial \mathbf{r}}{\partial s_v} = \mathbf{e}_v, \qquad \frac{\partial \mathbf{r}}{\partial s_w} = \mathbf{e}_w \tag{8.14}$$

Hence from (8.13) and (8.14) we have

$$\frac{\partial \mathbf{r}}{\partial u} = \frac{ds_u}{du} \mathbf{e}_u = h_u \mathbf{e}_u, \qquad \frac{\partial \mathbf{r}}{\partial v} = \frac{ds_v}{dv} \mathbf{e}_v = h_v \mathbf{e}_v, \qquad \frac{\partial \mathbf{r}}{\partial w} = \frac{ds_w}{dw} \mathbf{e}_w = h_w \mathbf{e}_w \tag{8.15}$$

So

$$h_u = \frac{ds_u}{du}, \qquad h_v = \frac{ds_v}{dv}, \qquad h_w = \frac{ds_w}{dw} \tag{8.16}$$

or

$$ds_u = h_u\,du, \qquad ds_v = h_v\,dv, \qquad ds_w = h_w\,dw \tag{8.10}$$

Thus the differential coordinates du, dv, dw must be multiplied by h_u, h_v, h_w to obtain arc lengths.

(8.11): Because of (8.15) we have

$$d\mathbf{r} = \frac{\partial \mathbf{r}}{\partial u}\,du + \frac{\partial \mathbf{r}}{\partial v}\,dv + \frac{\partial \mathbf{r}}{\partial w}\,dw = h_u\,du\,\mathbf{e}_u + h_v\,dv\,\mathbf{e}_v + h_w\,dw\,\mathbf{e}_w \tag{8.17}$$

Hence the arc length ds between the points (u, v, w) and $(u + du, v + dv, w + dw)$ is

$$ds^2 = d\mathbf{r} \cdot d\mathbf{r} = (h_u\,du)^2 + (h_v\,dv)^2 + (h_w\,dw)^2 \tag{8.11}$$

since $\mathbf{e}_u$, $\mathbf{e}_v$, $\mathbf{e}_w$ forms an orthonormal basis.

(8.12): The differential element of volume in this space is a parallelepiped, three sides of which are the vectors

$$\frac{\partial \mathbf{r}}{\partial u}\,du = h_u\,du\,\mathbf{e}_u, \qquad \frac{\partial \mathbf{r}}{\partial v}\,dv = h_v\,dv\,\mathbf{e}_v, \qquad \frac{\partial \mathbf{r}}{\partial w}\,dw = h_w\,dw\,\mathbf{e}_w$$

Hence the volume element dV is the triple scalar product of these (see Example 1-13):

$$\begin{aligned}
dV &= (h_u\,du\,\mathbf{e}_u) \cdot [(h_v\,dv\,\mathbf{e}_v) \times (h_w\,dw\,\mathbf{e}_w)] \\
&= [\mathbf{e}_u \mathbf{e}_v \mathbf{e}_w]\, h_u h_v h_w\,du\,dv\,dw \\
&= h_u h_v h_w\,du\,dv\,dw
\end{aligned} \tag{8.12}$$

B. Cylindrical coordinate system

The **cylindrical coordinate system** is defined by the transformation

CYLINDRICAL COORDINATE SYSTEM
$$x = u\cos v, \qquad y = u\sin v, \qquad z = w \tag{8.18}$$

The ranges of u, v, and w are given by $u \geqslant 0$, $0 \leqslant v < 2\pi$, and $-\infty < w < \infty$.

By definition (8.2), the Jacobian of the transformation is

$$J = \frac{\partial(x, y, z)}{\partial(u, v, w)} = \begin{vmatrix} \cos v & -u\sin v & 0 \\ \sin v & u\cos v & 0 \\ 0 & 0 & 1 \end{vmatrix} = u(\cos^2 v + \sin^2 v) = u \tag{8.19}$$

which is not zero except on the z-axis. The inverse transformation equations are

$$u = (x^2 + y^2)^{1/2}, \qquad v = \tan^{-1}\left(\frac{y}{x}\right), \qquad w = z \tag{8.20}$$

EXAMPLE 8-4: (a) Determine the scale factors h_u, h_v, h_w and the unit vectors $\mathbf{e}_u$, $\mathbf{e}_v$, $\mathbf{e}_w$ for the cylindrical coordinate system. (b) Prove that the cylindrical coordinate system is an orthogonal system.

Solution:

(a) Since the position vector is

$$\mathbf{r} = u\cos v\,\mathbf{i} + u\sin v\,\mathbf{j} + w\mathbf{k}$$

its partial derivatives are

$$\frac{\partial \mathbf{r}}{\partial u} = \cos v \mathbf{i} + \sin v \mathbf{j}, \qquad \frac{\partial \mathbf{r}}{\partial v} = -u \sin v \mathbf{i} + u \cos v \mathbf{j}, \qquad \frac{\partial \mathbf{r}}{\partial w} = \mathbf{k}$$

Hence the scale factors are

$$h_u = \left| \frac{\partial \mathbf{r}}{\partial u} \right| = (\cos^2 v + \sin^2 v)^{1/2} = 1$$

$$h_v = \left| \frac{\partial \mathbf{r}}{\partial v} \right| = [(-u \sin v)^2 + (u \cos v)^2]^{1/2} = u$$

$$h_w = \left| \frac{\partial \mathbf{r}}{\partial w} \right| = 1$$

and the base vectors are

$$\mathbf{e}_u = \frac{1}{h_u} \frac{\partial \mathbf{r}}{\partial u} = \cos v \mathbf{i} + \sin v \mathbf{j}$$

$$\mathbf{e}_v = \frac{1}{h_v} \frac{\partial \mathbf{r}}{\partial v} = -\sin v \mathbf{i} + \cos v \mathbf{j}$$

$$\mathbf{e}_w = \frac{1}{h_w} \frac{\partial \mathbf{r}}{\partial w} = \mathbf{k}$$

(b) Performing the various dot products among $\mathbf{e}_u$, $\mathbf{e}_v$, and $\mathbf{e}_w$,

$$\mathbf{e}_u \cdot \mathbf{e}_v = (\cos v)(-\sin v) + (\sin v)(\cos v) = 0$$
$$\mathbf{e}_v \cdot \mathbf{e}_w = (-\sin v)(0) + (\cos v)(0) + (0)(1) = 0$$
$$\mathbf{e}_w \cdot \mathbf{e}_u = (0)(\cos v) + (0)(\sin v) + (1)(0) = 0$$

Hence $\mathbf{e}_u$, $\mathbf{e}_v$, $\mathbf{e}_w$ are mutually orthogonal.

C. Spherical coordinate system

The **spherical coordinate system** is defined by the transformation

$$x = u \sin v \cos w, \qquad y = u \sin v \sin w, \qquad z = u \cos v \qquad (8.21)$$

where $u \geqslant 0, 0 \leqslant v < \pi, 0 \leqslant w < 2\pi$.

By definition (8.2), the Jacobian of the transformation is

$$J = \frac{\partial(x, y, z)}{\partial(u, v, w)} = \begin{vmatrix} \sin v \cos w & u \cos v \cos w & -u \sin v \sin w \\ \sin v \sin w & u \cos v \sin w & u \sin v \cos w \\ \cos v & -u \sin v & 0 \end{vmatrix}$$

$$= u^2 \sin v \qquad (8.22)$$

which is nonzero everywhere but on the z-axis. The inverse transformation is

$$u = (x^2 + y^2 + z^2)^{1/2}, \qquad v = \tan^{-1}\left[\frac{(x^2 + y^2)^{1/2}}{z}\right], \qquad w = \tan^{-1}\left(\frac{y}{x}\right)$$

$$(8.23)$$

EXAMPLE 8-5: (a) Find the scale factors h_u, h_v, h_w and the unit vectors $\mathbf{e}_u, \mathbf{e}_v, \mathbf{e}_w$ for the spherical coordinate system. (b) Prove that the spherical coordinate system is an orthogonal system.

Solution:
(a) Since the position vector is

$$\mathbf{r} = u \sin v \cos w \mathbf{i} + u \sin v \sin w \mathbf{j} + u \cos v \mathbf{k}$$

its partial derivatives are

$$\frac{\partial \mathbf{r}}{\partial u} = \sin v \cos w \mathbf{i} + \sin v \sin w \mathbf{j} + \cos v \mathbf{k}$$

$$\frac{\partial \mathbf{r}}{\partial v} = u \cos v \cos w \mathbf{i} + u \cos v \sin w \mathbf{j} - u \sin v \mathbf{k}$$

$$\frac{\partial \mathbf{r}}{\partial w} = -u \sin v \sin w \mathbf{i} + u \sin v \cos w \mathbf{j}$$

Then the scale factors are

$$h_u = \left| \frac{\partial \mathbf{r}}{\partial u} \right| = [(\sin v \cos w)^2 + (\sin v \sin w)^2 + (\cos v)^2]^{1/2} = 1$$

$$h_v = \left| \frac{\partial \mathbf{r}}{\partial v} \right| = u[(\cos v \cos w)^2 + (\cos v \sin w)^2 + (-\sin v)^2]^{1/2} = u$$

$$h_w = \left| \frac{\partial \mathbf{r}}{\partial w} \right| = u[(-\sin v \sin w)^2 + (\sin v \cos w)^2]^{1/2} = u \sin v$$

and the base vectors are

$$\mathbf{e}_u = \frac{1}{h_u} \frac{\partial \mathbf{r}}{\partial u} = \sin v \cos w \mathbf{i} + \sin v \sin w \mathbf{j} + \cos v \mathbf{k}$$

$$\mathbf{e}_v = \frac{1}{h_v} \frac{\partial \mathbf{r}}{\partial v} = \cos v \cos w \mathbf{i} + \cos v \sin w \mathbf{j} - \sin v \mathbf{k}$$

$$\mathbf{e}_w = \frac{1}{h_w} \frac{\partial \mathbf{r}}{\partial w} = -\sin w \mathbf{i} + \cos w \mathbf{j}$$

(b) Performing the various dot products among $\mathbf{e}_u$, $\mathbf{e}_v$, and $\mathbf{e}_w$,

$$\mathbf{e}_u \cdot \mathbf{e}_v = \sin v \cos v \cos^2 w + \sin v \cos v \sin^2 w - \cos v \sin v = 0$$

$$\mathbf{e}_v \cdot \mathbf{e}_w = -\cos v \cos w \sin w + \cos v \sin w \cos w = 0$$

$$\mathbf{e}_w \cdot \mathbf{e}_u = -\sin w \sin v \cos w + \cos w \sin v \sin w = 0$$

Thus $\mathbf{e}_u, \mathbf{e}_v, \mathbf{e}_w$ are mutually orthogonal.

8-3. Gradient, Divergence, and Curl in Orthogonal Curvilinear Coordinates

If $\psi = \psi(u, v, w)$ is an arbitrary scalar function, the *gradient* of ψ in orthogonal curvilinear coordinates is

$$\nabla \psi = \frac{\partial \psi}{\partial u} \nabla u + \frac{\partial \psi}{\partial v} \nabla v + \frac{\partial \psi}{\partial w} \nabla w \qquad (8.24)$$

$$\nabla \psi = \frac{1}{h_u} \frac{\partial \psi}{\partial u} \mathbf{e}_u + \frac{1}{h_v} \frac{\partial \psi}{\partial v} \mathbf{e}_v + \frac{1}{h_w} \frac{\partial \psi}{\partial w} \mathbf{e}_w \qquad (8.25)$$

If $\mathbf{f} = f_u \mathbf{e}_u + f_v \mathbf{e}_v + f_w \mathbf{e}_w$ is a vector expressed with the basis $\mathbf{e}_u, \mathbf{e}_v, \mathbf{e}_w$, the *divergence* in the orthogonal curvilinear coordinates is

$$\nabla \cdot \mathbf{f} = \frac{1}{h_u h_v h_w} \left[\frac{\partial}{\partial u} (h_v h_w f_u) + \frac{\partial}{\partial v} (h_w h_u f_v) + \frac{\partial}{\partial w} (h_u h_v f_w) \right] \qquad (8.26)$$

The *Laplacian* of the scalar function ψ is

$$\nabla^2 \psi = \nabla \cdot (\nabla \psi) = \frac{1}{h_u h_v h_w} \left[\frac{\partial}{\partial u} \left(\frac{h_v h_w}{h_u} \frac{\partial \psi}{\partial u} \right) + \frac{\partial}{\partial v} \left(\frac{h_w h_u}{h_v} \frac{\partial \psi}{\partial v} \right) + \frac{\partial}{\partial w} \left(\frac{h_u h_v}{h_w} \frac{\partial \psi}{\partial w} \right) \right]$$

$$(8.27)$$

The *curl* of the vector function **f** is

$$\nabla \times \mathbf{f} = \frac{1}{h_v h_w}\left[\frac{\partial}{\partial v}(h_w f_w) - \frac{\partial}{\partial w}(h_v f_v)\right]\mathbf{e}_u + \frac{1}{h_w h_u}\left[\frac{\partial}{\partial w}(h_u f_u) - \frac{\partial}{\partial u}(h_w f_w)\right]\mathbf{e}_v$$

$$+ \frac{1}{h_u h_v}\left[\frac{\partial}{\partial u}(h_v f_v) - \frac{\partial}{\partial v}(h_u f_u)\right]\mathbf{e}_w \tag{8.28}$$

The curl of the vector function (8.28) can also be written in determinant form as

$$\nabla \times \mathbf{f} = \frac{1}{h_u h_v h_w}\begin{vmatrix} h_u\mathbf{e}_u & h_v\mathbf{e}_v & h_w\mathbf{e}_w \\ \dfrac{\partial}{\partial u} & \dfrac{\partial}{\partial v} & \dfrac{\partial}{\partial w} \\ h_u f_u & h_v f_v & h_w f_w \end{vmatrix} \tag{8.29}$$

EXAMPLE 8-6: Verify the definition of the gradient given in (8.24).

Solution: From definition (5.4) the gradient in the rectangular coordinates is

$$\nabla\psi = \frac{\partial\psi}{\partial x}\mathbf{i} + \frac{\partial\psi}{\partial y}\mathbf{j} + \frac{\partial\psi}{\partial z}\mathbf{k}$$

If $\psi = \psi(u, v, w) = \psi[u(x, y, z), v(x, y, z), w(x, y, z)]$, then using the chain rules of elementary calculus,

$$\frac{\partial\psi}{\partial x} = \frac{\partial\psi}{\partial u}\frac{\partial u}{\partial x} + \frac{\partial\psi}{\partial v}\frac{\partial v}{\partial x} + \frac{\partial\psi}{\partial w}\frac{\partial w}{\partial x}$$

$$\frac{\partial\psi}{\partial y} = \frac{\partial\psi}{\partial u}\frac{\partial u}{\partial y} + \frac{\partial\psi}{\partial v}\frac{\partial v}{\partial y} + \frac{\partial\psi}{\partial w}\frac{\partial w}{\partial y}$$

$$\frac{\partial\psi}{\partial z} = \frac{\partial\psi}{\partial u}\frac{\partial u}{\partial z} + \frac{\partial\psi}{\partial v}\frac{\partial v}{\partial z} + \frac{\partial\psi}{\partial w}\frac{\partial w}{\partial z}$$

And then by rearranging, we have

$$\nabla\psi = \frac{\partial\psi}{\partial u}\left(\frac{\partial u}{\partial x}\mathbf{i} + \frac{\partial u}{\partial y}\mathbf{j} + \frac{\partial u}{\partial z}\mathbf{k}\right) + \frac{\partial\psi}{\partial v}\left(\frac{\partial v}{\partial x}\mathbf{i} + \frac{\partial v}{\partial y}\mathbf{j} + \frac{\partial v}{\partial z}\mathbf{k}\right)$$

$$+ \frac{\partial\psi}{\partial w}\left(\frac{\partial w}{\partial x}\mathbf{i} + \frac{\partial w}{\partial y}\mathbf{j} + \frac{\partial w}{\partial z}\mathbf{k}\right)$$

$$= \frac{\partial\psi}{\partial u}\nabla u + \frac{\partial\psi}{\partial v}\nabla v + \frac{\partial\psi}{\partial w}\nabla w$$

EXAMPLE 8-7: Verify the form for gradient given in (8.25).

Solution: Substituting

$$\nabla u = \frac{1}{h_u}\mathbf{e}_u, \qquad \nabla v = \frac{1}{h_v}\mathbf{e}_v, \qquad \nabla w = \frac{1}{h_w}\mathbf{e}_w$$

in (8.24), we obtain

$$\nabla\psi = \frac{1}{h_u}\frac{\partial\psi}{\partial u}\mathbf{e}_u + \frac{1}{h_v}\frac{\partial\psi}{\partial v}\mathbf{e}_v + \frac{1}{h_w}\frac{\partial\psi}{\partial w}\mathbf{e}_w$$

(See Problem 8-4.)

8-4. Special Coordinate Systems

A. Rectangular Cartesian coordinates (x, y, z)

Since $u = x$, $v = y$, and $w = z$ for the rectangular coordinates, the scale factors are

$$h_u = h_x = 1, \qquad h_v = h_y = 1, \qquad h_w = h_x = 1 \qquad (8.30)$$

and the coordinate surfaces for these coordinates are

$$x = \text{const}, \qquad y = \text{const}, \qquad z = \text{const}$$

Let $\psi = \psi(x, y, z)$ be an arbitrary scalar function and let $\mathbf{f} = f_1\mathbf{i} + f_2\mathbf{j} + f_3\mathbf{k}$ be any vector-valued function. From the definitions given in Chapter 5, the gradient of ψ, the divergence and curl of $\mathbf{f}$, and the Laplacian of ψ are

$$\nabla\psi = \frac{\partial\psi}{\partial x}\mathbf{i} + \frac{\partial\psi}{\partial y}\mathbf{j} + \frac{\partial\psi}{\partial z}\mathbf{k} \qquad (8.31)$$

$$\nabla \cdot \mathbf{f} = \frac{\partial f_1}{\partial x} + \frac{\partial f_2}{\partial y} + \frac{\partial f_3}{\partial z} \qquad (8.32)$$

$$\nabla \times \mathbf{f} = \begin{vmatrix} \mathbf{i} & \mathbf{j} & \mathbf{k} \\ \dfrac{\partial}{\partial x} & \dfrac{\partial}{\partial y} & \dfrac{\partial}{\partial z} \\ f_1 & f_2 & f_3 \end{vmatrix} \qquad (8.33)$$

$$\nabla^2\psi = \frac{\partial^2\psi}{\partial x^2} + \frac{\partial^2\psi}{\partial y^2} + \frac{\partial^2\psi}{\partial z^2} \qquad (8.34)$$

B. Circular cylindrical coordinates (ρ, ϕ, z)

Since $u = \rho, v = \phi, w = z$ with $0 \leqslant \rho < \infty, 0 \leqslant \phi < 2\pi, -\infty < z < \infty$ for the circular cylindrical coordinates shown in Figure 8-2, the transformation from rectangular to cylindrical coordinates is

$$x = \rho\cos\phi, \qquad y = \rho\sin\phi, \qquad z = z \qquad (8.35)$$

The unit base vectors for this coordinate system, are $\mathbf{e}_\rho$, $\mathbf{e}_\phi$, and $\mathbf{k}$. The coordinate equation for the surface of a right circular cylinder having the z-axis as a common axis is

$$\rho = (x^2 + y^2)^{1/2} = \text{const}$$

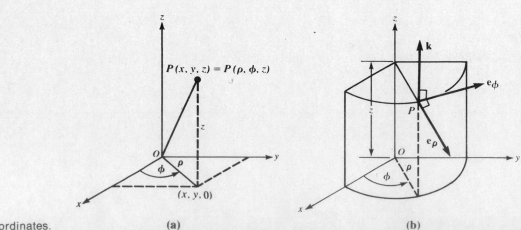

Figure 8-2
Cylindrical coordinates.

(a)

(b)

the equation for the half-plane through the z-axis is

$$\phi = \tan^{-1}\left(\frac{y}{x}\right) = \text{const}$$

and the equation for the plane parallel to the xy-plane is

$$z = \text{const}$$

From the results of Example 8-4 the scale factors are

$$h_u = h_\rho = 1, \qquad h_v = h_\phi = \rho, \qquad h_w = h_z = 1 \qquad (8.36)$$

Note: Some calculus textbooks may use the variables (r, θ, z) for circular cylindrical coordinates rather than (ρ, ϕ, z)

EXAMPLE 8-8: For circular cylindrical coordinates (ρ, ϕ, z), if $\psi = \psi(\rho, \phi, z)$, $\mathbf{f} = f_\rho \mathbf{e}_\rho + f_\phi \mathbf{e}_\phi + f_3 \mathbf{k}$, evaluate (a) $(ds)^2$, (b) dV, (c) $\nabla \psi$, (d) $\nabla \cdot \mathbf{f}$, (e) $\nabla \times \mathbf{f}$, and (f) $\nabla^2 \psi$.

Solution: Using (8.11) and (8.12) and the scale factors, we obtain

(a)
$$(ds)^2 = (d\rho)^2 + (\rho \, d\phi)^2 + (dz)^2 \qquad (8.37)$$

(b)
$$dV = \rho \, d\rho \, d\phi \, dz \qquad (8.38)$$

Now using (8.25) and (8.26) and the scale factors, we get

(c)
$$\nabla \psi = \frac{\partial \psi}{\partial \rho} \mathbf{e}_\rho + \frac{1}{\rho} \frac{\partial \psi}{\partial \phi} \mathbf{e}_\phi + \frac{\partial \psi}{\partial z} \mathbf{k} \qquad (8.39)$$

(d)
$$\nabla \cdot \mathbf{f} = \frac{1}{\rho} \frac{\partial}{\partial \rho} (\rho f_\rho) + \frac{1}{\rho} \frac{\partial f_\phi}{\partial \phi} + \frac{\partial f_3}{\partial z} \qquad (8.40)$$

and from (8.29) and (8.27) and the scale factors, we get

(e)
$$\nabla \times \mathbf{f} = \frac{1}{\rho} \begin{vmatrix} \mathbf{e}_\rho & \rho \mathbf{e}_\phi & \mathbf{k} \\ \dfrac{\partial}{\partial \rho} & \dfrac{\partial}{\partial \phi} & \dfrac{\partial}{\partial z} \\ f_\rho & \rho f_\phi & f_3 \end{vmatrix}$$

$$= \frac{1}{\rho}\left(\frac{\partial f_3}{\partial \phi} - \frac{\partial(\rho f_\phi)}{\partial z}\right)\mathbf{e}_\rho + \left(\frac{\partial f_\rho}{\partial z} - \frac{\partial f_3}{\partial \rho}\right)\mathbf{e}_\phi + \frac{1}{\rho}\left(\frac{\partial(\rho f_\phi)}{\partial \rho} - \frac{\partial f_\rho}{\partial \phi}\right)\mathbf{k} \qquad (8.41)$$

(f)
$$\nabla^2 \psi = \frac{1}{\rho} \frac{\partial}{\partial \rho}\left(\rho \frac{\partial \psi}{\partial \rho}\right) + \frac{1}{\rho^2} \frac{\partial^2 \psi}{\partial \phi^2} + \frac{\partial^2 \psi}{\partial z^2} \qquad (8.42)$$

EXAMPLE 8-9: Find the relationships between the unit base vectors $\mathbf{i}, \mathbf{j}, \mathbf{k}$ of the rectangular Cartesian coordinate system and the unit base vectors $\mathbf{e}_\rho, \mathbf{e}_\phi, \mathbf{k}$ of the circular cylindrical coordinate system.

Solution: From the results of Example 8-4 we have

$$\mathbf{e}_\rho = \cos \phi \, \mathbf{i} + \sin \phi \, \mathbf{j} \qquad (8.43a)$$

$$\mathbf{e}_\phi = -\sin \phi \, \mathbf{i} + \cos \phi \, \mathbf{j} \qquad (8.43b)$$

$$\mathbf{k} = \mathbf{k} \qquad (8.43c)$$

From (1.73) we have the relationships

$$\mathbf{e}_\rho = (\mathbf{e}_\rho \cdot \mathbf{i})\mathbf{i} + (\mathbf{e}_\rho \cdot \mathbf{j})\mathbf{j} + (\mathbf{e}_\rho \cdot \mathbf{k})\mathbf{k} \qquad (8.44a)$$

$$\mathbf{e}_\phi = (\mathbf{e}_\phi \cdot \mathbf{i})\mathbf{i} + (\mathbf{e}_\phi \cdot \mathbf{j})\mathbf{j} + (\mathbf{e}_\phi \cdot \mathbf{k})\mathbf{k} \qquad (8.44b)$$

Comparing sets (8.43) and (8.44), we have

$$\mathbf{e}_\rho \cdot \mathbf{i} = \cos \phi, \qquad \mathbf{e}_\rho \cdot \mathbf{j} = \sin \phi, \qquad \mathbf{e}_\rho \cdot \mathbf{k} = 0$$
$$\mathbf{e}_\phi \cdot \mathbf{i} = -\sin \phi, \qquad \mathbf{e}_\phi \cdot \mathbf{j} = \cos \phi, \qquad \mathbf{e}_\phi \cdot \mathbf{k} = 0 \qquad (8.45)$$

Table 8-1 shows the results of (8.45) in tabular form.

TABLE 8-1 Relationship between Rectangular and Cylindrical Base Vectors

$\cdot$	$\mathbf{i}$	$\mathbf{j}$	$\mathbf{k}$
$\mathbf{e}_\rho$	$\cos \phi$	$\sin \phi$	0
$\mathbf{e}_\phi$	$-\sin \phi$	$\cos \phi$	0
$\mathbf{k}$	0	0	1

Note: We can also obtain (8.45) from Figure 8-2b by applying the definition of the dot product (1.22).

Using (8.45), base vectors $\mathbf{i}, \mathbf{j}, \mathbf{k}$ can be expressed in terms of $\mathbf{e}_\rho, \mathbf{e}_\phi, \mathbf{k}$:

$$\mathbf{i} = (\mathbf{i} \cdot \mathbf{e}_\rho)\mathbf{e}_\rho + (\mathbf{i} \cdot \mathbf{e}_\phi)\mathbf{e}_\phi + (\mathbf{i} \cdot \mathbf{k})\mathbf{k} = \cos \phi \mathbf{e}_\rho - \sin \phi \mathbf{e}_\phi \qquad (8.46a)$$
$$\mathbf{j} = (\mathbf{j} \cdot \mathbf{e}_\rho)\mathbf{e}_\rho + (\mathbf{j} \cdot \mathbf{e}_\phi)\mathbf{e}_\phi + (\mathbf{j} \cdot \mathbf{k})\mathbf{k} = \sin \phi \mathbf{e}_\rho + \cos \phi \mathbf{e}_\phi \qquad (8.46b)$$
$$\mathbf{k} = \mathbf{k} \qquad (8.46c)$$

Using matrix notation, the transformations between $\mathbf{i}, \mathbf{j}, \mathbf{k}$ and $\mathbf{e}_\rho, \mathbf{e}_\phi, \mathbf{k}$ (8.43) and (8.46), can be written, respectively, as

$$\begin{bmatrix} \mathbf{e}_\rho \\ \mathbf{e}_\phi \\ \mathbf{k} \end{bmatrix} = \begin{bmatrix} \cos \phi & \sin \phi & 0 \\ -\sin \phi & \cos \phi & 0 \\ 0 & 0 & 1 \end{bmatrix} \begin{bmatrix} \mathbf{i} \\ \mathbf{j} \\ \mathbf{k} \end{bmatrix} \qquad (8.47)$$

$$\begin{bmatrix} \mathbf{i} \\ \mathbf{j} \\ \mathbf{k} \end{bmatrix} = \begin{bmatrix} \cos \phi & -\sin \phi & 0 \\ \sin \phi & \cos \phi & 0 \\ 0 & 0 & 1 \end{bmatrix} \begin{bmatrix} \mathbf{e}_\rho \\ \mathbf{e}_\phi \\ \mathbf{k} \end{bmatrix} \qquad (8.48)$$

Note: Unit vectors $\mathbf{e}_\rho$ and $\mathbf{e}_\phi$ vary in direction from point to point.

C. Spherical coordinates (r, θ, ϕ)

Since $u = r$, $v = \theta$, $w = \phi$ with $0 \leqslant r < \infty$, $0 \leqslant \theta \leqslant \pi$, $0 \leqslant \phi < 2\pi$, as shown in Figure 8-3, the transformation from the rectangular to the spherical coordinates is

$$x = r \sin \theta \cos \phi, \qquad y = r \sin \theta \sin \phi, \qquad z = r \cos \theta \qquad (8.49)$$

The coordinate equation for the surface of the concentric spheres centered at the origin is

$$r = (x^2 + y^2 + z^2)^{1/2} = \text{const}$$

The equation for circular cones centered on the z-axis and having vertices at the origin is

$$\theta = \cos^{-1}\left[\frac{z}{(x^2 + y^2 + z^2)^{1/2}}\right] = \text{const}$$

and the equation for the half-plane through the z-axis is

$$\phi = \tan^{-1}\left(\frac{y}{x}\right) = \text{const}$$

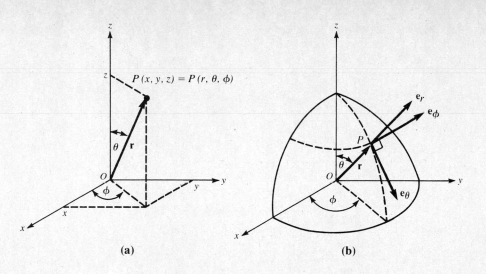

Figure 8-3
Spherical coordinates.

From the results of Example 8-5, the scale factors are

$$h_u = h_r = 1, \qquad h_v = h_\theta = r, \qquad h_w = h_\phi = r \sin \theta \qquad (8.50)$$

Note: Some calculus textbooks use (ρ, ϕ, θ) rather than (r, θ, ϕ) for spherical coordinates.

EXAMPLE 8-10: For spherical coordinates (r, θ, ϕ), if $\psi = \psi(r, \theta, \phi)$ and $\mathbf{f} = f_r \mathbf{e}_r + f_\theta \mathbf{e}_\theta + f_\phi \mathbf{e}_\phi$, evaluate (a) $(ds)^2$, (b) dV, (c) $\nabla \psi$, (d) $\nabla \cdot \mathbf{f}$, (e) $\nabla \times \mathbf{f}$, and (f) $\nabla^2 \psi$.

Solution: As in Example 8-8, we use (8.11), (8.12), (8.25), (8.26), (8.29), (8.27), and the scale factors:

(a)
$$(ds)^2 = (dr)^2 + (r \, d\theta)^2 + (r \sin \theta \, d\phi)^2 \qquad (8.51)$$

(b)
$$dV = r^2 \sin \theta \, dr \, d\theta \, d\phi \qquad (8.52)$$

(c)
$$\nabla \psi = \frac{\partial \psi}{\partial r} \mathbf{e}_r + \frac{1}{r} \frac{\partial \psi}{\partial \theta} \mathbf{e}_\theta + \frac{1}{r \sin \theta} \frac{\partial \psi}{\partial \phi} \mathbf{e}_\phi \qquad (8.53)$$

(d)
$$\nabla \cdot \mathbf{f} = \frac{1}{r^2 \sin \theta} \left[\sin \theta \frac{\partial}{\partial r} (r^2 f_r) + r \frac{\partial}{\partial \theta} (\sin \theta f_\theta) + r \frac{\partial f_\phi}{\partial \phi} \right]$$

$$= \frac{1}{r^2} \frac{\partial}{\partial r} (r^2 f_r) + \frac{1}{r \sin \theta} \frac{\partial}{\partial \theta} (\sin \theta f_\theta) + \frac{1}{r \sin \theta} \frac{\partial f_\phi}{\partial \phi} \qquad (8.54)$$

(e)
$$\nabla \times \mathbf{f} = \frac{1}{r^2 \sin \theta} \begin{vmatrix} \mathbf{e}_r & r \mathbf{e}_\theta & r \sin \theta \mathbf{e}_\phi \\ \dfrac{\partial}{\partial r} & \dfrac{\partial}{\partial \theta} & \dfrac{\partial}{\partial \phi} \\ f_r & r f_\theta & r \sin \theta f_\phi \end{vmatrix} \qquad (8.55a)$$

or equivalently

$$\nabla \times \mathbf{f} = \frac{1}{r \sin \theta} \left[\frac{\partial}{\partial \theta} (\sin \theta f_\phi) - \frac{\partial f_\theta}{\partial \phi} \right] \mathbf{e}_r + \frac{1}{r} \left[\frac{1}{\sin \theta} \frac{\partial f_r}{\partial \phi} - \frac{\partial}{\partial r} (r f_\phi) \right] \mathbf{e}_\theta$$

$$+ \frac{1}{r} \left[\frac{\partial}{\partial r} (r f_\theta) - \frac{\partial f_r}{\partial \theta} \right] \mathbf{e}_\phi \qquad (8.55b)$$

$$\nabla^2 \psi = \frac{1}{r^2 \sin \theta} \left[\sin \theta \frac{\partial}{\partial r} \left(r^2 \frac{\partial \psi}{\partial r} \right) + \frac{\partial}{\partial \theta} \left(\sin \theta \frac{\partial \psi}{\partial \theta} \right) + \frac{1}{\sin \theta} \frac{\partial^2 \psi}{\partial \phi^2} \right]$$

$$= \frac{1}{r^2} \frac{\partial}{\partial r} \left(r^2 \frac{\partial \psi}{\partial r} \right) + \frac{1}{r^2 \sin \theta} \frac{\partial}{\partial \theta} \left(\sin \theta \frac{\partial \psi}{\partial \theta} \right) + \frac{1}{r^2 \sin^2 \theta} \frac{\partial^2 \psi}{\partial \phi^2} \qquad (8.56)$$

EXAMPLE 8-11: Find the relationships between unit base vectors **i**, **j**, **k** of the rectangular Cartesian coordinate system and unit base vectors $\mathbf{e}_r$, $\mathbf{e}_\theta$, $\mathbf{e}_\phi$ of the spherical coordinate system.

Solution: From the results of Example 8-5, we have

$$\mathbf{e}_r = \sin\theta\cos\phi\,\mathbf{i} + \sin\theta\sin\phi\,\mathbf{j} + \cos\theta\,\mathbf{k}$$
$$\mathbf{e}_\theta = \cos\theta\cos\phi\,\mathbf{i} + \cos\theta\sin\phi\,\mathbf{j} - \sin\theta\,\mathbf{k} \quad (8.57)$$
$$\mathbf{e}_\phi = -\sin\phi\,\mathbf{i} + \cos\phi\,\mathbf{j}$$

From (1.73)

$$\mathbf{e}_r = (\mathbf{e}_r \cdot \mathbf{i})\mathbf{i} + (\mathbf{e}_r \cdot \mathbf{j})\mathbf{j} + (\mathbf{e}_r \cdot \mathbf{k})\mathbf{k}$$
$$\mathbf{e}_\theta = (\mathbf{e}_\theta \cdot \mathbf{i})\mathbf{i} + (\mathbf{e}_\theta \cdot \mathbf{j})\mathbf{j} + (\mathbf{e}_\theta \cdot \mathbf{k})\mathbf{k} \quad (8.58)$$
$$\mathbf{e}_\phi = (\mathbf{e}_\phi \cdot \mathbf{i})\mathbf{i} + (\mathbf{e}_\phi \cdot \mathbf{j})\mathbf{j} + (\mathbf{e}_\phi \cdot \mathbf{k})\mathbf{k}$$

Comparing (8.57) and (8.58), we obtain

$$\mathbf{e}_r \cdot \mathbf{i} = \sin\theta\cos\phi, \qquad \mathbf{e}_r \cdot \mathbf{j} = \sin\theta\sin\phi, \qquad \mathbf{e}_r \cdot \mathbf{k} = \cos\theta$$
$$\mathbf{e}_\theta \cdot \mathbf{i} = \cos\theta\cos\phi, \qquad \mathbf{e}_\theta \cdot \mathbf{j} = \cos\theta\sin\phi, \qquad \mathbf{e}_\theta \cdot \mathbf{k} = -\sin\theta \quad (8.59)$$
$$\mathbf{e}_\phi \cdot \mathbf{i} = -\sin\phi \qquad \mathbf{e}_\phi \cdot \mathbf{j} = \cos\phi \qquad \mathbf{e}_\phi \cdot \mathbf{k} = 0$$

The results of (8.59) are tabulated in Table 8-2. The relations (8.59) can also be obtained from Figure 8-3 by applying the definition of the dot product (1.22).

TABLE 8-2 Relationship between Rectangular and Spherical Base Vectors

$\cdot$	**i**	**j**	**k**
$\mathbf{e}_r$	$\sin\theta\cos\phi$	$\sin\theta\sin\phi$	$\cos\theta$
$\mathbf{e}_\theta$	$\cos\theta\cos\phi$	$\cos\theta\sin\phi$	$-\sin\theta$
$\mathbf{e}_\phi$	$-\sin\phi$	$\cos\phi$	0

Using set (8.59), we can now express **i**, **j**, **k** in terms of $\mathbf{e}_r$, $\mathbf{e}_\theta$, $\mathbf{e}_\phi$:

$$\mathbf{i} = (\mathbf{i} \cdot \mathbf{e}_r)\mathbf{e}_r + (\mathbf{i} \cdot \mathbf{e}_\theta)\mathbf{e}_\theta + (\mathbf{i} \cdot \mathbf{e}_\phi)\mathbf{e}_\phi$$
$$= \sin\theta\cos\phi\,\mathbf{e}_r + \cos\theta\cos\phi\,\mathbf{e}_\theta - \sin\phi\,\mathbf{e}_\phi \quad (8.60a)$$
$$\mathbf{j} = (\mathbf{j} \cdot \mathbf{e}_r)\mathbf{e}_r + (\mathbf{j} \cdot \mathbf{e}_\theta)\mathbf{e}_\theta + (\mathbf{j} \cdot \mathbf{e}_\phi)\mathbf{e}_\phi$$
$$= \sin\theta\sin\phi\,\mathbf{e}_r + \cos\theta\sin\phi\,\mathbf{e}_\theta + \cos\phi\,\mathbf{e}_\phi \quad (8.60b)$$
$$\mathbf{k} = (\mathbf{k} \cdot \mathbf{e}_r)\mathbf{e}_r + (\mathbf{k} \cdot \mathbf{e}_\theta)\mathbf{e}_\theta + (\mathbf{k} \cdot \mathbf{e}_\phi)\mathbf{e}_\phi$$
$$= \cos\theta\,\mathbf{e}_r - \sin\theta\,\mathbf{e}_\theta \quad (8.60c)$$

Note: The unit vectors $\mathbf{e}_r$, $\mathbf{e}_\theta$, $\mathbf{e}_\phi$ vary in direction from point to point as we move through the space, but the unit vectors **i**, **j**, **k** maintain their fixed directions.

In matrix notation the transformations between **i**, **j**, **k** and $\mathbf{e}_r$, $\mathbf{e}_\theta$, $\mathbf{e}_\phi$ expressed in sets (8.57) and (8.60) can be written, respectively, as

$$\begin{bmatrix} \mathbf{e}_r \\ \mathbf{e}_\theta \\ \mathbf{e}_\phi \end{bmatrix} = \begin{bmatrix} \sin\theta\cos\phi & \sin\theta\sin\phi & \cos\theta \\ \cos\theta\cos\phi & \cos\theta\sin\phi & -\sin\theta \\ -\sin\phi & \cos\phi & 0 \end{bmatrix} \begin{bmatrix} \mathbf{i} \\ \mathbf{j} \\ \mathbf{k} \end{bmatrix} \quad (8.61)$$

$$\begin{bmatrix} \mathbf{i} \\ \mathbf{j} \\ \mathbf{k} \end{bmatrix} = \begin{bmatrix} \sin\theta\cos\phi & \cos\theta\cos\phi & -\sin\phi \\ \sin\theta\sin\phi & \cos\theta\sin\phi & \cos\phi \\ \cos\theta & -\sin\theta & 0 \end{bmatrix} \begin{bmatrix} \mathbf{e}_r \\ \mathbf{e}_\theta \\ \mathbf{e}_\phi \end{bmatrix} \quad (8.62)$$

SUMMARY

1. The transformation between the rectangular coordinates (x, y, z) and the curvilinear coordinates (u, v, w) is defined by

$$x = x(u, v, w), \qquad y = y(u, v, w), \qquad z = z(u, v, w)$$

or

$$u = u(x, y, z), \qquad v = v(x, y, z), \qquad w = w(x, y, z)$$

2. The curvilinear coordinate surfaces are defined by

$$u(x, y, z) = c_1, \qquad v(x, y, z) = c_2, \qquad w(x, y, z) = c_3$$

where c_1, c_2, c_3 are constants. The curvilinear coordinate curve u-curve is defined by

$$x = x(u, v, w), \qquad y = y(u, v, w), \qquad z = z(u, v, w)$$

where v and w are constant, and the v- and w-curves are defined similarly.

3. There are two sets of base vectors at each point P in space:

   (a) The first set is the set of unit vectors $\mathbf{e}_u, \mathbf{e}_v, \mathbf{e}_w$ tangent at point P to the coordinate curves given by

   $$\mathbf{e}_u = \frac{1}{h_u} \frac{\partial \mathbf{r}}{\partial u}, \qquad \mathbf{e}_v = \frac{1}{h_v} \frac{\partial \mathbf{r}}{\partial v}, \qquad \mathbf{e}_w = \frac{1}{h_w} \frac{\partial \mathbf{r}}{\partial w}$$

   where $\mathbf{r}$ is the position vector of P and

   $$h_u = \left| \frac{\partial \mathbf{r}}{\partial u} \right|, \qquad h_v = \left| \frac{\partial \mathbf{r}}{\partial v} \right|, \qquad h_w = \left| \frac{\partial \mathbf{r}}{\partial w} \right|$$

   are called scale factors.

   (b) The second set is the set of unit vectors $\mathbf{e}_U, \mathbf{e}_V, \mathbf{e}_W$ normal at point P to the coordinate surfaces given by

   $$\mathbf{e}_U = \frac{1}{H_u} \nabla u, \qquad \mathbf{e}_V = \frac{1}{H_v} \nabla v, \qquad \mathbf{e}_W = \frac{1}{H_w} \nabla w$$

   where

   $$H_u = |\nabla u|, \qquad H_v = |\nabla v|, \qquad H_w = |\nabla w|$$

4. In an orthogonal curvilinear coordinate system, $\mathbf{e}_u, \mathbf{e}_v, \mathbf{e}_w$ are mutually orthogonal at every point; that is,

$$\mathbf{e}_u \cdot \mathbf{e}_v = \mathbf{e}_v \cdot \mathbf{e}_w = \mathbf{e}_w \cdot \mathbf{e}_u = 0$$

and

$$\mathbf{e}_u = \mathbf{e}_U, \qquad \mathbf{e}_v = \mathbf{e}_V, \qquad \mathbf{e}_w = \mathbf{e}_W$$

5. In an orthogonal curvilinear coordinate system

   (a) $\qquad ds_u = h_u\, du, \qquad ds_v = h_v\, dv, \qquad ds_w = h_w\, dw$

   (b) $\qquad (ds)^2 = (h_u\, du)^2 + (h_v\, dv)^2 + (h_w\, dw)^2$

   (c) $\qquad dV = h_u h_v h_w\, du\, dv\, dw$

   where ds represents the infinitesimal arc length between the points (u, v, w) and $(u + du, v + dv, w + dw)$, and dV is the volume element.

6. The cylindrical coordinate system (ρ, ϕ, z) is defined by the transformation

$$x = \rho \cos \phi, \qquad y = \rho \sin \phi, \qquad z = z$$
$$0 \leqslant \rho < \infty, \qquad 0 \leqslant \phi < 2\pi, \qquad -\infty < z < \infty$$

and the scale factors are

$$h_u = h_\rho = 1, \qquad h_v = h_\phi = \rho, \qquad h_w = h_z = 1$$

7. The transformations between base vectors **i, j, k** in rectangular coordinates and base vectors $\mathbf{e}_\rho, \mathbf{e}_\phi, \mathbf{k}$ in cylindrical coordinates are given by

$$\begin{bmatrix} \mathbf{e}_\rho \\ \mathbf{e}_\phi \\ \mathbf{k} \end{bmatrix} = \begin{bmatrix} \cos\phi & \sin\phi & 0 \\ -\sin\phi & \cos\phi & 0 \\ 0 & 0 & 1 \end{bmatrix} \begin{bmatrix} \mathbf{i} \\ \mathbf{j} \\ \mathbf{k} \end{bmatrix}$$

$$\begin{bmatrix} \mathbf{i} \\ \mathbf{j} \\ \mathbf{k} \end{bmatrix} = \begin{bmatrix} \cos\phi & -\sin\phi & 0 \\ \sin\phi & \cos\phi & 0 \\ 0 & 0 & 1 \end{bmatrix} \begin{bmatrix} \mathbf{e}_\rho \\ \mathbf{e}_\phi \\ \mathbf{k} \end{bmatrix}$$

8. The spherical coordinate system (r, θ, ϕ) is defined by the transformation

$$x = r\sin\theta\cos\phi, \qquad y = r\sin\theta\sin\phi, \qquad z = r\cos\theta$$
$$0 \leqslant r < \infty, \qquad 0 \leqslant \theta \leqslant \pi, \qquad 0 \leqslant \phi < 2\pi$$

and the scale factors are

$$h_u = h_r = 1, \qquad h_v = h_\theta = r, \qquad h_w = h_\phi = r\sin\theta$$

9. The transformations between base vectors **i, j, k** and base vectors $\mathbf{e}_r, \mathbf{e}_\theta, \mathbf{e}_\phi$ in spherical coordinates are given by

$$\begin{bmatrix} \mathbf{e}_r \\ \mathbf{e}_\theta \\ \mathbf{e}_\phi \end{bmatrix} = \begin{bmatrix} \sin\theta\cos\phi & \sin\theta\sin\phi & \cos\theta \\ \cos\theta\cos\phi & \cos\theta\sin\phi & -\sin\theta \\ -\sin\phi & \cos\phi & 0 \end{bmatrix} \begin{bmatrix} \mathbf{i} \\ \mathbf{j} \\ \mathbf{k} \end{bmatrix}$$

$$\begin{bmatrix} \mathbf{i} \\ \mathbf{j} \\ \mathbf{k} \end{bmatrix} = \begin{bmatrix} \sin\theta\cos\phi & \cos\theta\cos\phi & -\sin\phi \\ \sin\theta\sin\phi & \cos\theta\sin\phi & \cos\phi \\ \cos\theta & -\sin\theta & 0 \end{bmatrix} \begin{bmatrix} \mathbf{e}_r \\ \mathbf{e}_\theta \\ \mathbf{e}_\phi \end{bmatrix}$$

10. Gradient, divergence, and curl in the orthogonal curvilinear coordinates are given by

$$\nabla\psi = \frac{1}{h_u}\frac{\partial\psi}{\partial u}\mathbf{e}_u + \frac{1}{h_v}\frac{\partial\psi}{\partial v}\mathbf{e}_v + \frac{1}{h_w}\frac{\partial\psi}{\partial w}\mathbf{e}_w$$

$$\nabla\cdot\mathbf{f} = \frac{1}{h_u h_v h_w}\left[\frac{\partial}{\partial u}(h_v h_w f_u) + \frac{\partial}{\partial v}(h_w h_u f_v) + \frac{\partial}{\partial w}(h_u h_v f_w)\right]$$

$$\nabla\times\mathbf{f} = \frac{1}{h_u h_v h_w}\begin{vmatrix} h_u\mathbf{e}_u & h_v\mathbf{e}_v & h_w\mathbf{e}_w \\ \dfrac{\partial}{\partial u} & \dfrac{\partial}{\partial v} & \dfrac{\partial}{\partial w} \\ h_u f_u & h_v f_v & h_w f_w \end{vmatrix}$$

RAISE YOUR GRADES

Can you explain...?

☑ how the scale factors and the base vectors are defined in a curvilinear coordinate system

☑ how to find the scale factors and the base vectors in the cylindrical coordinate system

☑ how to find the scale factors and the base vectors in the spherical coordinate system

☑ what the major differences and the relationships are between the base vectors of the rectangular Cartesian coordinate system and the base vectors of the cylindrical and spherical coordinate systems
☑ how the gradient, divergence, and curl are defined in orthogonal curvilinear coordinates

SOLVED PROBLEMS

Curvilinear Coordinates

PROBLEM 8-1 Show that $\partial\mathbf{r}/\partial u$, $\partial\mathbf{r}/\partial v$, $\partial\mathbf{r}/\partial w$ and ∇u, ∇v, ∇w are reciprocal sets of vectors.

Solution: Since

$$u = u(x, y, z) = u[x(u,v,w), y(u,v,w), z(u,v,w)] \tag{a}$$

differentiating with respect to u yields

$$\frac{\partial u}{\partial x}\frac{\partial x}{\partial u} + \frac{\partial u}{\partial y}\frac{\partial y}{\partial u} + \frac{\partial u}{\partial z}\frac{\partial z}{\partial u} = 1 \tag{b}$$

Since $\dfrac{\partial\mathbf{r}}{\partial u} = \dfrac{\partial x}{\partial u}\mathbf{i} + \dfrac{\partial y}{\partial u}\mathbf{j} + \dfrac{\partial z}{\partial u}\mathbf{k}$ and $\nabla u = \dfrac{\partial u}{\partial x}\mathbf{i} + \dfrac{\partial u}{\partial y}\mathbf{j} + \dfrac{\partial u}{\partial z}\mathbf{k}$, we can write eq. (b) as

$$\frac{\partial\mathbf{r}}{\partial u}\cdot\nabla u = 1 \tag{c}$$

Similarly,

$$\frac{\partial\mathbf{r}}{\partial v}\cdot\nabla v = 1, \qquad \frac{\partial\mathbf{r}}{\partial w}\cdot\nabla w = 1 \tag{d}$$

Now differentiating eq. (a) with respect to v,

$$\frac{\partial u}{\partial x}\frac{\partial x}{\partial v} + \frac{\partial u}{\partial y}\frac{\partial y}{\partial v} + \frac{\partial u}{\partial z}\frac{\partial z}{\partial v} = 0$$

or

$$\frac{\partial\mathbf{r}}{\partial v}\cdot\nabla u = 0 \tag{e}$$

Similarly,

$$\frac{\partial\mathbf{r}}{\partial w}\cdot\nabla u = 0, \quad \frac{\partial\mathbf{r}}{\partial u}\cdot\nabla v = 0, \quad \frac{\partial\mathbf{r}}{\partial w}\cdot\nabla v = 0, \quad \frac{\partial\mathbf{r}}{\partial u}\cdot\nabla w = 0, \quad \frac{\partial\mathbf{r}}{\partial v}\cdot\nabla w = 0 \tag{f}$$

If $u = q_1$, $v = q_2$, and $w = q_3$, then eqs. (e) and (f) can be summarized as

$$\frac{\partial\mathbf{r}}{\partial q_m}\cdot(\nabla q_n) = \delta_{mn} = \begin{cases} 1 & \text{if} \quad m = n \\ 0 & \text{if} \quad m \neq n \end{cases} \tag{g}$$

where δ_{mn} is the Kronecker delta. Hence by the definition of reciprocal sets of vectors (see Problem 1-19), we conclude that $\partial\mathbf{r}/\partial u$, $\partial\mathbf{r}/\partial v$, $\partial\mathbf{r}/\partial w$ and ∇u, ∇v, ∇w are reciprocal sets of vectors.

PROBLEM 8-2 Show that

$$\frac{\partial \mathbf{r}}{\partial u} = \frac{1}{\alpha} \nabla v \times \nabla w, \qquad \frac{\partial \mathbf{r}}{\partial v} = \frac{1}{\alpha} \nabla w \times \nabla u, \qquad \frac{\partial \mathbf{r}}{\partial w} = \frac{1}{\alpha} \nabla u \times \nabla v \qquad \text{(a)}$$

$$\nabla u = \frac{1}{\beta}\left(\frac{\partial \mathbf{r}}{\partial v} \times \frac{\partial \mathbf{r}}{\partial w}\right), \qquad \nabla v = \frac{1}{\beta}\left(\frac{\partial \mathbf{r}}{\partial w} \times \frac{\partial \mathbf{r}}{\partial u}\right), \qquad \nabla w = \frac{1}{\beta}\left(\frac{\partial \mathbf{r}}{\partial u} \times \frac{\partial \mathbf{r}}{\partial v}\right) \qquad \text{(b)}$$

where

$$\alpha = [\nabla u \, \nabla v \, \nabla w] = \nabla u \cdot \nabla v \times \nabla w \qquad \text{(c)}$$

$$\beta = \left[\frac{\partial \mathbf{r}}{\partial u} \frac{\partial \mathbf{r}}{\partial v} \frac{\partial \mathbf{r}}{\partial w}\right] = \frac{\partial \mathbf{r}}{\partial u} \cdot \frac{\partial \mathbf{r}}{\partial v} \times \frac{\partial \mathbf{r}}{\partial w} = \frac{1}{\alpha} \qquad \text{(d)}$$

Solution: Since the vectors $\partial \mathbf{r}/\partial u$, $\partial \mathbf{r}/\partial v$, $\partial \mathbf{r}/\partial w$ and ∇u, ∇v, ∇w form reciprocal sets of vectors, eqs. (a)–(d) follow directly from the results of Problems 1-19, 1-20, and 1-21.

PROBLEM 8-3 If a system of curvilinear coordinates is orthogonal, show that

$$\mathbf{e}_u = \mathbf{e}_U, \qquad \mathbf{e}_v = \mathbf{e}_V, \qquad \mathbf{e}_w = \mathbf{e}_W \qquad \text{(a)}$$

Solution: From Problem 8-1, we know that $\mathbf{e}_u, \mathbf{e}_v, \mathbf{e}_w$ and $\mathbf{e}_U, \mathbf{e}_V, \mathbf{e}_W$ are reciprocal sets of vectors. Now the orthonormal system $\mathbf{e}_u, \mathbf{e}_v, \mathbf{e}_w$ satisfies

$$\mathbf{e}_u = \mathbf{e}_v \times \mathbf{e}_w, \qquad \mathbf{e}_v = \mathbf{e}_w \times \mathbf{e}_u, \qquad \mathbf{e}_w = \mathbf{e}_u \times \mathbf{e}_v$$

Hence by Example 8-2 and Problem 8-2, we have the result (a).

PROBLEM 8-4 Show that in an orthogonal curvilinear coordinate system

$$h_u = \frac{1}{H_u}, \qquad h_v = \frac{1}{H_v}, \qquad h_w = \frac{1}{H_w} \qquad \text{(a)}$$

$$\mathbf{e}_u = h_u \nabla u, \qquad \mathbf{e}_v = h_v \nabla v, \qquad \mathbf{e}_w = h_w \nabla w \qquad \text{(b)}$$

$$[\nabla u \, \nabla v \, \nabla w] = \nabla u \cdot \nabla v \times \nabla w = \frac{1}{h_u h_v h_w} \qquad \text{(c)}$$

Solution:
(a) By (8.4) and (8.6), and because of eq. (c) in Problem 8-1, we have

$$\mathbf{e}_u \cdot \mathbf{e}_U = \left(\frac{1}{h_u}\frac{\partial \mathbf{r}}{\partial u}\right) \cdot \left(\frac{1}{H_u}\nabla u\right) = \frac{1}{h_u H_u}\left(\frac{\partial \mathbf{r}}{\partial u} \cdot \nabla u\right) = \frac{1}{h_u H_u} \qquad \text{(d)}$$

But in an orthogonal system, by the result of Problem 8-3, we have

$$\frac{1}{h_u H_u} = \mathbf{e}_u \cdot \mathbf{e}_U = \mathbf{e}_u \cdot \mathbf{e}_u = 1$$

Hence $h_u = 1/H_u$. By similar reasoning, we can determine that $h_v = 1/H_v$ and $h_w = 1/H_w$.
(b) By combining the results of Problem 8-3 with (8.6) and eq. (a) of Problem 8-4, we have

$$\mathbf{e}_u = \mathbf{e}_U = \frac{1}{H_u}\nabla u = h_u \nabla u$$

$$\mathbf{e}_v = \mathbf{e}_V = \frac{1}{H_v}\nabla v = h_v \nabla v$$

$$\mathbf{e}_w = \mathbf{e}_W = \frac{1}{H_w}\nabla w = h_w \nabla w$$

(c) By eq. (b) of Problem 8-4 we have

$$\nabla u = \frac{1}{h_u}\,\mathbf{e}_u, \qquad \nabla v = \frac{1}{h_v}\,\mathbf{e}_v, \qquad \nabla w = \frac{1}{h_w}\,\mathbf{e}_w \tag{e}$$

And because of condition (8.9) we can write

$$[\nabla u\,\nabla v\,\nabla w] = \nabla u \cdot \nabla v \times \nabla w$$

$$= \frac{1}{h_u h_v h_w}\,\mathbf{e}_u \cdot \mathbf{e}_v \times \mathbf{e}_w$$

$$= \frac{1}{h_u h_v h_w}$$

Orthogonal Curvilinear Coordinates

PROBLEM 8-5 Find H_u, H_v, H_w and the vectors $\mathbf{e}_U$, $\mathbf{e}_V$, $\mathbf{e}_W$ for the cylindrical coordinate system.

Solution: From the inverse transformation (8.20) we have

$$\nabla u = \frac{1}{(x^2+y^2)^{1/2}}\,(x\mathbf{i}+y\mathbf{j}), \qquad \nabla v = \frac{1}{(x^2+y^2)}\,(-y\mathbf{i}+x\mathbf{j}), \qquad \nabla w = \mathbf{k}$$

Hence the scale factors are

$$H_u = |\nabla u| = 1 = \frac{1}{h_u}$$

$$H_v = |\nabla v| = \frac{1}{(x^2+y^2)^{1/2}} = \frac{1}{u} = \frac{1}{h_v}$$

$$H_w = |\nabla w| = 1 = \frac{1}{h_w}$$

and the base vectors are

$$\mathbf{e}_U = \frac{1}{H_u}\nabla u = \frac{1}{(x^2+y^2)^{1/2}}\,(x\mathbf{i}+y\mathbf{j})$$

$$= \frac{1}{u}\,(u\cos v\,\mathbf{i} + u\sin v\,\mathbf{j})$$

$$= \cos v\,\mathbf{i} + \sin v\,\mathbf{j}$$
$$= \mathbf{e}_u$$

$$\mathbf{e}_V = \frac{1}{H_v}\nabla v = (x^2+y^2)^{1/2}\frac{1}{(x^2+y^2)}\,(-y\mathbf{i}+x\mathbf{j})$$

$$= \frac{1}{(x^2+y^2)^{1/2}}\,(-y\mathbf{i}+x\mathbf{j})$$

$$= \frac{1}{u}\,(-u\sin v\,\mathbf{i} + u\cos v\,\mathbf{j})$$

$$= -\sin v\,\mathbf{i} + \cos v\,\mathbf{j}$$
$$= \mathbf{e}_v$$

$$\mathbf{e}_W = \frac{1}{H_w}\nabla w = \mathbf{k} = \mathbf{e}_w$$

PROBLEM 8-6 Determine H_u, H_v, H_w and $\mathbf{e}_U$, $\mathbf{e}_V$, $\mathbf{e}_W$ for the spherical coordinate system.

Solution: From (8.23) we have

$$\nabla u = \frac{1}{(x^2 + y^2 + z^2)^{1/2}}(x\mathbf{i} + y\mathbf{j} + z\mathbf{k})$$

$$\nabla v = \frac{1}{(x^2 + y^2)^{1/2}(x^2 + y^2 + z^2)}[zx\mathbf{i} + zy\mathbf{j} - (x^2 + y^2)\mathbf{k}]$$

$$\nabla w = \frac{1}{(x^2 + y^2)}(-y\mathbf{i} + x\mathbf{j})$$

Hence the scale factors are

$$H_u = |\nabla u| = 1 = \frac{1}{h_u}$$

$$H_u = |\nabla v| = \frac{1}{(x^2 + y^2 + z^2)^{1/2}} = \frac{1}{u} = \frac{1}{h_v}$$

$$H_w = |\nabla w| = \frac{1}{(x^2 + y^2)^{1/2}} = \frac{1}{u \sin v} = \frac{1}{h_w}$$

and the base vectors are

$$\mathbf{e}_U = \frac{1}{H_u}\nabla u = \frac{1}{(x^2 + y^2 + z^2)^{1/2}}(x\mathbf{i} + y\mathbf{j} + z\mathbf{k})$$

$$= \sin v \cos w\mathbf{i} + \sin v \sin w\mathbf{j} + \cos v\mathbf{k}$$

$$= \mathbf{e}_u$$

$$\mathbf{e}_V = \frac{1}{H_v}\nabla v = \frac{1}{(x^2 + y^2)^{1/2}(x^2 + y^2 + z^2)^{1/2}}[zx\mathbf{i} + zy\mathbf{j} - (x^2 + y^2)\mathbf{k}]$$

$$= \cos v \cos w\mathbf{i} + \cos v \sin w\mathbf{j} - \sin v\mathbf{k}$$

$$= \mathbf{e}_v$$

$$\mathbf{e}_W = \frac{1}{H_w}\nabla w = \frac{1}{(x^2 + y^2)^{1/2}}(-y\mathbf{i} + x\mathbf{j})$$

$$= -\sin w\mathbf{i} + \cos w\mathbf{j}$$

$$= \mathbf{e}_w$$

PROBLEM 8-7   Using the relation

$$(ds)^2 = (dx)^2 + (dy)^2 + (dz)^2 = (h_u\,du)^2 + (h_v\,dv)^2 + (h_w\,dw)^2 \tag{a}$$

find the scale factors h_u, h_v, h_w in cylindrical coordinates and in spherical coordinates.

Solution: From the definition of cylindrical coordinates in (8.18) we find

$$dx = \cos v\,du - u \sin v\,dv$$
$$dy = \sin v\,du + u \cos v\,dv$$
$$dz = dw$$

Thus

$$(dx)^2 + (dy)^2 + (dz)^2 = \cos^2 v(du)^2 + u^2\sin^2 v(dv)^2 - 2u \sin v \cos v\,du\,dv$$
$$+ \sin^2 v(du)^2 + u^2\cos^2 v(dv)^2$$
$$+ 2u \sin v \cos v\,du\,dv + (dw)^2$$
$$= (\sin^2 v + \cos^2 v)(du)^2 + u^2(\sin^2 v + \cos^2 v)(dv)^2 + (dw)^2$$
$$= (du)^2 + (u\,dv)^2 + (dw)^2$$
$$= (h_u\,du)^2 + (h_v\,dv)^2 + (h_w\,dw)^2$$

Hence in cylindrical coordinates

$$h_u = 1, \qquad h_v = u, \qquad h_w = 1$$

From the definition of spherical coordinates in (8.21) we find

$$dx = \sin v \cos w \, du + u \cos v \cos w \, dv - u \sin v \sin w \, dw$$
$$dy = \sin v \sin w \, du + u \cos v \sin w \, dv + u \sin v \cos w \, dw$$
$$dz = \cos v \, du - u \sin v \, dv$$

Thus

$$(dx)^2 + (dy)^2 + (dz)^2$$
$$= \sin^2 v \cos^2 w (du)^2 + u^2 \cos^2 v \cos^2 w (dv)^2 + u^2 \sin^2 v \sin^2 w (dw)^2$$
$$\quad + 2u \sin v \cos v \cos^2 w \, du \, dv - 2u^2 \sin v \cos v \sin w \cos w \, dv \, dw$$
$$\quad - 2u \sin^2 v \sin w \cos w \, dw \, du$$
$$\quad + \sin^2 v \sin^2 w (du)^2 + u^2 \cos^2 v \sin^2 w (dv)^2 + u^2 \sin^2 v \cos^2 w (dw)^2$$
$$\quad + 2u \sin v \cos v \sin^2 w \, du \, dv + 2u^2 \sin v \cos v \sin w \cos w \, dv \, dw$$
$$\quad + 2u \sin^2 v \sin w \cos w \, dw \, du$$
$$\quad + \cos^2 v (du)^2 + u^2 \sin^2 v (dv)^2 - 2u \sin v \cos v \, du \, dv$$
$$= [\sin^2 v (\cos^2 w + \sin^2 w) + \cos^2 v](du)^2$$
$$\quad + u^2 [\cos^2 v (\cos^2 w + \sin^2 w) + \sin^2 v](dv)^2$$
$$\quad + u^2 \sin^2 v (\sin^2 w + \cos^2 w)(dw)^2$$
$$= (du)^2 + (u \, dv)^2 + (u \sin v \, dw)^2$$
$$= (h_u \, du)^2 + (h_v \, dv)^2 + (h_w \, dw)^2$$

Hence in spherical coordinates

$$h_u = 1, \qquad h_v = u, \qquad h_w = u \sin v$$

Note: Relation (a) of this problem is true only for orthogonal coordinates.

PROBLEM 8-8 Verify the expression for divergence given in (8.26).

Solution: We know that vectors $\mathbf{e}_u$, $\mathbf{e}_v$, $\mathbf{e}_w$ form a right-handed orthonormal basis. Using eq. (b) of Problem 8-4, we have

$$\mathbf{e}_u = \mathbf{e}_v \times \mathbf{e}_w = h_v h_w \nabla v \times \nabla w$$
$$\mathbf{e}_v = \mathbf{e}_w \times \mathbf{e}_u = h_w h_u \nabla w \times \nabla u \qquad \text{(a)}$$
$$\mathbf{e}_w = \mathbf{e}_u \times \mathbf{e}_v = h_u h_v \nabla u \times \nabla v$$

Thus

$$\mathbf{f} = f_u \mathbf{e}_u + f_v \mathbf{e}_v + f_w \mathbf{e}_w$$
$$= h_v h_w f_u \nabla v \times \nabla w + h_w h_u f_v \nabla w \times \nabla u + h_u h_v f_w \nabla u \times \nabla v \qquad \text{(b)}$$

Since by (5.41),

$$\nabla \cdot (\phi \mathbf{f}) = \phi \nabla \cdot \mathbf{f} + \mathbf{f} \cdot \nabla \phi$$

we have

$$\nabla \cdot (f_u \mathbf{e}_u) = \nabla \cdot (h_v h_w f_u \nabla v \times \nabla w)$$
$$= h_v h_w f_u \nabla \cdot (\nabla v \times \nabla w) + (\nabla v \times \nabla w) \cdot \nabla (h_v h_w f_u) \qquad \text{(c)}$$

Again, by identity (5.43),

$$\nabla \cdot (\mathbf{f} \times \mathbf{g}) = \mathbf{g} \cdot (\nabla \times \mathbf{f}) - \mathbf{f} \cdot (\nabla \times \mathbf{g})$$

and using $\nabla \times (\nabla \phi) = \mathbf{0}$ from (5.31), we have

$$\nabla \cdot (\nabla v \times \nabla w) = \nabla w \cdot (\nabla \times \nabla v) - \nabla v \cdot (\nabla \times \nabla w) = 0$$

Next, by (8.25) and eq. (a) of this problem, we have

$$\mathbf{V}(h_v h_w f_u) = \frac{1}{h_u}\frac{\partial}{\partial u}(h_v h_w f_u)\mathbf{e}_u + \frac{1}{h_v}\frac{\partial}{\partial v}(h_v h_w f_u)\mathbf{e}_v + \frac{1}{h_w}\frac{\partial}{\partial w}(h_v h_w f_u)\mathbf{e}_w$$

and

$$\mathbf{V}v \times \mathbf{V}w = \frac{1}{h_v h_w}\mathbf{e}_u$$

Since $\mathbf{e}_u \cdot \mathbf{e}_u = 1$ and $\mathbf{e}_u \cdot \mathbf{e}_v = \mathbf{e}_u \cdot \mathbf{e}_w = 0$, taking the dot product yields

$$(\mathbf{V}v \times \mathbf{V}w) \cdot \mathbf{V}(h_v h_w f_u) = \frac{1}{h_u h_v h_w}\frac{\partial}{\partial u}(h_v h_w f_u)$$

Thus from eq. (c) of this problem,

$$\mathbf{V} \cdot (f_u \mathbf{e}_u) = \frac{1}{h_u h_v h_w}\frac{\partial}{\partial u}(h_v h_w f_u) \tag{d}$$

By a similar set of operations (or by cyclic changes of indices), we arrive at

$$\mathbf{V} \cdot (f_v \mathbf{e}_v) = \frac{1}{h_u h_v h_w}\frac{\partial}{\partial v}(h_w h_u f_v) \tag{e}$$

$$\mathbf{\Delta} \cdot (f_w \mathbf{e}_w) = \frac{1}{h_u h_v h_w}\frac{\partial}{\partial w}(h_u h_v f_w) \tag{f}$$

Adding eqs. (d), (e), and (f), we obtain (8.26).

PROBLEM 8-9   Verify the formula for the Laplacian of the scalar function (8.27).

Solution: Applying (8.26) to the vector $\mathbf{V}\psi$ given by (8.25), that is, substituting

$$f_u = \frac{1}{h_u}\frac{\partial \psi}{\partial u}, \qquad f_v = \frac{1}{h_v}\frac{\partial \psi}{\partial v}, \qquad f_w = \frac{1}{h_w}\frac{\partial \psi}{\partial w}$$

into (8.26), we obtain (8.27).

PROBLEM 8-10   Verify the formula for the curl of a vector function (8.28).

Solution: Using eq. (b) of Problem 8-4, we have

$$\mathbf{f} = f_u \mathbf{e}_u + f_v \mathbf{e}_v + f_w \mathbf{e}_w$$

$$= h_u f_u \mathbf{V}u + h_v f_v \mathbf{V}v + h_w f_w \mathbf{V}w \tag{a}$$

Since by identity (5.42),

$$\mathbf{V} \times (\phi \mathbf{f}) = \phi \mathbf{V} \times \mathbf{f} + (\mathbf{V}\phi) \times \mathbf{f}$$

and because $\mathbf{V} \times \mathbf{V}u = \mathbf{0}$, then

$$\mathbf{V} \times (f_u \mathbf{e}_u) = \mathbf{V} \times (h_u f_u \mathbf{V}u)$$

$$= (h_u f_u)\mathbf{V} \times \mathbf{V}u + \mathbf{V}(h_u f_u) \times \mathbf{V}u$$

$$= \mathbf{V}(h_u f_u) \times \mathbf{V}u \tag{b}$$

Now, using (8.25) and eq. (e) of Problem 8-4, we have

$$\mathbf{V}(h_u f_u) \times \mathbf{V}u = \frac{1}{h_u^2}\frac{\partial}{\partial u}(h_u f_u)\mathbf{e}_u \times \mathbf{e}_u + \frac{1}{h_u h_v}\frac{\partial}{\partial v}(h_u f_u)\mathbf{e}_v \times \mathbf{e}_u$$

$$+ \frac{1}{h_w h_u}\frac{\partial}{\partial w}(h_u f_u)\mathbf{e}_w \times \mathbf{e}_u$$

Because $\mathbf{e}_u, \mathbf{e}_v, \mathbf{e}_w$ form a right-handed orthonormal basis,

$$\mathbf{e}_u \times \mathbf{e}_u = \mathbf{0}, \qquad \mathbf{e}_v \times \mathbf{e}_u = -\mathbf{e}_w, \qquad \mathbf{e}_w \times \mathbf{e}_u = \mathbf{e}_v$$

and so

$$\nabla \times (f_u \mathbf{e}_u) = \nabla(h_u f_u) \times \nabla u$$

$$= \frac{1}{h_w h_u} \frac{\partial}{\partial w}(h_u f_u)\mathbf{e}_v - \frac{1}{h_u h_v}\frac{\partial}{\partial v}(h_u f_u)\mathbf{e}_w \qquad \text{(c)}$$

By similar reasoning (or by cyclic changes in indices), we obtain

$$\nabla \times (f_v \mathbf{e}_v) = \frac{1}{h_u h_v}\frac{\partial}{\partial u}(h_v f_v)\mathbf{e}_w - \frac{1}{h_v h_w}\frac{\partial}{\partial w}(h_v f_v)\mathbf{e}_u \qquad \text{(d)}$$

$$\nabla \times (f_w \mathbf{e}_w) = \frac{1}{h_v h_w}\frac{\partial}{\partial v}(h_w f_w)\mathbf{e}_u - \frac{1}{h_w h_u}\frac{\partial}{\partial u}(h_w f_w)\mathbf{e}_v \qquad \text{(e)}$$

Adding eqs. (c), (d), and (e), we obtain (8.28).

PROBLEM 8-11   Using the definition of divergence given by (6.30), that is,

$$\nabla \cdot \mathbf{f} = \lim_{\Delta V \to 0} \frac{1}{\Delta V} \oiint_S \mathbf{f} \cdot d\mathbf{S}$$

derive the expression (8.26) for the divergence in orthogonal curvilinear coordinates.

Solution: Let the point $P(u,v,w)$ be at the center of the curvilinear volume element dV, as shown in Figure 8-4. By (8.12)

$$dV = h_u h_v h_w \, du \, dv \, dw$$

If $\mathbf{f} = f_u \mathbf{e}_u + f_v \mathbf{e}_v + f_w \mathbf{e}_w$, the flux for the outward normal through surface $ABCD$ (for constant u) is

$$-f_u h_v h_w \, dv \, dw + \frac{1}{2}\frac{\partial}{\partial u}(f_u h_v h_w)du\,dv\,dw \qquad \text{(a)}$$

and through surface $EFGH$, the flux is

$$f_u h_v h_w \, dv \, dw + \frac{1}{2}\frac{\partial}{\partial u}(f_u h_v h_w)du\,dv\,dw \qquad \text{(b)}$$

(Ignore infinitesimals of higher order.) Adding eqs. (a) and (b), the net outflow of the flux through the two surfaces for constant u is

$$\frac{\partial}{\partial u}(f_u h_v h_w)du\,dv\,dw \qquad \text{(c)}$$

Computing the similar results for the other two pairs of surfaces and adding these together produces

$$\oiint_S \mathbf{f}\cdot d\mathbf{S} = \left[\frac{\partial}{\partial u}(f_u h_v h_w) + \frac{\partial}{\partial v}(f_v h_w h_u) + \frac{\partial}{\partial w}(f_w h_u h_v)\right]du\,dv\,dw \qquad \text{(d)}$$

Dividing eq. (d) by dV yields

$$\nabla \cdot \mathbf{f} = \frac{1}{h_u h_v h_w}\left[\frac{\partial}{\partial u}(f_u h_v h_w) + \frac{\partial}{\partial v}(f_v h_w h_u) + \frac{\partial}{\partial w}(f_w h_u h_v)\right]$$

PROBLEM 8-12   Using the definition of curl given by (6.33),

$$\mathbf{n}\cdot(\nabla\times\mathbf{f}) = \lim_{\Delta S \to 0}\frac{1}{\Delta S}\oint_C \mathbf{f}\cdot d\mathbf{r}$$

derive expression (8.28) for the curl of $\mathbf{f}$ in orthogonal curvilinear coordinates.

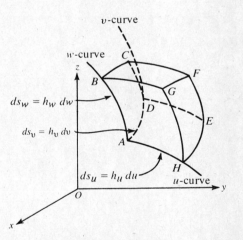

Figure 8-4
Curvilinear volume element.

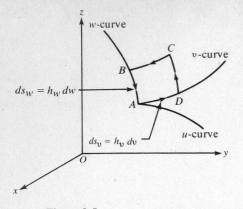

Figure 8-5
Curvilinear surface element.

Solution: Let $\mathbf{f} = f_u\mathbf{e}_u + f_v\mathbf{e}_v + f_w\mathbf{e}_w$ and calculate $\mathbf{e}_u \cdot (\nabla \times \mathbf{f})$ first. Consider a closed curve $C_u(ADCB)$ lying in the surface $u = $ constant, as shown in Figure 8-5. The surface element dS enclosed by C_u is

$$dS = h_v h_w \, dv \, dw$$

The circulation around the closed curve C_u in the positive direction is

$$\oint_{C_u} \mathbf{f} \cdot d\mathbf{r} = \int_A^D \mathbf{f} \cdot d\mathbf{r} + \int_D^C \mathbf{f} \cdot d\mathbf{r} + \int_C^B \mathbf{f} \cdot d\mathbf{r} + \int_B^A \mathbf{f} \cdot d\mathbf{r} \tag{a}$$

Ignoring infinitesimals of higher order, we obtain

$$\int_A^D \mathbf{f} \cdot d\mathbf{r} = f_v h_v \, dv$$

$$\int_D^C \mathbf{f} \cdot d\mathbf{r} = \left[f_w h_w + \frac{\partial}{\partial v}(f_w h_w)dv \right]dw$$

$$\int_C^B \mathbf{f} \cdot d\mathbf{r} = -\left[f_v h_v + \frac{\partial}{\partial w}(f_v h_v)dw \right]dv$$

$$\int_B^A \mathbf{f} \cdot d\mathbf{r} = -f_w h_w \, dw$$

Adding these results yields

$$\oint_{C_u} \mathbf{f} \cdot d\mathbf{r} = \left[\frac{\partial}{\partial v}(f_w h_w) - \frac{\partial}{\partial w}(f_v h_v) \right]dv \, dw \tag{b}$$

Dividing eq. (b) by dS gives

$$\mathbf{e}_u \cdot (\nabla \times \mathbf{f}) = \frac{1}{h_v h_w}\left[\frac{\partial}{\partial v}(f_w h_w) - \frac{\partial}{\partial w}(f_v h_v) \right] \tag{c}$$

By cyclic permutation of the indices we obtain the remaining two components of $\nabla \times \mathbf{f}$. Combining these components, we get

$$\nabla \times \mathbf{f} = \frac{1}{h_v h_w}\left[\frac{\partial}{\partial v}(f_w h_w) - \frac{\partial}{\partial w}(f_v h_v) \right]\mathbf{e}_u$$

$$+ \frac{1}{h_w h_u}\left[\frac{\partial}{\partial w}(f_u h_u) - \frac{\partial}{\partial u}(f_w h_w) \right]\mathbf{e}_v$$

$$+ \frac{1}{h_u h_v}\left[\frac{\partial}{\partial u}(f_v h_v) - \frac{\partial}{\partial v}(f_u h_u) \right]\mathbf{e}_w$$

which is the desired expression for curl $\mathbf{f}$.

Special Coordinate Systems

PROBLEM 8-13 If

$$\mathbf{f} = f_1\mathbf{i} + f_2\mathbf{j} + f_3\mathbf{k} = f_\rho\mathbf{e}_\rho + f_\phi\mathbf{e}_\phi + f_3\mathbf{k} \tag{a}$$

find the relationship between f_1, f_2, f_3 and f_ρ, f_ϕ, f_3.

Solution: Using (8.46), we have

$$\mathbf{f} = f_1\mathbf{i} + f_2\mathbf{j} + f_3\mathbf{k}$$

$$= f_1(\cos\phi\,\mathbf{e}_\rho - \sin\phi\,\mathbf{e}_\phi) + f_2(\sin\phi\,\mathbf{e}_\rho + \cos\phi\,\mathbf{e}_\phi) + f_3\mathbf{k}$$

$$= (f_1\cos\phi + f_2\sin\phi)\mathbf{e}_\rho + (-f_1\sin\phi + f_2\cos\phi)\mathbf{e}_\phi + f_3\mathbf{k} \tag{b}$$

Since $\mathbf{f} = f_\rho\mathbf{e}_\rho + f_\phi\mathbf{e}_\phi + f_3\mathbf{k}$, comparing coefficients yields

$$f_\rho = f_1\cos\phi + f_2\sin\phi \tag{c.1}$$

$$f_\phi = -f_1 \sin \phi + f_2 \cos \phi \tag{c.2}$$

$$f_3 = f_3 \tag{c.3}$$

Similarly, using (8.43), we have

$$\mathbf{f} = f_\rho \mathbf{e}_\rho + f_\phi \mathbf{e}_\phi + f_3 \mathbf{k}$$

$$= f_\rho(\cos \phi \mathbf{i} + \sin \phi \mathbf{j}) + f_\phi(-\sin \phi \mathbf{i} + \cos \phi \mathbf{j}) + f_3 \mathbf{k}$$

$$= (f_\rho \cos \phi - f_\phi \sin \phi)\mathbf{i} + (f_\rho \sin \phi + f_\phi \cos \phi)\mathbf{j} + f_3 \mathbf{k}$$

Again, because $\mathbf{f} = f_1 \mathbf{i} + f_2 \mathbf{j} + f_3 \mathbf{k}$, comparing coefficients gives

$$f_1 = f_\rho \cos \phi - f_\phi \sin \phi \tag{d.1}$$

$$f_2 = f_\rho \sin \phi + f_\phi \cos \phi \tag{d.2}$$

$$f_3 = f_3 \tag{d.3}$$

In matrix notation, sets (c) and (d) of this problem can be expressed, respectively, as

$$\begin{bmatrix} f_\rho \\ f_\phi \\ f_3 \end{bmatrix} = \begin{bmatrix} \cos \phi & \sin \phi & 0 \\ -\sin \phi & \cos \phi & 0 \\ 0 & 0 & 1 \end{bmatrix} \begin{bmatrix} f_1 \\ f_2 \\ f_3 \end{bmatrix} \tag{e}$$

$$\begin{bmatrix} f_1 \\ f_2 \\ f_3 \end{bmatrix} = \begin{bmatrix} \cos \phi & -\sin \phi & 0 \\ \sin \phi & \cos \phi & 0 \\ 0 & 0 & 1 \end{bmatrix} \begin{bmatrix} f_\rho \\ f_\phi \\ f_3 \end{bmatrix} \tag{f}$$

Comparing (8.47) and eq. (e) of this problem with (8.48) and eq. (f) of this problem, we observe that the transformations of the base vectors and the components of a vector have the same matrix.

PROBLEM 8-14 Transform $\mathbf{f} = x\mathbf{i} + y\mathbf{j} + z\mathbf{k}$ into circular cylindrical coordinates.

Solution: Since $f_1 = x, f_2 = y$, and $f_3 = z$, from eq. set (c) of Problem 8-13 we have

$$f_\rho = f_1 \cos \phi + f_2 \sin \phi = x \cos \phi + y \sin \phi$$

$$f_\phi = -f_1 \sin \phi + f_2 \cos \phi = -x \sin \phi + y \cos \phi$$

$$f_3 = f_3 = z$$

Substituting $x = \rho \cos \phi$ and $y = \rho \sin \phi$, we obtain

$$f_\rho = \rho \cos^2 \phi + \rho \sin^2 \phi = \rho(\cos^2 \phi + \sin^2 \phi) = \rho$$

$$f_\phi = -\rho \cos \phi \sin \phi + \rho \sin \phi \cos \phi = 0$$

$$f_3 = z$$

Hence

$$\mathbf{f} = \rho \mathbf{e}_\rho + z\mathbf{k}$$

Alternate Solution: Using (8.46), we have

$$\mathbf{f} = x\mathbf{i} + y\mathbf{j} + z\mathbf{k}$$

$$= x(\cos \phi \mathbf{e}_\rho - \sin \phi \mathbf{e}_\phi) + y(\sin \phi \mathbf{e}_\rho + \cos \phi \mathbf{e}_\phi) + z\mathbf{k}$$

$$= (x \cos \phi + y \sin \phi)\mathbf{e}_\rho + (-x \sin \phi + y \cos \phi)\mathbf{e}_\phi + z\mathbf{k}$$

Substituting $x = \rho \cos \phi$ and $y = \rho \sin \phi$, we obtain

$$\mathbf{f} = \rho(\cos^2 \phi + \sin^2 \phi)\mathbf{e}_\rho + (-\rho \cos \phi \sin \phi + \rho \sin \phi \cos \phi)\mathbf{e}_\phi + z\mathbf{k}$$

$$= \rho \mathbf{e}_\rho + z\mathbf{k}$$

PROBLEM 8-15 Transform $\mathbf{f} = (1/\rho)\mathbf{e}_\rho$ into rectangular Cartesian coordinates.

Solution: Since $f_\rho = 1/\rho$, $f_\phi = 0$, and $f_3 = 0$, from set (d) of Problem 8-13 we have

$$f_1 = f_\rho \cos \phi - f_\phi \sin \phi = \frac{1}{\rho} \cos \phi$$

$$f_2 = f_\rho \sin \phi + f_\phi \cos \phi = \frac{1}{\rho} \sin \phi$$

$$f_3 = f_3 = 0$$

Substituting

$$\rho = (x^2 + y^2)^{1/2}, \qquad \cos \phi = \frac{x}{\rho} = \frac{x}{(x^2 + y^2)^{1/2}}, \qquad \sin \phi = \frac{y}{\rho} = \frac{y}{(x^2 + y^2)^{1/2}}$$

we obtain

$$f_1 = \frac{x}{\rho^2} = \frac{x}{x^2 + y^2} \qquad \text{and} \qquad f_2 = \frac{y}{\rho^2} = \frac{y}{x^2 + y^2}$$

Hence

$$\mathbf{f} = \frac{x}{(x^2 + y^2)} \mathbf{i} + \frac{y}{(x^2 + y^2)} \mathbf{j}$$

Alternate Solution: Using set (8.43), we have

$$\mathbf{f} = \frac{1}{\rho} \mathbf{e}_\rho = \frac{1}{\rho} (\cos \phi \mathbf{i} + \sin \phi \mathbf{j}) = \frac{\cos \phi}{\rho} \mathbf{i} + \frac{\sin \phi}{\rho} \mathbf{j}$$

Then substituting $\cos \phi = x/\rho$, $\sin \phi = y/\rho$, and $\rho^2 = x^2 + y^2$, we obtain

$$\mathbf{f} = \frac{x}{\rho^2} \mathbf{i} + \frac{y}{\rho^2} \mathbf{j} = \frac{x}{x^2 + y^2} \mathbf{i} + \frac{y}{x^2 + y^2} \mathbf{j}$$

PROBLEM 8-16 Using the expressions for spherical coordinates given in (8.53) through (8.56), derive the following:

(a) $$\nabla f(r) = f'(r)\mathbf{e}_r$$
(b) $$\nabla \cdot [f(r)\mathbf{r}] = 3f(r) + rf'(r)$$
(c) $$\nabla \times [f(r)\mathbf{r}] = \mathbf{0}$$

(d) $$\nabla^2 f(r) = \frac{2}{r} f'(r) + f''(r)$$

Solution:
(a) From (8.53) we have

$$\nabla f(r) = \frac{\partial f(r)}{\partial r} \mathbf{e}_r = f'(r)\mathbf{e}_r$$

(b) Since $\mathbf{r} = r\mathbf{e}_r$ and $f(r)\mathbf{r} = rf(r)\mathbf{e}_r$, from (8.54) we have

$$\nabla \cdot [f(r)\mathbf{r}] = \nabla \cdot [rf(r)\mathbf{e}_r]$$

$$= \frac{1}{r^2} \frac{\partial}{\partial r} [r^3 f(r)]$$

$$= \frac{1}{r^2} [3r^2 f(r) + r^3 f'(r)]$$

$$= 3f(r) + rf'(r)$$

(c) From (8.55a) we have

$$\mathbf{V} \times [f(r)\mathbf{r}] = \mathbf{V} \times [rf(r)\mathbf{e}_r]$$

$$= \frac{1}{r^2 \sin \theta} \begin{vmatrix} \mathbf{e}_r & r\mathbf{e}_\theta & r\sin\theta\mathbf{e}_\phi \\ \dfrac{\partial}{\partial r} & \dfrac{\partial}{\partial \theta} & \dfrac{\partial}{\partial \phi} \\ rf(r) & 0 & 0 \end{vmatrix}$$

$$= \mathbf{0}$$

(d) From (8.56) we have

$$\mathbf{V}^2 f(r) = \frac{1}{r^2}\frac{\partial}{\partial r}\left(r^2 \frac{\partial f(r)}{\partial r}\right)$$

$$= \frac{1}{r^2}\frac{d}{dr}\left[r^2 f'(r)\right]$$

$$= \frac{1}{r^2}\left[2rf'(r) + r^2 f''(r)\right]$$

$$= \frac{2}{r}f'(r) + f''(r)$$

PROBLEM 8-17   Find the relationships between f_1, f_2, f_3 and f_r, f_θ, f_ϕ if

$$\mathbf{f} = f_1\mathbf{i} + f_2\mathbf{j} + f_3\mathbf{k} = f_r\mathbf{e}_r + f_\theta\mathbf{e}_\theta + f_\phi\mathbf{e}_\phi$$

Solution: Using set (8.60), we have

$$\begin{aligned}
\mathbf{f} &= f_1\mathbf{i} + f_2\mathbf{j} + f_3\mathbf{k} \\
&= f_1(\sin\theta\cos\phi\,\mathbf{e}_r + \cos\theta\cos\phi\,\mathbf{e}_\theta - \sin\phi\,\mathbf{e}_\phi) \\
&\quad + f_2(\sin\theta\sin\phi\,\mathbf{e}_r + \cos\theta\sin\phi\,\mathbf{e}_\theta + \cos\phi\,\mathbf{e}_\phi) + f_3(\cos\theta\,\mathbf{e}_r - \sin\theta\,\mathbf{e}_\theta) \\
&= (f_1\sin\theta\cos\phi + f_2\sin\theta\sin\phi + f_3\cos\theta)\mathbf{e}_r \\
&\quad + (f_1\cos\theta\cos\phi + f_2\cos\theta\sin\phi - f_3\sin\theta)\mathbf{e}_\theta \\
&\quad + (-f_1\sin\phi + f_2\cos\phi)\mathbf{e}_\phi
\end{aligned}$$
(a)

Since $\mathbf{f} = f_r\mathbf{e}_r + f_\theta\mathbf{e}_\theta + f_\phi\mathbf{e}_\phi$, comparing coefficients yields

$$f_r = f_1\sin\theta\cos\phi + f_2\sin\theta\sin\phi + f_3\cos\theta \tag{b.1}$$

$$f_\theta = f_1\cos\theta\cos\phi + f_2\cos\theta\sin\phi - f_3\sin\theta \tag{b.2}$$

$$f_\phi = -f_1\sin\phi + f_2\cos\phi \tag{b.3}$$

Similarly, using set (8.57), we have

$$\begin{aligned}
\mathbf{f} &= f_r\mathbf{e}_r + f_\theta\mathbf{e}_\theta + f_\phi\mathbf{e}_\phi \\
&= f_r(\sin\theta\cos\phi\,\mathbf{i} + \sin\theta\sin\phi\,\mathbf{j} + \cos\theta\,\mathbf{k}) \\
&\quad + f_\theta(\cos\theta\cos\phi\,\mathbf{i} + \cos\theta\sin\phi\,\mathbf{j} - \sin\theta\,\mathbf{k}) + f_\phi(-\sin\phi\,\mathbf{i} + \cos\phi\,\mathbf{j}) \\
&= (f_r\sin\theta\cos\phi + f_\theta\cos\theta\cos\phi - f_\phi\sin\phi)\mathbf{i} \\
&\quad + (f_r\sin\theta\sin\phi + f_\theta\cos\theta\sin\phi + f_\phi\cos\phi)\mathbf{j} + (f_r\cos\theta - f_\theta\sin\theta)\mathbf{k}
\end{aligned}$$

Again, because $\mathbf{f} = f_1\mathbf{i} + f_2\mathbf{j} + f_3\mathbf{k}$, comparing coefficients yields

$$f_1 = f_r\sin\theta\cos\phi + f_\theta\cos\theta\cos\phi - f_\phi\sin\phi \tag{c.1}$$

$$f_2 = f_r\sin\theta\sin\phi + f_\theta\cos\theta\sin\phi + f_\phi\cos\phi \tag{c.2}$$

$$f_3 = f_r\cos\theta - f_\theta\sin\theta \tag{c.3}$$

In matrix notation the transformations sets (b) and (c) just given can be expressed, respectively, as

$$
\begin{bmatrix} f_r \\ f_\theta \\ f_\phi \end{bmatrix} = \begin{bmatrix} \sin\theta\cos\phi & \sin\theta\sin\phi & \cos\theta \\ \cos\theta\cos\phi & \cos\theta\sin\phi & -\sin\theta \\ -\sin\phi & \cos\phi & 0 \end{bmatrix} \begin{bmatrix} f_1 \\ f_2 \\ f_3 \end{bmatrix} \tag{d}
$$

$$
\begin{bmatrix} f_1 \\ f_2 \\ f_3 \end{bmatrix} = \begin{bmatrix} \sin\theta\cos\phi & \cos\theta\cos\phi & -\sin\phi \\ \sin\theta\sin\phi & \cos\theta\sin\phi & \cos\phi \\ \cos\theta & -\sin\theta & 0 \end{bmatrix} \begin{bmatrix} f_r \\ f_\theta \\ f_\phi \end{bmatrix} \tag{e}
$$

[Again, note that the transformations of the base vectors and the components of a vector have the same matrix.]

PROBLEM 8-18 Transform $\mathbf{f} = x\mathbf{i} + y\mathbf{j} + z\mathbf{k}$ to spherical coordinates.

Solution: Since $f_1 = x$, $f_2 = y$, $f_3 = z$, from eq. (b) or eq. (d) of Problem 8-17, we have

$$
f_r = x\sin\theta\cos\phi + y\sin\theta\sin\phi + z\cos\theta
$$
$$
f_\theta = x\cos\theta\cos\phi + y\cos\theta\sin\phi - z\sin\theta
$$
$$
f_\phi = -x\sin\phi + y\cos\phi
$$

Substituting the spherical coordinate transformation equations $x = r\sin\theta\cos\phi$, $y = r\sin\theta\sin\phi$, and $z = r\cos\theta$, we obtain

$$
\begin{aligned}
f_r &= r\sin^2\theta\cos^2\phi + r\sin^2\theta\sin^2\phi + r\cos^2\theta \\
&= r\sin^2\theta(\cos^2\phi + \sin^2\phi) + r\cos^2\theta = r(\sin^2\theta + \cos^2\theta) \\
&= r \\
f_\theta &= r\sin\theta\cos\theta\cos^2\phi + r\sin\theta\cos\theta\sin^2\phi - r\cos\theta\sin\theta \\
&= r\sin\theta\cos\theta(\cos^2\phi + \sin^2\phi - 1) \\
&= 0 \\
f_\phi &= -r\sin\theta\sin\phi\cos\phi + r\sin\theta\sin\phi\cos\phi \\
&= 0
\end{aligned}
$$

Hence $\mathbf{f} = f_r\mathbf{e}_r = r\mathbf{e}_r$.

PROBLEM 8-19 Transform $\mathbf{f} = \dfrac{1}{r\sin\theta}\,\mathbf{e}_\phi$ into rectangular coordinates.

Solution: Using (8.57c), we have

$$
\mathbf{f} = \frac{1}{r\sin\theta}\,\mathbf{e}_\phi = \frac{1}{r\sin\theta}(-\sin\phi\mathbf{i} + \cos\phi\mathbf{j}) = -\frac{\sin\phi}{r\sin\theta}\,\mathbf{i} + \frac{\cos\phi}{r\sin\theta}\,\mathbf{j}
$$

Now from $x = r\sin\theta\cos\phi$, $y = r\sin\theta\sin\phi$, and $z = r\cos\theta$,

$$
\sin\phi = \frac{y}{r\sin\theta}, \qquad \cos\phi = \frac{x}{r\sin\theta}, \qquad x^2 + y^2 = r^2\sin^2\theta
$$

Hence

$$
\mathbf{f} = -\frac{y}{r^2\sin^2\theta}\,\mathbf{i} + \frac{x}{r^2\sin^2\theta}\,\mathbf{j} = -\frac{y}{x^2 + y^2}\,\mathbf{i} + \frac{x}{x^2 + y^2}\,\mathbf{j}
$$

PROBLEM 8-20 Find the relationships between $\mathbf{e}_\rho$, $\mathbf{e}_\phi$, $\mathbf{k}$ of the circular cylindrical coordinate system and $\mathbf{e}_r$, $\mathbf{e}_\theta$, $\mathbf{e}_\phi$ of the spherical coordinate system.

Curvilinear Orthogonal Coordinates **195**

Solution: Substituting set (8.46) into set (8.57), we have

$$\mathbf{e}_r = \sin\theta\cos\phi(\cos\phi\mathbf{e}_\rho - \sin\phi\mathbf{e}_\phi) + \sin\theta\sin\phi(\sin\phi\mathbf{e}_\rho + \cos\phi\mathbf{e}_\phi) + \cos\theta\mathbf{k}$$

$$= \sin\theta(\cos^2\phi + \sin^2\phi)\mathbf{e}_\rho + (-\sin\theta\cos\phi\sin\phi + \sin\theta\sin\phi\cos\phi)\mathbf{e}_\phi + \cos\theta\mathbf{k}$$

$$= \sin\theta\mathbf{e}_\rho + \cos\theta\mathbf{k} \tag{a.1}$$

$$\mathbf{e}_\theta = \cos\theta\cos\phi(\cos\phi\mathbf{e}_\rho - \sin\phi\mathbf{e}_\phi) + \cos\theta\sin\phi(\sin\phi\mathbf{e}_\rho + \cos\phi\mathbf{e}_\phi) - \sin\theta\mathbf{k}$$

$$= \cos\theta(\cos^2\phi + \sin^2\phi)\mathbf{e}_\rho$$
$$+ (-\cos\theta\cos\phi\sin\phi + \cos\theta\sin\phi\cos\phi)\mathbf{e}_\phi - \sin\theta\mathbf{k}$$

$$= \cos\theta\mathbf{e}_\rho - \sin\theta\mathbf{k} \tag{a.2}$$

$$\mathbf{e}_\phi = -\sin\phi(\cos\phi\mathbf{e}_\rho - \sin\phi\mathbf{e}_\phi) + \cos\phi(\sin\phi\mathbf{e}_\rho + \cos\phi\mathbf{e}_\phi)$$

$$= (-\sin\phi\cos\phi + \cos\phi\sin\phi)\mathbf{e}_\rho + (\sin^2\phi + \cos^2\phi)\mathbf{e}_\phi$$

$$= \mathbf{e}_\phi \tag{a.3}$$

Similarly, by substituting set (8.60) into set (8.43), we obtain

$$\mathbf{e}_\rho = \sin\theta\mathbf{e}_r + \cos\theta\mathbf{e}_\theta \tag{b.1}$$

$$\mathbf{e}_\phi = \mathbf{e}_\phi \tag{b.2}$$

$$\mathbf{k} = \cos\theta\mathbf{e}_r - \sin\theta\mathbf{e}_\theta \tag{b.3}$$

Using matrix notation, sets (a) and (b) of this problem can be expressed, respectively, as

$$\begin{bmatrix} \mathbf{e}_r \\ \mathbf{e}_\theta \\ \mathbf{e}_\phi \end{bmatrix} = \begin{bmatrix} \sin\theta & 0 & \cos\theta \\ \cos\theta & 0 & -\sin\theta \\ 0 & 1 & 0 \end{bmatrix} \begin{bmatrix} \mathbf{e}_\rho \\ \mathbf{e}_\phi \\ \mathbf{k} \end{bmatrix} \tag{c}$$

$$\begin{bmatrix} \mathbf{e}_\rho \\ \mathbf{e}_\phi \\ \mathbf{k} \end{bmatrix} = \begin{bmatrix} \sin\theta & \cos\theta & 0 \\ 0 & 0 & 1 \\ \cos\theta & -\sin\theta & 0 \end{bmatrix} \begin{bmatrix} \mathbf{e}_r \\ \mathbf{e}_\theta \\ \mathbf{e}_\phi \end{bmatrix} \tag{d}$$

This pair of matrix equations can also be obtained by simple matrix product operations using the pair (8.61), (8.48) and the pair (8.47), (8.62), respectively.

Supplementary Exercises

PROBLEM 8-21 Show that the Jacobian

$$J\left(\frac{x, y, z}{u, v, w}\right) = h_u h_v h_w$$

and thus that the volume element dV given by

$$dV = J\left(\frac{x, y, z}{u, v, w}\right) du\,dv\,dw$$

is $h_u h_v h_w\,du\,dv\,dw$, in agreement with (8.12).

PROBLEM 8-22 Using curvilinear coordinates and the vector identity (5.45),

$$\nabla \cdot \nabla\mathbf{f} = \nabla^2\mathbf{f} = \nabla(\nabla \cdot \mathbf{f}) - \nabla \times (\nabla \times \mathbf{f})$$

obtain the curvilinear form of the vector Laplacian $\nabla^2\mathbf{f}$.

PROBLEM 8-23   Show that the volume-integral definition of the operator $\mathbf{V}$ given by (6.34)

$$\mathbf{V}\{\ \} = \lim_{\Delta V \to 0} \frac{1}{\Delta V} \oiint_{S} \mathbf{n}\{\ \}\, dS$$

leads to the expression

$$\mathbf{V}\{\ \} = \frac{1}{h_u h_v h_w} \left[\frac{\partial}{\partial u}(h_v h_w \mathbf{e}_u\{\ \}) + \frac{\partial}{\partial v}(h_w h_u \mathbf{e}_v\{\ \}) + \frac{\partial}{\partial w}(h_u h_v \mathbf{e}_w\{\ \}) \right]$$

where $\{\ \}$ is any scalar function.

PROBLEM 8-24   Using the form of the $\mathbf{V}$ operator given in Problem 8.23, find $\mathbf{V}\psi$, $\mathbf{V} \cdot \mathbf{f}$, and $\mathbf{V} \times \mathbf{f}$. Also show that they reduce to (8.25), (8.26), and (8.28), respectively.

PROBLEM 8-25   In circular cylindrical coordinates $(\rho, \phi\, z)$, if vector field $\mathbf{f}$ is

$$\mathbf{f}(\rho, \phi) = f_\rho(\rho, \phi)\mathbf{e}_\rho + f_\phi(\rho, \phi)\mathbf{e}_\phi$$

show that $\mathbf{V} \times \mathbf{f}$ has only a z-component.

PROBLEM 8-26   Transform $\mathbf{f} = (x/y)\mathbf{i}$ to circular cylindrical coordinates.

Answer: $\cos\phi \cot\phi\, \mathbf{e}_\rho - \cos\phi\, \mathbf{e}_\phi$

PROBLEM 8-27   Transform $\mathbf{f} = \rho\mathbf{e}_\rho + \rho\mathbf{e}_\phi$ to rectangular coordinates.

Answer: $(x - y)\mathbf{i} + (x + y)\mathbf{j}$

PROBLEM 8-28   Show that the following three forms (in spherical coordinates) of $\mathbf{V}^2\psi(r)$ are equivalent:

(a)

$$\frac{1}{r^2}\frac{d}{dr}\left[r^2 \frac{d\psi(r)}{dr} \right]$$

(b)

$$\frac{1}{r}\frac{d^2}{dr^2}[r\psi(r)]$$

(c)

$$\frac{d^2\psi(r)}{dr^2} + \frac{2}{r}\frac{d\psi(r)}{dr}$$

PROBLEM 8-29   Transform $\mathbf{f} = (1/r)\mathbf{e}_r$ to rectangular coordinates.

Answer: $\dfrac{x}{x^2 + y^2 + z^2}\mathbf{i} + \dfrac{y}{x^2 + y^2 + z^2}\mathbf{j} + \dfrac{z}{x^2 + y^2 + z^2}\mathbf{k}$

PROBLEM 8-30   For the **elliptic cylindrical coordinates** (u, v, z) the transformation between the coordinates (x, y, z) and (u, v, z) is given by $x = a\cosh u \cos v$, $y = a\sinh u \sin v$, $z = z$, where a is a constant.

(a) Show that the families of coordinate surfaces are

    (1) elliptic cylinders, $u = \text{const}$, $0 \leqslant u < \infty$
    (2) hyperbolic cylinders, $v = \text{const}$, $0 \leqslant v < 2\pi$
    (3) planes parallel to the xy-plane, $z = \text{const}$, $-\infty < z < \infty$

(b) Is this system orthogonal?
(c) Find the scale factors.

Answer: (b) Yes   (c) $h_u = a(\sinh^2 u + \sin^2 v)^{1/2}$, $h_v = a(\sinh^2 u + \sin^2 v)^{1/2}$, $h_z = 1$

PROBLEM 8-31 For the **parabolic cylindrical coordinates** (ξ, η, z) the transformation between the coordinates (x, y, z) and (ξ, η, z) is given by $x = \xi\eta, \; y = \frac{1}{2}(\eta^2 - \xi^2), z = z$.

(a) Show that the families of coordinate surfaces are

(1) parabolic cylinders, $\xi = $ const, $-\infty < \xi < \infty$
(2) parabolic cylinders, $\eta = $ const, $0 \leqslant \eta < \infty$
(3) plane-parallel to the xy-plane, $z = $ const, $-\infty < z < \infty$

(b) Find the scale factors.

Answer: (b) $h_\xi = (\xi^2 + \eta^2)^{1/2}, h_\eta = (\xi^2 + \eta^2)^{1/2}, h_z = 1$

PROBLEM 8-32 For the **prolate spheroidal coordinates** (u, v, ϕ) the transformation between the coordinates (x, y, z) and (u, v, ϕ) is $x = a \sinh u \sin v \cos \phi$, $y = a \sinh u \sin v \sin \phi, \; z = a \cosh u \cos v$, where a is a constant.

(a) Show that the families of coordinate surfaces are

(1) prolate spheroids, $u = $ const, $0 \leqslant u < \infty$
(2) hyperboloids or two sheets, $v = $ const, $0 \leqslant v \leqslant \pi$
(3) half-plane through the z-axis, $\phi = $ const, $0 \leqslant \phi < 2\pi$

(b) Find the scale factors.

Answer: (b) $h_u = a(\sinh^2 u + \sin^2 v)^{1/2} = a(\cosh^2 u - \cos^2 v)^{1/2}$

$\qquad h_v = a(\sinh^2 u + \sin^2 v)^{1/2}$

$\qquad h_\phi = a \sinh u \sin v$

PROBLEM 8-33 For the **oblate spheroidal coordinates** (u, v, ϕ) the transformation between the coordinates (x, y, z) and (u, v, ϕ) is $x = a \cosh u \cos v \cos \phi$, $y = a \cosh u \cos v \sin \phi, \; z = a \sinh u \sin v$, where a is a constant.

(a) Show that the families of coordinate surfaces are

(1) oblate spheroids, $u = $ const, $0 \leqslant u < \infty$
(2) hyperboloids of one sheet, $v = $ const, $-\pi/2 \leqslant v \leqslant \pi/2$
(3) half-planes through the z-axis, $\phi = $ const, $0 \leqslant \phi < 2\pi$

(b) Find the scale factors.

Answers: (b) $h_u = a(\sinh^2 u + \sin^2 v)^{1/2} = a(\cosh^2 u - \cos^2 v)^{1/2}$

$\qquad h_v = a(\sinh^2 u + \sin^2 v)^{1/2}$

$\qquad h_\phi = a \cosh u \cos v$

PROBLEM 8-34 For **paraboloidal coordinates** (ξ, η, ϕ) the transformation equations are given by $x = \xi\eta \cos \phi, \; y = \xi\eta \sin \phi, \; z = \frac{1}{2}(\eta^2 - \xi^2)$.

(a) Show that the families of coordinate surfaces are

(1) paraboloids about the positive z-axis, $\xi = $ const, $0 \leqslant \xi < \infty$
(2) paraboloids about the negative z-axis, $\eta = $ const, $0 \leqslant \eta < \infty$
(3) half-planes through the z-axis, $\phi = $ const, $0 \leqslant \phi < 2\pi$

(b) Find the scale factors.
(c) Show that $\mathbf{e}_\xi \times \mathbf{e}_\eta = -\mathbf{e}_\phi$.

Answer: (b) $h_\xi = (\xi^2 + \eta^2)^{1/2}, h_\eta = (\xi^2 + \eta^2)^{1/2}, h_\phi = \xi\eta$

PROBLEM 8-35 Verify that in cylindrical coordinates

$$\nabla \ln \rho = \nabla \times (\mathbf{k}\phi)$$

PROBLEM 8-36   Verify the following relations in spherical coordinates: $\nabla(1/r) = \nabla \times (\cos\theta\,\nabla\phi)$; and $\nabla\phi = \nabla \times (r\,\nabla\theta/\sin\theta)$.

PROBLEM 8-37   Using the matrix multiplication method, derive eqs (c) and (d) of Problem 8.20 from eqs. (8.47) and (8.48) and eqs. (8.61) and (8.62).

PROBLEM 8-38   Express $\partial/\partial x$, $\partial/\partial y$, and $\partial/\partial z$ in spherical coordinates.

Answer:
$$\frac{\partial}{\partial x} = \sin\theta\cos\phi\,\frac{\partial}{\partial r} + \cos\theta\cos\phi\,\frac{1}{r}\frac{\partial}{\partial\theta} - \frac{\sin\phi}{r\sin\theta}\frac{\partial}{\partial\phi}$$

$$\frac{\partial}{\partial y} = \sin\theta\sin\phi\,\frac{\partial}{\partial r} + \cos\theta\sin\phi\,\frac{1}{r}\frac{\partial}{\partial\theta} + \frac{\cos\phi}{r\sin\theta}\frac{\partial}{\partial\phi}$$

$$\frac{\partial}{\partial z} = \cos\theta\,\frac{\partial}{\partial r} - \sin\theta\,\frac{1}{r}\frac{\partial}{\partial\theta}$$

PROBLEM 8-39   In Section 4-5, the set of transformations

$$x = x(u,v), \qquad y = y(u,v), \qquad z = z(u,v)$$

is interpreted as parametric equations of a surface S is space. They can be considered as a special case of (8.1), in which w is a constant; the surface S then corresponds to a surface $w(x,y,z) = \text{const}$. The curvilinear coordinates on S are u and v. Let $\mathbf{r} = x\mathbf{i} + y\mathbf{j} + z\mathbf{k}$.

(a) Show that the curves $u = \text{const}$ and $v = \text{const}$ intersect at right angles, so that the coordinates are orthogonal, *if and only if*

$$\frac{\partial x}{\partial u}\frac{\partial x}{\partial v} + \frac{\partial y}{\partial u}\frac{\partial y}{\partial v} + \frac{\partial z}{\partial u}\frac{\partial z}{\partial v} = 0$$

(b) Show that the element of an arc on a curve $u = u(t)$, $v = v(t)$ on S is

$$ds^2 = E\,du^2 + 2F\,du\,dv + G\,dv^2$$

where

$$E = |\partial\mathbf{r}/\partial u|^2, \qquad F = (\partial\mathbf{r}/\partial u)\cdot(\partial\mathbf{r}/\partial v), \qquad G = |\partial\mathbf{r}/\partial v|^2$$

(c) Show that the coordinates are othogonal when

$$ds^2 = E\,du^2 + G\,dv^2$$

(d) Show that the area of the surface S is

$$S = \iint\limits_{R_{uv}} \sqrt{EG - F^2}\,du\,dv$$

9 APPLICATIONS TO MECHANICS

THIS CHAPTER IS ABOUT

☑ **Displacement, Velocity, and Acceleration Vectors**
☑ **Angular Velocity and Angular Acceleration**
☑ **Force and Moment**
☑ **Work and Energy**

9-1. Displacement, Velocity, and Acceleration Vectors

A. Definitions

Let the position vector $\mathbf{r}(t)$ of a moving particle in space be dependent on time t. If the initial point of the particle has a position vector $\mathbf{r}(t_0)$ at time t_0, the **displacement vector** of the particle for any arbitrary time t is

DISPLACEMENT
VECTOR
$$\mathbf{r}(t) - \mathbf{r}(t_0) \tag{9.1}$$

The **velocity vector** $\mathbf{v}(t)$ and the **acceleration vector** $\mathbf{a}(t)$ of the particle are defined by the equations

VELOCITY VECTOR
$$\mathbf{v}(t) = \frac{d\mathbf{r}}{dt} = \mathbf{r}'(t) \tag{9.2}$$

ACCELERATION
VECTOR
$$\mathbf{a}(t) = \frac{d\mathbf{v}(t)}{dt} = \frac{d^2\mathbf{r}}{dt^2} = \mathbf{r}''(t) \tag{9.3}$$

The **speed** $v(t)$ of a particle at time t is the magnitude of the velocity vector $\mathbf{v}(t)$. If s is the arc length that measures the distance of the particle from its starting point on a path C along a curve, then from (4.37)

$$v(t) = |\mathbf{v}(t)| = |\mathbf{r}'(t)| = \frac{ds}{dt} \tag{9.4}$$

The position vector $\mathbf{r}$, velocity vector $\mathbf{v}$, and acceleration vector $\mathbf{a}$ in rectangular coordinates are given by

$$\mathbf{r}(t) = x(t)\mathbf{i} + y(t)\mathbf{j} + z(t)\mathbf{k} \tag{9.5}$$

$$\mathbf{v}(t) = \mathbf{r}'(t) = x'(t)\mathbf{i} + y'(t)\mathbf{j} + z'(t)\mathbf{k} \tag{9.6}$$

$$\mathbf{a}(t) = \mathbf{v}'(t) = \mathbf{r}''(t) = x''(t)\mathbf{i} + y''(t)\mathbf{j} + z''(t)\mathbf{k} \tag{9.7}$$

EXAMPLE 9-1: Verify (9.5), (9.6), and (9.7).

Solution: In rectangular coordinates the position vector is

$$\mathbf{r}(t) = x(t)\mathbf{i} + y(t)\mathbf{j} + z(t)\mathbf{k}$$

Since the unit vectors $\mathbf{i}, \mathbf{j}$, and $\mathbf{k}$ are constant in time,

$$\mathbf{v} = \mathbf{r}'(t) = x'(t)\mathbf{i} + y'(t)\mathbf{j} + z'(t)\mathbf{k}$$

$$\mathbf{a} = \mathbf{v}'(t) = \mathbf{r}''(t) = x''(t)\mathbf{i} + y''(t)\mathbf{j} + z''(t)\mathbf{k}$$

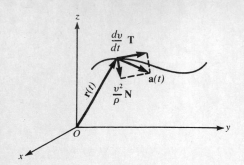

Figure 9-1
Centripetal acceleration.

B. Centripetal acceleration

The acceleration vector $\mathbf{a}(t)$ can be expressed as

$$\mathbf{a}(t) = \frac{dv}{dt}\mathbf{T} + \kappa v^2 \mathbf{N} = \frac{dv}{dt}\mathbf{T} + \frac{v^2}{\rho}\mathbf{N} \tag{9.8}$$

where κ and $\rho = 1/\kappa$ are the curvature and radius of curvature of the path C, $\mathbf{T}$ is the unit tangent vector to path C at $\mathbf{r}(t)$, and $\mathbf{N}$ is the unit principal normal vector (see the discussion of curvature in Section 4-6).

Equation (9.8) shows that the acceleration vector of the moving particle can be resolved into two components: a tangential component of magnitude dv/dt and a normal component of magnitude $v^2\kappa = v^2/\rho$. The normal component, called the **centripetal acceleration**, is caused by the fact that the velocity vector changes direction (see Figure 9-1).

EXAMPLE 9-2: Derive expression (9.8) for the acceleration vector.

Solution: From Section 4-4, we know that the velocity vector $\mathbf{v}(t) = \mathbf{r}'(t)$ is tangent to path C at the terminal point of the position vector $\mathbf{r}(t)$. If $v(t)$ is the speed of the particle, the velocity vector is

$$\mathbf{v}(t) = v(t)\frac{\mathbf{r}'(t)}{|\mathbf{r}'(t)|} = v(t)\mathbf{T} \tag{9.9}$$

Differentiating (9.9) with respect to time t and then using the Frenet–Serret formula (4.72a) and (9.4), the acceleration vector is

$$\mathbf{a}(t) = \mathbf{v}'(t) = \frac{d}{dt}[v(t)\mathbf{T}]$$

$$= \frac{dv}{dt}\mathbf{T} + v(t)\frac{d\mathbf{T}}{dt}$$

$$= \frac{dv}{dt}\mathbf{T} + v\frac{d\mathbf{T}}{ds}\frac{ds}{dt}$$

$$= \frac{dv}{dt}\mathbf{T} + v^2\kappa\mathbf{N} = \frac{dv}{dt}\mathbf{T} + \frac{v^2}{\rho}\mathbf{N}$$

9-2. Angular Velocity and Angular Acceleration

Let a particle P rotate at a fixed distance R about a fixed line L. As the particle moves in the circular path, the rate of change of the angular position $\theta(t)$ is called the **angular speed** ω, that is,

ANGULAR SPEED
$$\omega = \frac{d\theta(t)}{dt} \tag{9.10}$$

The **angular velocity** ω of the particle is a vector of magnitude ω. Its direction along L is that of a right-hand screw advance when the screw is turned in the same sense as the rotation of the particle (see Figure 9-2).

The **angular acceleration** of the particle $\boldsymbol{\alpha}(t)$ is

ANGULAR ACCELERATION
$$\boldsymbol{\alpha}(t) = \frac{d\omega}{dt} \tag{9.11}$$

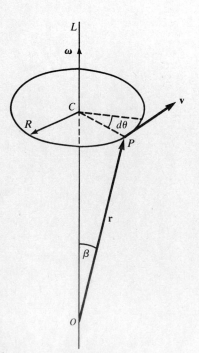

Figure 9-2
Angular velocity.

EXAMPLE 9-3: Show that the velocity vector $\mathbf{v}$ of a particle rotating about a fixed axis L can be written as

$$\mathbf{v} = \omega \times \mathbf{r} \tag{9.12}$$

where **r** specifies the position of the particle from L. Note that the origin O can be any point on L.

Solution: For motion in a circle of radius R the speed v of the particle is

$$v = R \frac{d\theta}{dt} = R\omega \tag{9.13}$$

where θ is the angular position and ω is the angular speed of the particle. Referring to Figure 9-2, if β is the angle between ω and **r**, then

$$|\omega \times \mathbf{r}| = \omega r \sin \beta = \omega R = v$$

The velocity vector **v** lies in the plane of the circle and is perpendicular to **r**. Since $\omega \times \mathbf{r}$ is perpendicular to the plane spanned by **r** and ω, it is perpendicular to the radius CP, in other words, **v** and $\omega \times \mathbf{r}$ are parallel. Thus

$$\mathbf{v} = \omega \times \mathbf{r}$$

9-3. Force and Moment

A. Newton's laws of motion

The motion of a particle in space is governed by **Newton's laws of motion**:

1. A particle remains in its state of rest or in uniform rectilinear motion unless acted upon by a force.
2. The time rate of change of the product of the mass and velocity of a particle is proportional to the total force acting upon the particle.
3. To every action or force, there is always an equal and opposite reaction or force.

> *Note:* The third law applies only to central forces, which are those exerted by two particles on each other and directed along the line connecting them.

Newton's second law of motion can be written as

NEWTON'S SECOND LAW OF MOTION
$$\mathbf{f} = \frac{d}{dt}(m\mathbf{v}) \tag{9.14}$$

If the mass m is a constant and **a** is the acceleration vector, then Newton's second law can be rewritten as

$$\mathbf{f} = m\frac{d\mathbf{v}}{dt} = m\frac{d^2\mathbf{r}}{dt^2} = m\mathbf{a} \tag{9.15}$$

A particle is said to be *free* if it experiences no force.

B. Momentum and moment

The product of mass m and velocity **v** is called the **linear momentum p** of the particle and is represented by

LINEAR MOMENTUM
$$\mathbf{p} = m\mathbf{v} \tag{9.16}$$

The angular momentum **L** of a particle with respect to a given origin is

ANGULAR MOMENTUM
$$\mathbf{L} = \mathbf{r} \times \mathbf{p} \tag{9.17}$$

where **r** is the position vector of the particle relative to the origin and **p** is the linear momentum.

The **moment** or **torque** τ of a force **f** acting on a particle relative to the origin is

TORQUE
$$\tau = \mathbf{r} \times \mathbf{f} \tag{9.18}$$

Using the equation for linear momentum (9.16), Newton's second law of motion (9.14) can be expressed as

$$\mathbf{f} = \frac{d\mathbf{p}}{dt} \tag{9.19}$$

Torque τ can also be expressed as

$$\tau = \frac{d\mathbf{L}}{dt} \tag{9.20}$$

EXAMPLE 9-4: Verify the expression (9.20) for torque.

Solution: From the definition of torque (9.18) and Newton's second law of motion (9.19)

$$\tau = \mathbf{r} \times \frac{d\mathbf{p}}{dt} \tag{9.21}$$

Differentiating (9.17) with respect to t, we have

$$\frac{d\mathbf{L}}{dt} = \frac{d}{dt}(\mathbf{r} \times \mathbf{p}) = \frac{d\mathbf{r}}{dt} \times \mathbf{p} + \mathbf{r} \times \frac{d\mathbf{p}}{dt}$$

Since $d\mathbf{p}/dt = \mathbf{f}$ and

$$\frac{d\mathbf{r}}{dt} \times \mathbf{p} = \mathbf{v} \times (m\mathbf{v}) = m(\mathbf{v} \times \mathbf{v}) = 0$$

we obtain

$$\frac{d\mathbf{L}}{dt} = \mathbf{r} \times \mathbf{f} = \tau$$

That is, the torque is equal to the time rate of change of angular momentum. *Note:* This derivation does not assume a constant mass.

9-4. Work and Energy

A. Work

The work W done by a force $\mathbf{f}$ to move a particle along a curve C from point A to point B is defined by the line integral

$$W_{AB} = \int_{\mathbf{r}_A}^{\mathbf{r}_B} \mathbf{f} \cdot d\mathbf{r} \tag{9.22}$$

where $\mathbf{r}_A$ and $\mathbf{r}_B$ are the position vectors of points A and B, respectively.

The work done by $\mathbf{f}$ is generally dependent on path C. If $\mathbf{f}$ is perpendicular to the direction of motion, $W = 0$, or, in other words, no work is done.

B. Potential energy

A force $\mathbf{f}$ is said to be **conservative** if work W done by that force around any closed path is zero:

CONSERVATIVE FORCE
$$\oint \mathbf{f} \cdot d\mathbf{r} = 0 \tag{9.23}$$

If $\mathbf{f}$ is a conservative force, then

$$\nabla \times \mathbf{f} = 0 \tag{9.24}$$

and **f** can be expressed as the gradient of a scalar function $-V$, or

**POTENTIAL
ENERGY**

$$\mathbf{f} = -\nabla V \qquad (9.25)$$

where V is called the **potential** or **potential energy**. The region is simply connected if V is single-valued. Since any constant can be added to V in (9.25) without affecting the result, the zero level of potential energy is arbitrary. Hence potential energy has no *absolute* meaning, and only *differences* of potential energy have physical meaning.

Specifically, potential energy V is the work done by a conservative force **f** in moving a particle from point A to point B. That is, the work done by a conservative force in moving a particle is the difference in potential energies of the particle at the two points. Symbolically, if **f** is a conservative force, the work done in moving the particle from point A to point B is

$$W_{AB} = \int_{\mathbf{r}_A}^{\mathbf{r}_B} \mathbf{f} \cdot d\mathbf{r} = V_A - V_B \qquad (9.26)$$

where V_A and V_B are the potential energies at points A and B, respectively.

EXAMPLE 9-5: Verify (9.24) and (9.25).

Solution: The condition (9.24) was established in Example 7-15. For convenience the solution is repeated here. Since **f** is conservative, we have from definition (9.23)

$$\oint_C \mathbf{f} \cdot d\mathbf{r} = 0$$

for any closed curve C. Now assume that $\nabla \times \mathbf{f} \neq \mathbf{0}$ at some point P on C. If we assume that $\nabla \times \mathbf{f}$ is continuous, there will be some region about P where $\nabla \times \mathbf{f} \neq \mathbf{0}$. Choose a small plane surface S in this region and a unit normal vector **n** to S parallel to $\nabla \times \mathbf{f}$, that is, $\nabla \times \mathbf{f} = a\mathbf{n}$, where $a > 0$. Let C be the boundary of S. Then by Stokes' theorem (7.26)

$$\oint_C \mathbf{f} \cdot d\mathbf{r} = \iint_S \nabla \times \mathbf{f} \cdot d\mathbf{S} = \iint_S a\mathbf{n} \cdot \mathbf{n}\, dS = a \iint_S dS = aS > 0$$

which contradicts the fact that **f** is conservative. Hence $\nabla \times \mathbf{f} = \mathbf{0}$.

Equation (9.25) was established in Problem 7-14. The solution to Problem 7-14 is repeated here. If $\mathbf{f} = -\nabla V$ for a scalar function V, then by Stokes' theorem (7.26) we have

$$\oint_C \mathbf{f} \cdot d\mathbf{r} = \oint_C (-\nabla V) \cdot d\mathbf{r}$$

$$= \iint_S \nabla \times (-\nabla V) \cdot d\mathbf{S}$$

where S is the surface enclosed by C. But from (5.31) we have $\nabla \times (-\nabla V) = \mathbf{0}$; hence

$$\oint_C \mathbf{f} \cdot d\mathbf{r} = 0$$

Thus we see that **f** is conservative.

C. Kinetic energy

The **kinetic energy** T of a particle is the scalar quantity defined as

KINETIC ENERGY $$T = \tfrac{1}{2}m\mathbf{v} \cdot \mathbf{v} = \tfrac{1}{2}mv^2 \qquad (9.27)$$

SUMMARY

1. If $\mathbf{r}(t)$ is the position vector of a moving particle in space, the velocity vector $\mathbf{v}(t)$ and the acceleration vector $\mathbf{a}(t)$ of the particle are defined by

$$\mathbf{v}(t) = \frac{d\mathbf{r}}{dt} = \mathbf{r}'(t)$$

$$\mathbf{a}(t) = \frac{d\mathbf{v}(t)}{dt} = \frac{d^2\mathbf{r}}{dt^2} = \mathbf{r}''(t)$$

2. If a particle rotates about a fixed line in a circular path, the angular velocity ω of the particle is a vector of magnitude $\omega = d\theta(t)/dt$ [where $\theta(t)$ is its angular position], and its direction along the fixed line is that of a right-hand screw advance when the screw is turned in the same sense as the rotation of the particle.

3. Newton's second law of motion can be expressed as

$$\mathbf{f} = \frac{d}{dt}(m\mathbf{v}) = \frac{d\mathbf{p}}{dt}$$

where $\mathbf{f}$ is the force acting upon the particle and $\mathbf{p}$ is the linear momentum of the particle, which is the product of mass m and velocity $\mathbf{v}$.

4. Torque τ can be expressed as

$$\tau = \mathbf{r} \times \mathbf{f} = \frac{d\mathbf{L}}{dt} = \frac{d}{dt}(\mathbf{r} \times \mathbf{p})$$

where $\mathbf{L}$ is the angular momentum of a particle with respect to an origin.

5. The work W_{AB} done by a force $\mathbf{f}$ to move a particle along a curve C from point A to point B is defined by

$$W_{AB} = \int_{\mathbf{r}_A}^{\mathbf{r}_B} \mathbf{f} \cdot d\mathbf{r}$$

where $\mathbf{r}_A$ and $\mathbf{r}_B$ are the position vectors of points A and B.

6. If a force $\mathbf{f}$ can be expressed as $\mathbf{f} = -\nabla V$, then V is called the potential energy and the force $\mathbf{f}$ is said to be conservative.

7. The kinetic energy T of a particle is defined by

$$T = \tfrac{1}{2}m\mathbf{v} \cdot \mathbf{v} = \tfrac{1}{2}mv^2$$

RAISE YOUR GRADES

Can you explain . . . ?

☑ how to evaluate the velocity and acceleration vectors of a particle
☑ how to evaluate the angular velocity and angular acceleration of a particle
☑ Newton's laws of motion
☑ how to evaluate the work done by a force $\mathbf{f}$ in moving a particle along a curve

SOLVED PROBLEMS

Displacement, Velocity, and Acceleration Vectors

PROBLEM 9-1 Show that the position velocity, and acceleration vectors in cylindrical coordinates (Figure 9-3) are

$$\mathbf{r}(t) = \rho \mathbf{e}_\rho + z \mathbf{k} \tag{a}$$

$$\mathbf{v}(t) = \frac{d\rho}{dt} \mathbf{e}_\rho + \rho \frac{d\phi}{dt} \mathbf{e}_\phi + \frac{dz}{dt} \mathbf{k} \tag{b}$$

$$\mathbf{a}(t) = \left[\frac{d^2\rho}{dt^2} - \rho \left(\frac{d\phi}{dt} \right)^2 \right] \mathbf{e}_\rho + \left(\rho \frac{d^2\phi}{dt^2} + 2 \frac{d\rho}{dt} \frac{d\phi}{dt} \right) \mathbf{e}_\phi + \frac{d^2z}{dt^2} \mathbf{k} \tag{c}$$

Solution: Referring to Figure 9-3 and Section 8-4, the position vector in cylindrical coordinates is given by

$$\mathbf{r}(t) = \rho \mathbf{e}_\rho + z \mathbf{k}$$

Since unit vector $\mathbf{k}$ is constant in time and the unit vector $\mathbf{e}_\rho$ changes direction in time, the velocity vector is

$$\mathbf{v}(t) = \frac{d\mathbf{r}(t)}{dt} = \frac{d\rho}{dt} \mathbf{e}_\rho + \rho \frac{d\mathbf{e}_\rho}{dt} + \frac{dz}{dt} \mathbf{k} \tag{d}$$

But from (8.43a) we have

$$\mathbf{e}_\rho = \cos \phi \, \mathbf{i} + \sin \phi \, \mathbf{j}$$

and hence, because of (8.43b), its derivative is

$$\frac{d\mathbf{e}_\rho}{dt} = -\sin \phi \frac{d\phi}{dt} \mathbf{i} + \cos \phi \frac{d\phi}{dt} \mathbf{j}$$

$$= \frac{d\phi}{dt} (-\sin \phi \, \mathbf{i} + \cos \phi \, \mathbf{j})$$

$$= \frac{d\phi}{dt} \mathbf{e}_\phi \tag{e}$$

So the velocity vector in cylindrical coordinates is

$$\mathbf{v}(t) = \frac{d\rho}{dt} \mathbf{e}_\rho + \rho \frac{d\phi}{dt} \mathbf{e}_\phi + \frac{dz}{dt} \mathbf{k}$$

Similarly, $\mathbf{e}_\phi$ also changes direction in time, and thus

$$\mathbf{a}(t) = \frac{d\mathbf{v}(t)}{dt} = \frac{d^2\rho}{dt^2} \mathbf{e}_\rho + \frac{d\rho}{dt} \frac{d\mathbf{e}_\rho}{dt} + \frac{d}{dt} \left(\rho \frac{d\phi}{dt} \right) \mathbf{e}_\phi + \rho \frac{d\phi}{dt} \frac{d\mathbf{e}_\phi}{dt} + \frac{d^2z}{dt^2} \mathbf{k} \tag{f}$$

From (8.43b), $\mathbf{e}_\phi$ in terms of $\mathbf{i}$ and $\mathbf{j}$ is

$$\mathbf{e}_\phi = -\sin \phi \, \mathbf{i} + \cos \phi \, \mathbf{j}$$

and hence, because of (8.43a), its derivative is

$$\frac{d\mathbf{e}_\phi}{dt} = -\cos \phi \frac{d\phi}{dt} \mathbf{i} - \sin \phi \frac{d\phi}{dt} \mathbf{j}$$

$$= -\frac{d\phi}{dt} (\cos \phi \, \mathbf{i} + \sin \phi \, \mathbf{j})$$

$$= -\frac{d\phi}{dt} \mathbf{e}_\rho \tag{g}$$

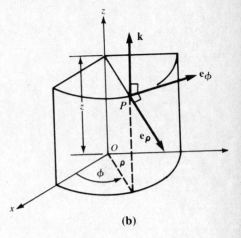

(b)

Figure 9-3
Cylindrical coordinates.

Substituting eqs. (e) and (g) into eq. (f) of this problem and rearranging the terms, we obtain the expression for the acceleration vector in cylindrical coordinates:

$$\mathbf{a}(t) = \left[\frac{d^2\rho}{dt^2} - \rho\left(\frac{d\phi}{dt}\right)^2\right]\mathbf{e}_\rho + \left(\rho\frac{d^2\phi}{dt^2} + 2\frac{d\rho}{dt}\frac{d\phi}{dt}\right)\mathbf{e}_\phi + \frac{d^2z}{dt^2}\mathbf{k}$$

PROBLEM 9-2 Find the tangential component a_t and normal component a_n of the acceleration vector $\mathbf{a}(t)$ if the velocity and acceleration vectors of a particle are known.

Solution: Let the velocity and acceleration vectors be (see (9.9))

$$\mathbf{v} = v(t)\mathbf{T} \tag{a}$$

$$\mathbf{a} = a_t\mathbf{T} + a_n\mathbf{N} \tag{b}$$

Since the velocity vector is in the direction of the tangent vector and perpendicular to the normal vector so that $\mathbf{v} \cdot \mathbf{T} = v$ and $\mathbf{v} \cdot \mathbf{N} = 0$, we take the dot product of both sides of eq. (b) with $\mathbf{v}$ and obtain

$$a_t = \frac{\mathbf{a} \cdot \mathbf{v}}{v} \tag{c}$$

Since $\mathbf{v} \times \mathbf{T} = \mathbf{0}$ and $|\mathbf{v} \times \mathbf{N}| = v$, by taking the cross product of both sides of eq. (b) with $\mathbf{v}$ and then taking the magnitude of both sides, we obtain

$$a_n = \frac{|\mathbf{a} \times \mathbf{v}|}{v} \tag{d}$$

Hence the acceleration vector in eq. (b) becomes

$$\mathbf{a} = \frac{\mathbf{a} \cdot \mathbf{v}}{v}\mathbf{T} + \frac{|\mathbf{a} \times \mathbf{v}|}{v}\mathbf{N} \tag{e}$$

Comparing eq. (b) of this problem with (9.8), we obtain the component expressions

$$a_t = \frac{dv}{dt} = \frac{\mathbf{a} \cdot \mathbf{v}}{v}$$

$$a_n = \frac{v^2}{\rho} = \frac{|\mathbf{a} \times \mathbf{v}|}{v}$$

PROBLEM 9-3 The position vector of a particle moving in a plane at any time t can be given by

$$\mathbf{r} = (a\cos\omega t)\mathbf{i} + (b\sin\omega t)\mathbf{j} \tag{a}$$

Show that its acceleration is always directed toward the origin.

Solution: Taking the derivative of eq. (a), the velocity vector is

$$\mathbf{v} = \mathbf{r}'(t) = -\omega a\sin\omega t\mathbf{i} + \omega b\cos\omega t\mathbf{j}$$

Then the acceleration vector is

$$\mathbf{a} = \mathbf{v}'(t) = \mathbf{r}''(t) = -\omega^2 a\cos\omega t\mathbf{i} - \omega^2 b\sin\omega t\mathbf{j}$$
$$= -\omega^2(a\cos\omega t\mathbf{i} + b\sin\omega t\mathbf{j})$$
$$= -\omega^2\mathbf{r} \tag{b}$$

Hence its acceleration is always in the radial direction, and because of the negative sign it is toward the origin. This is the centripetal acceleration caused by rotation.

PROBLEM 9-4 Verify **Kepler's second law of planetary motion:** If the acceleration of a particle moving in a plane is directed toward the origin, then

$$\rho^2\frac{d\phi}{dt} = \text{const} \tag{a}$$

Equivalently, the radius vector from the origin to the particle sweeps out equal areas in equal intervals of time.

Solution: Since the polar coordinates of a plane are merely cylindrical coordinates with $z = 0$, from eq. (c) of Problem 9-1 the acceleration vector **a** in plane polar coordinates can be expressed as

$$\mathbf{a} = \left[\frac{d^2\rho}{dt^2} - \rho\left(\frac{d\phi}{dt}\right)^2\right]\mathbf{e}_\rho + \left(\rho\frac{d^2\phi}{dt^2} + 2\frac{d\rho}{dt}\frac{d\phi}{dt}\right)\mathbf{e}_\phi$$

$$= \left[\frac{d^2\rho}{dt} - \rho\left(\frac{d\phi}{dt}\right)^2\right]\mathbf{e}_\rho + \frac{1}{\rho}\frac{d}{dt}\left(\rho^2\frac{d\phi}{dt}\right)\mathbf{e}_\phi \qquad (b)$$

Since the acceleration vector is directed toward the origin, the component of the acceleration in the $\mathbf{e}_\phi$ direction is zero; that is,

$$\frac{1}{\rho}\frac{d}{dt}\left(\rho^2\frac{d\phi}{dt}\right) = 0 \qquad (c)$$

Since the derivative of $\rho^2\,d\phi/dt$ is zero, we conclude that

$$\rho^2\frac{d\phi}{dt} = \text{const}$$

Also, from eq. (a) of Problem 9-1 (with $z = 0$), the radius vector in plane polar coordinates is

$$\mathbf{r} = \rho\mathbf{e}_\rho \qquad (d)$$

and from eq. (e) of Problem 9-1, its differential is

$$d\mathbf{r} = d\rho\,\mathbf{e}_\rho + \rho\,d\mathbf{e}_\rho = d\rho\,\mathbf{e}_\rho + \rho\,d\phi\,\mathbf{e}_\phi \qquad (e)$$

Now if A is the area swept out by position vector **r** at any time t after the initial t_0 value, then from the result of Problem 7-21

$$\oint_C \mathbf{r} \times d\mathbf{r} = 2\iint_S d\mathbf{S}$$

we have

$$|\mathbf{r} \times d\mathbf{r}| = 2\,dA \qquad (f)$$

From eqs. (d) and (e) we have

$$\mathbf{r} \times d\mathbf{r} = \rho\mathbf{e}_\rho \times (d\rho\,\mathbf{e}_\rho + \rho\,d\phi\,\mathbf{e}_\phi) = \rho^2\,d\phi\,\mathbf{k} \qquad (g)$$

since $\mathbf{e}_\rho \times \mathbf{e}_\rho = 0$ and $\mathbf{e}_\rho \times \mathbf{e}_\phi = \mathbf{k}$. Hence the radial area dA swept out during time dt is

$$dA = \tfrac{1}{2}\rho^2\,d\phi$$

and consequently, from eq. (a) of this problem

$$\frac{dA}{dt} = \frac{1}{2}\rho^2\frac{d\phi}{dt} = \text{const} \qquad (h)$$

This equation shows that equal areas are swept out in equal intervals of time.

Angular Velocity and Angular Acceleration

PROBLEM 9-5 If $\alpha = d\omega/dt$ is the angular acceleration of a particle rotating about a fixed axis L, show that its acceleration vector **a** can be expressed as

$$\mathbf{a} = \boldsymbol{\omega} \times \mathbf{v} + \boldsymbol{\alpha} \times \mathbf{r} \qquad (a)$$

or

$$\mathbf{a} = -\omega^2\mathbf{r} + \boldsymbol{\alpha} \times \mathbf{r} \qquad (b)$$

Solution:

(a) Differentiating the velocity vector (9.12), we obtain

$$\mathbf{a} = \frac{d\mathbf{v}}{dt} = \frac{d}{dt}(\boldsymbol{\omega} \times \mathbf{r})$$

$$= \boldsymbol{\omega} \times \frac{d\mathbf{r}}{dt} + \frac{d\boldsymbol{\omega}}{dt} \times \mathbf{r}$$

$$= \boldsymbol{\omega} \times \mathbf{v} + \boldsymbol{\alpha} \times \mathbf{r}$$

(b) Next since $\mathbf{v} = \boldsymbol{\omega} \times \mathbf{r}$, and using the definition of the vector triple product (1.59), we have

$$\mathbf{a} = \boldsymbol{\omega} \times (\boldsymbol{\omega} \times \mathbf{r}) + \boldsymbol{\alpha} \times \mathbf{r}$$
$$= (\boldsymbol{\omega} \cdot \mathbf{r})\boldsymbol{\omega} - \omega^2\mathbf{r} + \boldsymbol{\alpha} \times \mathbf{r} \tag{c}$$

If, in addition to choosing the origin O on the axis L (see Figure 9-2), we let O be in the plane of motion, then $\boldsymbol{\omega} \perp \mathbf{r}$ or $\boldsymbol{\omega} \cdot \mathbf{r} = 0$. Consequently

$$\mathbf{a} = -\omega^2\mathbf{r} + \boldsymbol{\alpha} \times \mathbf{r}$$

Note: $\boldsymbol{\alpha} \times \mathbf{r}$ is the tangential acceleration and $\boldsymbol{\omega} \times (\boldsymbol{\omega} \times \mathbf{r})$ is the centripetal acceleration.

PROBLEM 9-6 Show that when a particle rotates with a constant angular velocity, the curl of the velocity field is twice the angular velocity of the particle.

Solution: Since $\mathbf{v} = \boldsymbol{\omega} \times \mathbf{r}$, we can apply the result of Problem 5-23, obtaining the curl of the velocity field:

$$\nabla \times \mathbf{v} = \nabla \times (\boldsymbol{\omega} \times \mathbf{r})$$
$$= \boldsymbol{\omega}(\nabla \cdot \mathbf{r}) - \mathbf{r}(\nabla \cdot \boldsymbol{\omega}) + (\mathbf{r} \cdot \nabla)\boldsymbol{\omega} - (\boldsymbol{\omega} \cdot \nabla)\mathbf{r}$$

Since $\boldsymbol{\omega}$ is a constant vector, $\nabla \cdot \boldsymbol{\omega} = 0$ and $(\mathbf{r} \cdot \nabla)\boldsymbol{\omega} = 0$. From (5.17) and from the result of Problem 5-14, we have

$$\nabla \cdot \mathbf{r} = 3 \quad \text{and} \quad (\boldsymbol{\omega} \cdot \nabla)\mathbf{r} = \boldsymbol{\omega}$$

Hence the curl of $\mathbf{v}$ is

$$\nabla \times \mathbf{v} = \boldsymbol{\omega}(\nabla \cdot \mathbf{r}) - (\boldsymbol{\omega} \cdot \nabla)\mathbf{r} = 3\boldsymbol{\omega} - \boldsymbol{\omega} = 2\boldsymbol{\omega}$$

which indicates that the curl of the velocity field is twice the angular velocity of the particle.

Force and Moment

PROBLEM 9-7 The force exerted by the sun on a planet is given by

$$\mathbf{f} = -\left(Gm\frac{M}{r^3}\right)\mathbf{r} \tag{a}$$

where G is the gravitational constant, m is the mass of the planet, M is the mass of the sun, and $\mathbf{r}$ is the position vector from the sun at the origin to the planet.

(a) Using Newton's second law (9.15), show that

$$\frac{d}{dt}(\mathbf{r} \times \mathbf{v}) = \mathbf{r} \times \left(-\frac{GM}{r^3}\mathbf{r}\right) = \mathbf{0} \tag{b}$$

and hence, for some constant vector $\mathbf{h}$,

$$\mathbf{r} \times \mathbf{v} = \mathbf{r} \times \frac{d\mathbf{r}}{dt} = \mathbf{h} \tag{c}$$

(b) Show that $dA/dt = \frac{1}{2}h$, where A is the area swept out by **r**, thus proving Kepler's second law of planetary motion (see Problem 9-4).

Solution:

(a) From Newton's second law of motion (9.15) the force vector is $\mathbf{f} = -(GmM/r^3)\mathbf{r} = m(d\mathbf{v}/dt)$, and consequently

$$\frac{d\mathbf{v}}{dt} = -\frac{GM}{r^3}\mathbf{r} \qquad\qquad (d)$$

Since $d\mathbf{r}/dt = \mathbf{v}$ and $\mathbf{v} \times \mathbf{v} = \mathbf{0}$, we have

$$\frac{d}{dt}(\mathbf{r} \times \mathbf{v}) = \mathbf{r} \times \frac{d\mathbf{v}}{dt} + \frac{d\mathbf{r}}{dt} \times \mathbf{v} = \mathbf{r} \times \frac{d\mathbf{v}}{dt} \qquad\qquad (e)$$

Substituting the value of $d\mathbf{v}/dt$ from eq. (d) into eq. (e) and using $\mathbf{r} \times \mathbf{r} = \mathbf{0}$, we obtain

$$\frac{d}{dt}(\mathbf{r} \times \mathbf{v}) = \mathbf{r} \times \frac{d\mathbf{v}}{dt} = \mathbf{r} \times \left(-\frac{GM}{r^3}\mathbf{r}\right) = \mathbf{0} \qquad\qquad (f)$$

Since the derivative of $\mathbf{r} \times \mathbf{v}$ is zero, we conclude that $\mathbf{r} \times \mathbf{v}$ is a constant vector denoted by **h**; thus we can write

$$\mathbf{r} \times \mathbf{v} = \mathbf{r} \times \frac{d\mathbf{r}}{dt} = \mathbf{h}$$

(b) If A is the area swept out by position vector **r** at any time t after starting time t_0, then from the result of Problem 7-21,

$$\oint_C \mathbf{r} \times d\mathbf{r} = 2\iint_S d\mathbf{S}$$

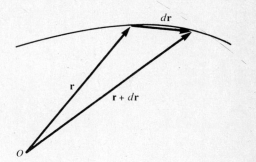

we have $|\mathbf{r} \times d\mathbf{r}| = 2\,dA$, where dA is the sector area (see Figure 9-4). Hence if h is a constant, then

$$2\frac{dA}{dt} = h \quad \text{or} \quad \frac{dA}{dt} = \frac{1}{2}h$$

that is, equal areas are swept out by r in equal intervals of time.

Figure 9-4
Kepler's second law of planetary motion.

PROBLEM 9-8 Show that the linear momentum of a free particle is conserved.

Solution: If the force **f** acting on a particle is zero, then Newton's second law of motion (9.19) becomes

$$\frac{d\mathbf{p}}{dt} = \mathbf{0}$$

where **p** is the linear momentum. But this expression implies that **p** is a constant vector in time. Therefore the linear momentum of a free particle is conserved.

PROBLEM 9-9 Show that the angular momentum of a particle is conserved if there is no torque acting on it.

Solution: If the torque τ acting on a particle is zero, then (9.20) for torque reduces to

$$\frac{d\mathbf{L}}{dt} = \mathbf{0}$$

which implies that the angular momentum **L** is a constant vector in time. Hence the angular momentum of a particle is conserved if there is no torque.

Work and Energy

PROBLEM 9-10   Verify the formula (9.26) for equating work and potential energy change.

Solution: Since **f** is conservative, we have from the expression for potential energy (9.25)

$$W_{AB} = \int_{\mathbf{r}_A}^{\mathbf{r}_B} \mathbf{f} \cdot d\mathbf{r} = - \int_{\mathbf{r}_A}^{\mathbf{r}_B} \nabla V \cdot d\mathbf{r}$$

Now from the result of Problem 5-4 we have

$$\nabla V \cdot d\mathbf{r} = dV$$

Hence the work done in moving the particle from point A to point B is

$$W_{AB} = \int_{\mathbf{r}_A}^{\mathbf{r}_B} \mathbf{f} \cdot d\mathbf{r} = - \int_{V=V_A}^{V=V_B} dV = -(V_B - V_A) = V_A - V_B$$

PROBLEM 9-11   (a) Show that a force (such as the gravitational force in Problem 9-7) that obeys the inverse square law,

$$\mathbf{f} = - \frac{k}{r^3} \mathbf{r} = - \frac{k}{r^2} \mathbf{e}_r$$

where k is a constant, is a conservative force. (b) Find the potential energy of the particle acted upon by this force.

Solution:
(a) From the result of Problem 5-19, that is, $\nabla \times [f(r)\mathbf{r}] = [f'(r)/r]\mathbf{r} \times \mathbf{r} = \mathbf{0}$, we have

$$\nabla \times \mathbf{f} = \nabla \times \left(- \frac{k}{r^3} \mathbf{r} \right) = \mathbf{0}$$

Hence the force **f** is conservative.
(b) The work done is

$$W_{AB} = \int_{\mathbf{r}_A}^{\mathbf{r}_B} \mathbf{f} \cdot d\mathbf{r} = \int_{\mathbf{r}_A}^{\mathbf{r}_B} \left(- \frac{k}{r^3} \mathbf{r} \right) \cdot d\mathbf{r}$$

$$= - \int_{\mathbf{r}_A}^{\mathbf{r}_B} \frac{k}{r^2} \, dr$$

$$= \frac{k}{r} \bigg|_{r_A}^{r_B}$$

$$= k \left(\frac{1}{r_B} - \frac{1}{r_A} \right)$$

$$= V_A - V_B$$

When $r_B = \infty$, it is usually understood that $V_B = 0$. With this datum, the gravitational potential becomes

$$V = - \frac{k}{r}$$

PROBLEM 9-12   Hooke's law states that an elastic force **f** is directly proportional to the displacement vector **r** of the particle upon which it acts. This is expressed as

$$\mathbf{f} = -k\mathbf{r}$$

where k is a constant. (a) Verify that the elastic force is a conservative force, and (b) find the potential energy of a particle acted upon by this force.

Solution:
(a) From the curl of **r** given by (5.29) the curl of **f** is

$$\mathbf{V} \times \mathbf{f} = \mathbf{V} \times (-k\mathbf{r}) = -k\mathbf{V} \times \mathbf{r} = 0$$

Hence the elastic force **f** is conservative.
(b) The work done in moving the particle from point A to point B is

$$W_{AB} = \int_{r_A}^{r_B} \mathbf{f} \cdot d\mathbf{r} = \int_{r_A}^{r_B} (-k\mathbf{r}) \cdot d\mathbf{r}$$

$$= -k \int_{r_A}^{r_B} r \, dr$$

$$= -\frac{k}{2} r^2 \Big|_{r_A}^{r_B}$$

$$= \frac{k}{2} r_A^2 - \frac{k}{2} r_B^2$$

$$= V_A - V_B$$

We take $V = 0$ when $r = 0$. Thus the potential energy of a particle becomes

$$V = \frac{k}{2} r^2$$

PROBLEM 9-13   Show that for a constant mass the work done in going from point A to point B is

$$W_{AB} = \frac{1}{2} m\mathbf{v}_B \cdot \mathbf{v}_B - \frac{1}{2} m\mathbf{v}_A \cdot \mathbf{v}_A = \frac{1}{2} m(v_B^2 - v_A^2) \qquad (a)$$

where $\mathbf{v}_A$ and $\mathbf{v}_B$ are the velocities of the particle at points A and B, respectively.

Solution: Since $\mathbf{f} = m(d\mathbf{v}/dt)$ and $d\mathbf{r} = (d\mathbf{r}/dt)dt = \mathbf{v} \, dt$, the work done is

$$W_{AB} = \int_{r_A}^{r_B} \mathbf{f} \cdot d\mathbf{r} = \int_{t_A}^{t_B} m \frac{d\mathbf{v}}{dt} \cdot \mathbf{v} \, dt = \int_{v_A}^{v_B} m\mathbf{v} \cdot d\mathbf{v}$$

$$= \frac{m}{2} \int_{v=v_A}^{v=v_B} d(\mathbf{v} \cdot \mathbf{v})$$

$$= \frac{m}{2} (\mathbf{v} \cdot \mathbf{v}) \Big|_{v_A}^{v_B}$$

$$= \frac{m}{2} (\mathbf{v}_B \cdot \mathbf{v}_B) - \frac{m}{2} (\mathbf{v}_A \cdot \mathbf{v}_A)$$

$$= \frac{m}{2} (v_B^2 - v_A^2)$$

Note: Using the equation for kinetic energy (9.27), eq. (a) can be rewritten as

$$W_{AB} = T_B - T_A \qquad (b)$$

which shows that the work done in going from point A to point B is equal to the difference between the kinetic energy of the particle at point B and that at point A.

PROBLEM 9-14   Prove the principle of conservation of energy: If a force **f** acting on a particle is conservative, the sum of the kinetic energy T and potential energy V of the particle is a constant:

$$T + V = E$$

where E is the **total mechanical energy** of the particle.

Solution: From (9.26) the work done by a conservative force to move a particle from point A to point B is

$$W_{AB} = V_A - V_B \tag{a}$$

where V_A and V_B are the potential energies of the particle at points A and B. Now, from eq. (b) of Problem 9-13, if T_A and T_B are the kinetic energies of the particle at points A and B, the work done is

$$W_{AB} = T_B - T_A \tag{b}$$

Hence equating (a) and (b), we obtain

$$T_A + V_A = T_B + V_B$$

So the total mechanical energy $T + V$ is the same at points A and B; thus it is a constant and we have

$$T_A + V_A = T_B + V_B = E$$

where E is a constant.

Supplementary Exercises

PROBLEM 9-15   Express $\mathbf{r}$, $\mathbf{v}$, and $\mathbf{a}$ in spherical coordinates.

Answer:

$$\mathbf{r} = r\mathbf{e}_r$$

$$\mathbf{v} = (dr/dt)\mathbf{e}_r + r(d\theta/dt)\mathbf{e}_\theta + r\sin\theta(d\phi/dt)\mathbf{e}_\phi$$

$$\mathbf{a} = \left[\frac{d^2 r}{dt^2} - r\left(\frac{d\theta}{dt}\right)^2 - r\sin^2\theta\left(\frac{d\phi}{dt}\right)^2\right]\mathbf{e}_r + \left[2\frac{dr}{dt}\frac{d\theta}{dt} + r\frac{d^2\theta}{dt^2} - r\sin\theta\cos\theta\left(\frac{d\phi}{dt}\right)^2\right]\mathbf{e}_\theta$$

$$+ \left(2\sin\theta\frac{dr}{dt}\frac{d\phi}{dt} + 2r\cos\theta\frac{d\theta}{dt}\frac{d\phi}{dt} + r\sin\theta\frac{d^2\phi}{dt^2}\right)\mathbf{e}_\phi$$

PROBLEM 9-16   A particle moves in the xy-plane with velocity $\mathbf{v}$ and acceleration $\mathbf{a}$. Show that $|\mathbf{v} \times \mathbf{a}| = v^3/\rho$ and the radius of curvature ρ can be evaluated by the formula

$$\rho = \frac{\left[\left(\dfrac{dx}{dt}\right)^2 + \left(\dfrac{dy}{dt}\right)^2\right]^{3/2}}{\left(\dfrac{dx}{dt}\dfrac{d^2 y}{dt^2} - \dfrac{dy}{dt}\dfrac{d^2 x}{dt^2}\right)}$$

[*Hint*: Write $\mathbf{v} = v\mathbf{T}$ and use (9.4).]

PROBLEM 9-17   A particle P moves on a straight line from the center of a disk toward the edge. The disk rotates counterclockwise with constant angular speed ω. Assuming that the disk is in the xy-plane, find the acceleration $\mathbf{a}$ of P. [*Hint*: Write $\mathbf{r} = t\mathbf{u}$, where $\mathbf{u}$ is a unit vector rotating with the disk.]

Answer: $2\dfrac{d\mathbf{u}}{dt} - \omega^2 t\mathbf{u}$

PROBLEM 9-18   A particle is attracted toward the origin with a force $\mathbf{f} = f(r)\mathbf{e}_r$; that is, the force is a central force. Show that the angular momentum of the particle is constant.

PROBLEM 9-19 The force acting upon a moving particle of mass m, which carries a charge q moving through an electromagnetic field is

$$\mathbf{f} = q(\mathbf{E} + \mathbf{v} \times \mathbf{B})$$

where $\mathbf{E}$ is the electric field vector, $\mathbf{B}$ is the magnetic induction vector, and $\mathbf{v}$ is the velocity of the charged particle. This equation is known as the **Lorentz force equation**. If there is no electric field, that is, $\mathbf{E} = \mathbf{0}$, and if the particle enters the magnetic field in a direction perpendicular to that of $\mathbf{B}$, show that the particle moves in a circular path, and find the radius of the circle.

Answer: mv/qB.

PROBLEM 9-20 Show that the moment of a force $\mathbf{f}$ applied at P relative to the origin O is unchanged if $\mathbf{f}$ slides along its line of action.

PROBLEM 9-21 Let $\mathbf{f} = -(k/r^2)\mathbf{r}$ be the central force that moves a particle P along the helix C, which is defined by

$$\mathbf{r}(\phi) = (a \cos \phi)\mathbf{i} + (a \sin \phi)\mathbf{j} + b\phi\mathbf{k}$$

Find the work done by $\mathbf{f}$ in moving P from the point $\phi = 0$ to the point $\phi = \pi$.

Answer: $k\left(\dfrac{1}{\sqrt{a^2 + b^2\pi^2}} - \dfrac{1}{a}\right)$

PROBLEM 9-22 The force of gravity near a point on the earth's surface (with the z-axis pointing up) can be expressed as $\mathbf{f} = -mg\mathbf{k}$, where m is the mass of the particle and $\mathbf{g} = -g\mathbf{k}$ is the acceleration of gravity. Find the potential energy of the particle at height z.

Answer: mgz

10 APPLICATIONS TO FLUID MECHANICS

THIS CHAPTER IS ABOUT

- ☑ **Equation of Continuity**
- ☑ **Equation of Motion**
- ☑ **Fluid Statics**
- ☑ **Steady Flow and Streamlines**
- ☑ **Irrotational Flow and Velocity Potential**
- ☑ **Vortex Flow and Circulation**
- ☑ **Equation of Energy**

10-1. Equation of Continuity

A. Definitions

- A **fluid** is a substance which deforms continuously under an applied shear stress.
- **Pressure** is the intensity of distributed force due to the action of fluids and is measured as force per unit area.
- **Density** is mass per unit volume.

The state of a moving fluid is completely determined if the distribution of the fluid velocity **v**, the pressure p, and the density ρ are given. All these quantities are, in general, functions of the space coordinates x, y, z and of time t; that is,

$$\mathbf{v}(\mathbf{r}, t) = \mathbf{v}(x, y, z, t) = \text{velocity of the fluid at position } \mathbf{r} \text{ and time } t$$

$$p(\mathbf{r}, t) = p(x, y, z, t) = \text{pressure of the fluid at position } \mathbf{r} \text{ and time } t$$

$$\rho(\mathbf{r}, t) = \rho(x, y, z, t) = \text{density of the fluid at position } \mathbf{r} \text{ and time } t$$

A fluid is said to be **incompressible** if its density ρ is constant.

B. Equation of continuity

The mass in any region R enclosed by the surface S is defined as

$$\iiint_R \rho \, dV \tag{10.1}$$

Since $\mathbf{v} \cdot d\mathbf{S}$ measures the volume of fluid crossing a surface element per unit time and $\rho \mathbf{v} \cdot d\mathbf{S}$ measures its mass, the surface integral

$$\oiint_S \rho \mathbf{v} \cdot d\mathbf{S} \tag{10.2}$$

is a measure of the rate at which the fluid mass flux is leaving region R through its surface S.

The decrease per unit time in the mass of fluid in region R can be written as

$$-\frac{\partial}{\partial t} \iiint_R \rho \, dV \tag{10.3}$$

If ρ and $\mathbf{v}$ are the density and velocity of a moving fluid, the **equation of continuity** of fluid dynamics is

EQUATION OF CONTINUITY
$$\nabla \cdot (\rho \mathbf{v}) + \frac{\partial \rho}{\partial t} = 0 \qquad (10.4)$$

The **mass flux density** of a fluid is given by the vector

MASS FLUX DENSITY
$$\mathbf{J} = \rho \mathbf{v} \qquad (10.5)$$

It has the same direction as the motion of the fluid, and its magnitude equals the mass of fluid flowing in unit time through a unit area perpendicular to the velocity vector. The unit of mass flux density is mass per square length per unit of time. Using $\mathbf{J}$, the equation of continuity (10.4) can be written as

$$\nabla \cdot \mathbf{J} + \frac{\partial \rho}{\partial t} = 0 \qquad (10.6)$$

EXAMPLE 10-1: Derive the equation of continuity (10.4).

Solution: Equating (10.2) and (10.3), we have

$$\oiint_S \rho \mathbf{v} \cdot d\mathbf{S} = -\frac{\partial}{\partial t} \iiint_R \rho \, dV \qquad (10.7)$$

By the divergence theorem (7.2), the surface integral on the left can be transformed as

$$\oiint_S \rho \mathbf{v} \cdot d\mathbf{S} = \iiint_R \nabla \cdot (\rho \mathbf{v}) \, dV \qquad (10.8)$$

Substituting this and transposing, (10.7) can be rewritten as

$$\iiint_R \left[\nabla \cdot (\rho \mathbf{v}) + \frac{\partial \rho}{\partial t} \right] dV = 0 \qquad (10.9)$$

Since (10.9) holds for any volume, the integrand must vanish. We then have the equation of continuity:

$$\nabla \cdot (\rho \mathbf{v}) + \frac{\partial \rho}{\partial t} = 0$$

10-2. Equation of Motion

A. Euler's equation of motion

The **viscosity** of a fluid is a measure of its resistance to deformation. A **perfect** or **ideal fluid** has zero viscosity. Thus a shear stress cannot be imposed, and internal friction cannot be induced.

Euler's equation of motion of a fluid is

EULER'S EQUATION OF MOTION
$$\frac{d\mathbf{v}}{dt} = \mathbf{f} - \frac{1}{\rho} \nabla p \qquad (10.10)$$

or

$$\frac{\partial \mathbf{v}}{\partial t} + (\mathbf{v} \cdot \nabla)\mathbf{v} = \mathbf{f} - \frac{1}{\rho} \nabla p \qquad (10.11)$$

where $\mathbf{f}(\mathbf{r})$ is the external force per unit mass acting on the fluid.

B. Boundary conditions

For a moving fluid the equation of motion can be supplemented by boundary conditions that must be satisfied at the surfaces bounding the fluid. For an ideal fluid and a solid surface the necessary boundary condition is simply that the fluid cannot penetrate the solid surface. Hence if the surface is at rest,

$$v_n = 0 \tag{10.12}$$

where v_n is the normal component of the fluid velocity at the surface.

For a moving surface v_n must be equal to the normal component of the velocity of the surface.

EXAMPLE 10-2: Establish Euler's equation (10.10) from Newton's second law of motion (9.19).

Solution: Consider some region R enclosed by surface S inside the fluid. If $\mathbf{f}(\mathbf{r})$ is the external force per unit mass acting on the fluid, the resultant external force acting on R is

$$\mathbf{f}_e = \iiint_R \rho \mathbf{f}\, dV \tag{10.13}$$

If $p(\mathbf{r}, t)$ is the fluid pressure, the resultant force due to fluid pressure is

$$\mathbf{f}_p = -\oiint_S p\, d\mathbf{S}$$

By the gradient theorem (7.19) this surface integral can be converted into a volume integral, that is,

$$\mathbf{f}_p = -\oiint_S p\, d\mathbf{S} = -\iiint_R \nabla p\, dV \tag{10.14}$$

By Newton's second law of motion (9.19)

$$\frac{d}{dt} \iiint_R \rho \mathbf{v}\, dV = \mathbf{f}_e + \mathbf{f}_p \tag{10.15}$$

But the left-hand term of (10.15) is the rate of change of the total amount of momentum associated with the fluid that is inside surface S at any instant. If this integral is changed from volume to mass by writing $\rho\, dV = dm$, then dm is invariable as S moves with the fluid. Hence the total force acting on R is

$$\frac{d}{dt} \iiint_R \rho \mathbf{v}\, dV = \iiint_R \frac{d\mathbf{v}}{dt}\, dm = \iiint_R \rho \frac{d\mathbf{v}}{dt}\, dV \tag{10.16}$$

by resubstituting $dm = \rho\, dV$.

Combining (10.13), (10.14), and (10.16), we obtain

$$\iiint_R \rho \frac{d\mathbf{v}}{dt}\, dV = \iiint_R \rho \mathbf{f}\, dV - \iiint_R \nabla p\, dV$$

or, upon transposing and combining,

$$\iiint_R \left(\rho \frac{d\mathbf{v}}{dt} - \rho \mathbf{f} + \nabla p \right) dV = \mathbf{0} \tag{10.17}$$

Since R is an arbitrary region in the fluid, and this is true for all R, the integrand must be zero and we get

$$\rho \frac{d\mathbf{v}}{dt} - \rho\mathbf{f} + \nabla p = 0$$

hence we obtain Euler's equation (10.10):

$$\frac{d\mathbf{v}}{dt} = \mathbf{f} - \frac{1}{\rho}\nabla p$$

10-3. Fluid Statics

We now consider the case in which a fluid is at rest and the pressure is in equilibrium with the applied external force field.

EXAMPLE 10-3: Derive the equation of equilibrium of the fluid at rest.

Solution: The velocity of the fluid at rest is zero so that $d\mathbf{v}/dt = \mathbf{0}$, and so (10.10) becomes

EQUILIBRIUM EQUATION $\quad\quad \dfrac{1}{\rho}\nabla p = \mathbf{f}$ $\quad\quad\quad\quad$ (10.18)

which is the desired equilibrium equation.

Equation (10.18) shows that surfaces of constant pressure are everywhere perpendicular to the external force field.

10-4. Steady Flow and Streamlines

A. Steady flow

A **steady flow** is a flow in which the velocity is constant in time at any point in the fluid; that is, $\partial v/\partial t = 0$. The **equation of motion for steady flow** is

EQUATION OF MOTION FOR STEADY FLOW $\quad \dfrac{1}{2}\nabla(v^2) - \mathbf{v} \times (\nabla \times \mathbf{v}) = \mathbf{f} - \dfrac{1}{\rho}\nabla p$ $\quad$ (10.19)

where $\mathbf{v}$ is the constant velocity in time at any point in the fluid, p is the fluid pressure, $\mathbf{f}$ is the external force per unit mass acting on the fluid, and ρ is the density of the fluid. (For derivation of (10.19), see Problem 10-8.)

B. Bernoulli's equation

A **pathline** is the trajectory of a single particle of fluid. A **streamline** is the curve found at any instant of time such that the tangent to the curve, at any point on the curve, gives the direction of the velocity at that point. In steady flow the streamlines do not vary with time, and they coincide with the pathlines.

The streamlines are determined by the set of differential equations

$$\frac{dx}{v_1} = \frac{dy}{v_2} = \frac{dz}{v_3} \qquad\qquad (10.20)$$

where $d\mathbf{r} = [dx, dy, dz]$ and $\mathbf{v} = [v_1, v_2, v_3]$.

The general form of **Bernoulli's equation** states that when an incompressible ideal fluid whose flow is steady moves under the action of conservative forces $\mathbf{f}$,

BERNOULLI'S EQUATION $\quad\quad V + \dfrac{p}{\rho} + \dfrac{1}{2}v^2 = c$ $\quad\quad\quad$ (10.21)

along a streamline, where c is a constant and $\mathbf{f} = -\nabla V$. (For the derivation of (10.21) see Problem 10-9.)

10-5. Irrotational Flow and Velocity Potential

A. Vorticity

The **vorticity** or **vortex vector** is defined as

VORTICITY $$\mathbf{\Omega} = \mathbf{\nabla} \times \mathbf{v} \qquad (10.22)$$

where **v** is the flow velocity vector. An **irrotational**, or **potential flow** is one for which $\mathbf{\Omega} = \mathbf{0}$ everywhere. A **rotational**, or **vortex flow** is one in which $\mathbf{\Omega}$ is not everywhere zero. A **stagnation point** is a point at which the velocity is zero.

B. Velocity potential

As we know from Problem 7-23, any vector field having zero curl can be expressed as the gradient of some scalar function. If we apply this result to the velocity vector **v** in an irrotational flow, this scalar function is called the **velocity potential** ϕ and we can write

$$\mathbf{v} = \mathbf{\nabla}\phi \qquad (10.23)$$

For irrotational flow of an incompressible ideal fluid, the velocity potential ϕ satisfies *Laplace's equation*:

$$\mathbf{\nabla}^2\phi = 0 \qquad (10.24)$$

EXAMPLE 10-4: Verify Laplace's equation (10.24).

Solution: The result of Problem 10-2 shows that if the fluid is incompressible,

$$\mathbf{\nabla} \cdot \mathbf{v} = 0$$

Since $\mathbf{v} = \mathbf{\nabla}\phi$ for an irrotational flow, we have

$$\mathbf{\nabla} \cdot \mathbf{v} = \mathbf{\nabla} \cdot \mathbf{\nabla}\phi = \mathbf{\nabla}^2\phi = 0$$

C. Bernoulli's equation for irrotational flow

Bernoulli's equation for irrotational flow states that if the flow of an ideal incompressible fluid is both steady and irrotational, then

BERNOULLI'S EQUATION FOR IRROTATIONAL FLOW $$V + \frac{p}{\rho} + \frac{1}{2}v^2 = c \qquad (10.25)$$

where V is the potential function such that $\mathbf{f} = -\mathbf{\nabla}V$, p is the pressure, **v** is the velocity, ρ is the density, and c is a constant.

Note: To see the distinction between form (10.25) and the earlier form (10.21), we see that in the general case, the "constant" c in (10.21) is a constant along any given streamline but that it is different for different streamlines. For an irrotational flow, the value c in (10.25) is constant throughout the fluid.

10-6. Vortex Flow and Circulation

A. Vortex tube

A **vortex line** is a line that is everywhere parallel to the vorticity defined in (10.22) as $\mathbf{\Omega} = \mathbf{\nabla} \times \mathbf{v}$. A **vortex tube** is a surface A generated by the vortex lines that pass through a closed curve C. It is characterized by the property

$$\mathbf{n} \cdot \mathbf{\Omega} = \mathbf{n} \cdot \mathbf{\nabla} \times \mathbf{v} = 0 \text{ on } A \qquad (10.26)$$

where **n** is the unit normal vector to the vortex tube A. Vortex lines or vortex tubes may change with time changes because, in general, the vorticity vector $\mathbf{\Omega}$ depends on time.

The strength of a vortex tube is defined as

STRENGTH OF A VORTEX TUBE $\qquad \iint\limits_S \boldsymbol{\Omega} \cdot d\mathbf{S}$ $\qquad$ (10.27)

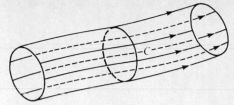

Figure 10-1
Vortex tube.

where S is the cross section of the vortex tube bounded by a simple closed curve C encircling the vortex tube (see Figure 10-1).

B. Circulation

The **circulation** of $\mathbf{v}$ (or simply "circulation") along a closed curve C is given by the line integral (see Section 6-1)

CIRCULATION $\qquad \oint_C \mathbf{v} \cdot d\mathbf{r}$ $\qquad$ (10.28)

EXAMPLE 10-5: Show that the strength of a vortex tube is equal to the circulation; that is,

$$\iint\limits_S \boldsymbol{\Omega} \cdot d\mathbf{S} = \oint_C \mathbf{v} \cdot d\mathbf{r} \qquad (10.29)$$

where S is the cross section of area bounded by C.

Solution: By applying Stokes' theorem (7.26), we have

$$\oint_C \mathbf{v} \cdot d\mathbf{r} = \iint\limits_S \boldsymbol{\nabla} \times \mathbf{v} \cdot d\mathbf{S} = \iint\limits_S \boldsymbol{\Omega} \cdot d\mathbf{S}$$

Note: The circulation may differ from zero in certain cases, even when the flow is irrotational; that is, $\boldsymbol{\Omega} = \mathbf{0}$ (see Problem 7-22).

C. Kelvin's theorem

Kelvin's theorem, or **the law of conservation of circulation**, states that the circulation of velocity is constant with respect to time for a closed curve C moving with the fluid particle if (a) the external force field acting on the fluid is conservative and (b) the fluid is a **barotropic fluid**. (A fluid is barotropic if its density is a function of pressure alone.) This is stated as

KELVIN'S THEOREM $\qquad K = \oint_C \mathbf{v} \cdot d\mathbf{r} = \text{const } t$ $\qquad$ (10.30)

EXAMPLE 10-6: Establish Kelvin's theorem (the law of conservation of circulation) in the form (10.30).

Solution: Let

$$K = \oint_C \mathbf{v} \cdot d\mathbf{r} \qquad (10.31)$$

The particles comprising the curve C change position with time, but in uniform flow the curve remains closed. At any time t the circulation around C is

$$K(t) = \oint_C \mathbf{v} \cdot \frac{d\mathbf{r}}{ds} \, ds = \oint_C \mathbf{v} \cdot \mathbf{T} \, ds \qquad (10.32)$$

where $\mathbf{T}$ is the unit vector tangent to C and s is the arc length variable.

Since both **v** and **T** vary with time,

$$\frac{dK}{dt} = \oint_C \frac{d}{dt}(\mathbf{v} \cdot \mathbf{T})\,ds = \oint_C \left(\frac{d\mathbf{v}}{dt} \cdot \mathbf{T} + \mathbf{v} \cdot \frac{d\mathbf{T}}{dt}\right)ds \qquad (10.33)$$

The variable s is the variable of integration along C at any instant of time and is therefore independent of the motion of the curve with time. Hence

$$\frac{d\mathbf{T}}{dt} = \frac{d}{dt}\left(\frac{d\mathbf{r}}{ds}\right) = \frac{d}{ds}\left(\frac{d\mathbf{r}}{dt}\right) = \frac{d\mathbf{v}}{ds} \qquad (10.34)$$

and (10.33) becomes

$$\frac{dK}{dt} = \oint_C \left(\frac{d\mathbf{v}}{dt} \cdot \mathbf{T} + \mathbf{v} \cdot \frac{d\mathbf{v}}{ds}\right)ds \qquad (10.35)$$

If the external force field $\mathbf{f}$ is conservative, it can be expressed as $\mathbf{f} = -\nabla V$, and Euler's equation (10.10) becomes

$$\frac{d\mathbf{v}}{dt} = -\nabla V - \frac{1}{\rho}\nabla p \qquad (10.36)$$

If ρ is a function of p alone and

$$\psi = V + \int \frac{dp}{\rho} \qquad (10.37)$$

then we obtain

$$\nabla\psi = \nabla V + \nabla\left(\int \frac{dp}{\rho}\right) \qquad (10.38)$$

Now from (5.15) we have

$$\nabla\left(\int \frac{dp}{\rho}\right) = \left(\frac{d}{dp}\int \frac{dp}{\rho}\right)\nabla p = \frac{1}{\rho}\nabla p \qquad (10.39)$$

Hence using (10.36),

$$\nabla\psi = \nabla V + \frac{1}{\rho}\nabla p = -\frac{d\mathbf{v}}{dt} \qquad (10.40)$$

Thus (10.35) reduces to

$$\frac{dK}{dt} = \oint_C \left(-\nabla\psi \cdot \mathbf{T} + \mathbf{v} \cdot \frac{d\mathbf{v}}{ds}\right)ds \qquad (10.41)$$

Since from identity (5.5) $\nabla\psi \cdot \mathbf{T} = \partial\psi/\partial s$,

$$\frac{dK}{dt} = -\oint_C \left[\frac{\partial}{\partial s}\left(\psi - \frac{1}{2}\mathbf{v} \cdot \mathbf{v}\right)\right]ds$$

$$= -\oint_C \left[\frac{\partial}{\partial s}\left(\psi - \frac{1}{2}v^2\right)\right]ds \qquad (10.42)$$

Since the quantity $\psi - \frac{1}{2}v^2$ has the same value at the common initial–terminal point of C, this integral vanishes, and we write

$$\frac{dK}{dt} = 0$$

By integrating this if we designate c as a constant, we obtain Kelvin's theorem:

$$K = \oint_C \mathbf{v} \cdot d\mathbf{r} = c$$

10-7. Equation of Energy

The **kinetic energy** T of a fluid in a simply connected region R with boundary S is

KINETIC ENERGY
$$T = \frac{1}{2} \iiint_R \rho v^2 \, dV = \frac{1}{2} \iiint_R \rho \mathbf{v} \cdot \mathbf{v} \, dV \qquad (10.43)$$

For the irrotational flow of an incompressible fluid in a region R in which the velocity potential ϕ is single-valued, the kinetic energy T of the fluid is

$$T = \frac{1}{2} \rho \oiint_S \phi \frac{\partial \phi}{\partial n} \, dS \qquad (10.44)$$

where S is the surface enclosing R and $\partial \phi / \partial n$ is the normal derivative of ϕ.

EXAMPLE 10-7: Verify the kinetic energy formula (10.44).

Solution: For irrotational flow we have $\mathbf{v} = \nabla \phi$. Hence from definition (10.43) the kinetic energy T is

$$T = \frac{1}{2} \rho \iiint_R |\nabla \phi|^2 \, dV \qquad (10.45)$$

Now if we set $\psi = \phi$ in Green's first theorem (7.10), we have

$$\iiint_R (\phi \nabla^2 \phi + |\nabla \phi|^2) \, dV = \oiint_S \phi \frac{\partial \phi}{\partial n} \, dS \qquad (10.46)$$

Thus (10.45) can be rewritten as

$$T = \frac{1}{2} \rho \oiint_S \phi \frac{\partial \phi}{\partial n} \, ds - \frac{1}{2} \rho \iiint_R \phi \nabla^2 \phi \, dV \qquad (10.47)$$

From (10.24), $\nabla^2 \phi = 0$; hence the kinetic energy is

$$T = \frac{1}{2} \rho \oiint_S \phi \frac{\partial \phi}{\partial n} \, dS$$

SUMMARY

1. The equation of continuity is

$$\nabla \cdot (\rho \mathbf{v}) + \frac{\partial \rho}{\partial t} = 0$$

   where ρ is the density of the fluid and $\mathbf{v}$ is the velocity of the fluid.
2. Euler's equation of motion is

$$\frac{d\mathbf{v}}{dt} = \mathbf{f} - \frac{1}{\rho} \nabla p$$

   where $\mathbf{f}$ is the external force per unit mass acting on the fluid and p is the pressure of the fluid.
3. An irrotational flow is one for which the vorticity is defined as $\Omega = \nabla \times \mathbf{v} = \mathbf{0}$ everywhere.
4. Kelvin's theorem states that

$$\oint_C \mathbf{v} \cdot d\mathbf{r} = \text{const}$$

   if the external force acting on the fluid is conservative and the fluid is barotropic.

RAISE YOUR GRADES

Can you explain...?

☑ how the equation of continuity is derived
☑ how Euler's equation of motion is derived from Newton's second law of motion
☑ what steady flow is
☑ what irrotational flow is

SOLVED PROBLEMS

Equation of Continuity

PROBLEM 10-1   Show that the equation of continuity (10.4) can be rewritten as

$$\rho \mathbf{V} \cdot \mathbf{v} + \mathbf{v} \cdot \mathbf{V}\rho + \frac{\partial \rho}{\partial t} = 0 \tag{a}$$

Solution: The equation of continuity (10.4) is

$$\mathbf{V} \cdot (\rho \mathbf{v}) + \frac{\partial \rho}{\partial t} = 0 \tag{b}$$

By (5.41), we have

$$\mathbf{V} \cdot (\rho \mathbf{v}) = \rho \mathbf{V} \cdot \mathbf{v} + \mathbf{v} \cdot \mathbf{V}\rho \tag{c}$$

Thus eq. (b) reduces to eq. (a).

PROBLEM 10-2   Show that the velocity vector of an incompressible fluid is solenoidal.

Solution: If the fluid is incompressible, then $\partial \rho / \partial t = 0$ and $\mathbf{V}\rho = \mathbf{0}$. Thus the equation of continuity expression (a) of Problem 10-1 reduces to

$$\rho \mathbf{V} \cdot \mathbf{v} = 0 \quad \text{or} \quad \mathbf{V} \cdot \mathbf{v} = 0$$

that is, $\mathbf{v}$ is solenoidal (see Problem 5-32).

Equation of Motion

PROBLEM 10-3   Derive Euler's equation of motion (10.11).

Solution: From (10.10) we have

$$\frac{d\mathbf{v}}{dt} = \mathbf{f} - \frac{1}{\rho}\mathbf{V}p \tag{a}$$

Since $\mathbf{v}(\mathbf{r}, t)$ depends on the time as well as the spatial coordinates, from the result of Problem 5-16 we have

$$d\mathbf{v} = \frac{\partial \mathbf{v}}{\partial t}\,dt + (d\mathbf{r} \cdot \mathbf{V})\mathbf{v}$$

or dividing both sides by dt,

$$\frac{d\mathbf{v}}{dt} = \frac{\partial \mathbf{v}}{\partial t} + (\mathbf{v} \cdot \nabla)\mathbf{v} \qquad \text{(b)}$$

Substituting eq. (b) into eq. (a) gives

$$\frac{\partial \mathbf{v}}{\partial t} + (\mathbf{v} \cdot \nabla)\mathbf{v} = \mathbf{f} - \frac{1}{\rho} \nabla p$$

PROBLEM 10-4   Show that Euler's equation (10.11) can be rewritten as

$$\frac{\partial \mathbf{v}}{\partial t} + \frac{1}{2} \nabla(v^2) - \mathbf{v} \times (\nabla \times \mathbf{v}) = \mathbf{f} - \frac{1}{\rho} \nabla p \qquad \text{(a)}$$

Solution: From Problem 5-23 we have the vector identity

$$\nabla(\mathbf{f} \cdot \mathbf{g}) = \mathbf{f} \times (\nabla \times \mathbf{g}) + \mathbf{g} \times (\nabla \times \mathbf{f}) + (\mathbf{f} \cdot \nabla)\mathbf{g} + (\mathbf{g} \cdot \nabla)\mathbf{f}$$

If we set $\mathbf{f} = \mathbf{g} = \mathbf{v}$, we obtain

$$\nabla(\mathbf{v} \cdot \mathbf{v}) = \nabla(v^2) = 2\mathbf{v} \times (\nabla \times \mathbf{v}) + 2(\mathbf{v} \cdot \nabla)\mathbf{v}$$

Upon simplification

$$(\mathbf{v} \cdot \nabla)\mathbf{v} = \tfrac{1}{2} \nabla(v^2) - \mathbf{v} \times (\nabla \times \mathbf{v}) \qquad \text{(b)}$$

Substituting eq. (b) into Euler's equation (10.11), we obtain

$$\frac{\partial \mathbf{v}}{\partial t} + \frac{1}{2} \nabla(v^2) - \mathbf{v} \times (\nabla \times \mathbf{v}) = \mathbf{f} - \frac{1}{\rho} \nabla p$$

PROBLEM 10-5   Derive the equation of motion for an incompressible fluid if the external force acting on the fluid is conservative.

Solution: If the external force field $\mathbf{f}$ is conservative, from (9.25) $\mathbf{f} = -\nabla V$, where V is a scalar potential function.

   If the fluid is incompressible, ρ is a constant and

$$\frac{1}{\rho} \nabla p = \nabla\left(\frac{p}{\rho}\right)$$

Substituting into eq. (a) of Problem 10-4, we obtain

$$\frac{\partial \mathbf{v}}{\partial t} - \mathbf{v} \times (\nabla \times \mathbf{v}) = -\nabla\left(V + \frac{p}{\rho} + \frac{1}{2} v^2\right) \qquad \text{(a)}$$

which is the desired equation.

Fluid Statics

PROBLEM 10-6   Show that if there is no external force, the pressure of a fluid is the same at every point in the fluid.

Solution: If $\mathbf{f} = \mathbf{0}$, then the equilibrium equation (10.18) becomes

$$\nabla p = \mathbf{0}$$

This equation implies that p is constant.

PROBLEM 10-7   Consider a fluid at rest in a uniform gravitational field. If the field is incompressible, show that

$$p = -\rho g z + c \qquad \text{(a)}$$

where c is a constant, g is the local gravitational constant, and z is the distance measured positively upward from the surface of the earth.

Solution: Taking the z-axis positively upward from the surface of the earth, we have

$$\mathbf{f} = -g\mathbf{k}$$

and from equilibrium equation (10.18), we have

$$\frac{\partial p}{\partial x} = \frac{\partial p}{\partial y} = 0 \quad \text{and} \quad \frac{\partial p}{\partial z} = -\rho g.$$

Integration of this pair of equations gives

$$p = -\rho g z + c$$

which indicates that the pressure decreases linearly with increasing height.

Steady Flow and Streamlines

PROBLEM 10-8   Derive the equation of motion for steady flow (10.19).

Solution: Since $\partial \mathbf{v}/\partial t = 0$ for steady flow, Euler's equation of motion (see Problem 10-4) reduces to the result (10.19).

PROBLEM 10-9   Verify Bernoulli's equation (10.21).

Solution: Since $\partial \mathbf{v}/\partial t = 0$ for steady flow, eq. (a) of Problem 10-5 becomes

$$\mathbf{v} \times (\mathbf{V} \times \mathbf{v}) = \mathbf{V}\left(V + \frac{p}{\rho} + \frac{1}{2}v^2 \right) \tag{a}$$

The vector $\mathbf{v} \times (\mathbf{V} \times \mathbf{v})$ is perpendicular to $\mathbf{v}$; therefore, if we dot product both sides of eq. (a) with $\mathbf{v}$, we obtain

$$\mathbf{v} \cdot \left[\mathbf{V}\left(V + \frac{p}{\rho} + \frac{1}{2}v^2 \right) \right] = 0 \tag{b}$$

Hence $\mathbf{V}(V + p/\rho + \frac{1}{2}v^2)$ is everywhere normal to the velocity field $\mathbf{v}$. So $\mathbf{v}$ is parallel to the surface and thus

$$V + \frac{p}{\rho} + \frac{1}{2}v^2 = c$$

which implies that $V + p/\rho + \frac{1}{2}v^2$ is constant along a streamline. The values of the constant are generally different for each streamline.

If v remains essentially constant, Bernoulli's equation (10.21) shows that the velocity is inversely proportional to the pressure.

PROBLEM 10-10   If $\mathbf{\Omega} = \mathbf{V} \times \mathbf{v}$ for steady flow of an incompressible fluid under the action of conservative forces, show that

$$(\mathbf{\Omega} \cdot \mathbf{V})\mathbf{v} - (\mathbf{v} \cdot \mathbf{V})\mathbf{\Omega} = 0 \tag{a}$$

Solution: If we take the curl of both sides of eq. (a) of Problem 10-9, we obtain

$$\mathbf{V} \times [\mathbf{v} \times (\mathbf{V} \times \mathbf{v})] = 0 \tag{b}$$

since by formula (5.31) $\mathbf{V} \times \mathbf{V}(V + p/\rho + \frac{1}{2}v^2) = \mathbf{0}$.

Substituting $\mathbf{V} \times \mathbf{v} = \mathbf{\Omega}$ in eq. (b) and using the result of Problem 5-23, we have

$$\mathbf{V} \times [\mathbf{v} \times \mathbf{\Omega}] = \mathbf{v}[\mathbf{V} \cdot \mathbf{\Omega}] - \mathbf{\Omega}(\mathbf{V} \cdot \mathbf{v}) + (\mathbf{\Omega} \cdot \mathbf{V})\mathbf{v} - (\mathbf{v} \cdot \mathbf{V})\mathbf{\Omega} = 0 \tag{c}$$

Since the fluid is incompressible, the result of Problem 10-2 gives us $\mathbf{V} \cdot \mathbf{v} = 0$. Also, from formula (5.32), $\mathbf{V} \cdot \mathbf{\Omega} = \mathbf{V} \cdot (\mathbf{V} \times \mathbf{v}) = 0$. Hence eq. (c) reduces to

$$(\mathbf{\Omega} \cdot \mathbf{V})\mathbf{v} - (\mathbf{v} \cdot \mathbf{V})\mathbf{\Omega} = 0$$

Irrotational Flow and Velocity Potential

PROBLEM 10-11 If a fluid rotates like a rigid body with constant angular velocity ω with respect to some axis, show that the vorticity $\boldsymbol{\Omega}$ is twice the angular velocity ω.

Solution: From Problem 9-6, we have

$$\mathbf{V} \times \mathbf{v} = 2\omega$$

Hence from the definition of vorticity in (10.22)

$$\boldsymbol{\Omega} = \mathbf{V} \times \mathbf{v} = 2\omega$$

PROBLEM 10-12 For an incompressible ideal fluid, which moves under the action of a conservative force field and whose flow is irrotational, show that

$$\frac{\partial \phi}{\partial t} + V + \frac{p}{\rho} + \frac{1}{2} v^2 = c(t) \tag{a}$$

where $c(t)$ is a function of time.

Solution: Substituting $\mathbf{v} = \mathbf{V}\phi$ and $\mathbf{V} \times \mathbf{v} = 0$ in eq. (a) of Problem 10-5,

$$\frac{\partial \mathbf{v}}{\partial t} - \mathbf{v} \times (\mathbf{V} \times \mathbf{v}) = -\mathbf{V}\left(V + \frac{p}{\rho} + \frac{1}{2} v^2\right)$$

we have

$$\frac{\partial}{\partial t} (\mathbf{V}\phi) = -\mathbf{V}\left(V + \frac{p}{\rho} + \frac{1}{2} v^2\right)$$

or upon simplification

$$\mathbf{V}\left(\frac{\partial \phi}{\partial t} + V + \frac{p}{\rho} + \frac{1}{2} v^2\right) = 0$$

Therefore, if $c(t)$ is a function of time, then

$$\frac{\partial \phi}{\partial t} + V + \frac{p}{\rho} + \frac{1}{2} v^2 = c(t)$$

PROBLEM 10-13 Verify Bernoulli's equation (10.25) for irrotational flow.

Solution: For steady flow, $c(t)$ is no longer a function of time and is a constant, and $\partial \phi / \partial t = 0$ in the result of Problem 10-12. Hence eq. (a) of Problem 10-12 reduces to Bernoulli's equation (10.25):

$$V + \frac{p}{\rho} + \frac{1}{2} v^2 = c$$

PROBLEM 10-14 If there is no action of conservative forces, show that in the steady flow of an incompressible fluid the greatest pressure occurs at the stagnation point.

Solution: If there is no action of external conservative forces, Bernoulli's equation (10.25) reduces to

$$\frac{p}{\rho} + \frac{1}{2} v^2 = c \tag{a}$$

where c is a constant. Solving for p, we obtain

$$p = c\rho - \frac{1}{2} \rho v^2 \tag{b}$$

Thus the pressure p is maximum at points where the velocity v is zero.

Vortex Flow and Circulation

PROBLEM 10-15 If a fluid is incompressible and the external force field is conservative, show that

$$\frac{d\mathbf{\Omega}}{dt} = (\mathbf{\Omega} \cdot \nabla)\mathbf{v} \tag{a}$$

Solution: Euler's equation of motion for this case is eq. (a) of Problem 10-5, that is,

$$\frac{\partial \mathbf{v}}{\partial t} - \mathbf{v} \times (\nabla \times \mathbf{v}) = -\nabla\left(V + \frac{p}{\rho} + \frac{1}{2}v^2\right) \tag{b}$$

Taking the curl of both sides of eq. (b), using $\nabla \times \mathbf{v} = \mathbf{\Omega}$ and denoting $\psi = V + p/\rho + \frac{1}{2}v^2$, we obtain

$$\frac{\partial \mathbf{\Omega}}{\partial t} - \nabla \times (\mathbf{v} \times \mathbf{\Omega}) = \mathbf{0} \tag{c}$$

since $\nabla \times \nabla\psi = \mathbf{0}$ for a scalar function ψ as derived in (5.31).

Now following the procedure established in Problem 10-10, we make use of the vector identity of Problem 5-23 and obtain

$$\nabla \times (\mathbf{v} \times \mathbf{\Omega}) = \mathbf{v}(\nabla \cdot \mathbf{\Omega}) - \mathbf{\Omega}(\nabla \cdot \mathbf{v}) + (\mathbf{\Omega} \cdot \nabla)\mathbf{v} - (\mathbf{v} \cdot \nabla)\mathbf{\Omega} \tag{d}$$

Again, since $\nabla \cdot \mathbf{\Omega} = \nabla \cdot (\nabla \times \mathbf{v}) = 0$ and $\nabla \cdot \mathbf{v} = 0$, we have

$$\nabla \times (\mathbf{v} \times \mathbf{\Omega}) = (\mathbf{\Omega} \cdot \nabla)\mathbf{v} - (\mathbf{v} \cdot \nabla)\mathbf{\Omega} \tag{e}$$

Substituting eq. (e) into eq. (c), we obtain

$$\frac{\partial \mathbf{\Omega}}{\partial t} + (\mathbf{v} \cdot \nabla)\mathbf{\Omega} = (\mathbf{\Omega} \cdot \nabla)\mathbf{v} \tag{f}$$

From the results of Problems 5-5 and 5-16, or eq. (b) of Problem 10-3, we have

$$\frac{\partial}{\partial t} + (\mathbf{v} \cdot \nabla) = \frac{d}{dt}$$

Hence substituting this in the left-hand side of eq. (f), we obtain

$$\frac{d\mathbf{\Omega}}{dt} = (\mathbf{\Omega} \cdot \nabla)\mathbf{v}$$

PROBLEM 10-16 Prove **Helmholtz's first theorem on vorticity**: In a fluid flow vortex tubes move with the fluid.

Solution: If $\mathbf{n}$ is a unit vector perpendicular to the vortex tube A, then on A the vortex tube is characterized by the property

$$\mathbf{n} \cdot \mathbf{\Omega} = \mathbf{n} \cdot \nabla \times \mathbf{v} = 0$$

If Γ is a simple closed curve lying on a vortex tube, but not encircling the tube (see Figure 10-2a), then

$$K = \oint_\Gamma \mathbf{v} \cdot d\mathbf{r} = \iint_{A_\Gamma} \mathbf{\Omega} \cdot d\mathbf{S} = \iint_{A_\Gamma} \mathbf{\Omega} \cdot \mathbf{n} \, dS = 0 \tag{a}$$

where A_Γ is the part of the vortex tube A enclosed by Γ. Assume that Γ is carried by the fluid motion into the simple closed curve Γ', and A' is the surface passing through Γ' and generated by the original vortex tube A (see Figure 10-2b). Then from Kelvin's theorem (10.30), the circulation remains zero for all time; that is,

$$K = \oint_{\Gamma'} \mathbf{v} \cdot d\mathbf{r} = \iint_{A_{\Gamma'}} \nabla \times \mathbf{v} \cdot d\mathbf{S} = \iint_{A_{\Gamma'}} \mathbf{\Omega} \cdot \mathbf{n} \, dS = 0 \tag{b}$$

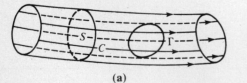

(a)

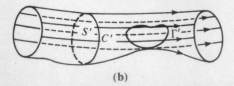

(b)

Figure 10-2

where $A_{\Gamma'}$ is the part of A' enclosed by Γ'. Since $A_{\Gamma'}$ is arbitrary, it follows that on A' the integrand in eq. (b) is

$$\boldsymbol{\Omega} \cdot \mathbf{n} = 0$$

Hence A' is a vortex tube. In other words, as time varies, the vortex tube maintains itself as a vortex tube. We can thus conclude that vortex tubes move with the fluid.

PROBLEM 10-17 Prove **Helmholtz's second theorem on vorticity:** The strength of a vortex tube remains constant.

Solution: The strength of a vortex tube is defined by (10.27) and is equated to the circulation in (10.29),

$$\iint_S \boldsymbol{\Omega} \cdot d\mathbf{S} = \iint_S \boldsymbol{\nabla} \times \mathbf{v} \cdot d\mathbf{S} = \oint_C \mathbf{v} \cdot d\mathbf{r}$$

where S is any surface cutting the vortex tube bounded by a closed curve C encircling the vortex tube (see Figure 10-2a).

If C' is the curve into which C is transformed by the fluid motion (see Figure 10-2b), then by Kelvin's theorem (10.30) the circulation remains constant in time and

$$K = \oint_C \mathbf{v} \cdot d\mathbf{r} = \oint_{C'} \mathbf{v} \cdot d\mathbf{r} = \iint_{S'} \boldsymbol{\Omega} \cdot d\mathbf{S} \tag{a}$$

where S' is the cross section of the displaced vortex tube bounded by C'. Hence the strength of a vortex tube is constant with respect to time.

Next consider a closed surface S consisting of a vortex tube A with cross sections S_1 and S_2 bounded by closed curves C_1 and C_2, respectively (Figure 10-3). Since $\boldsymbol{\Omega} = \boldsymbol{\nabla} \times \mathbf{v}$ is solenoidal, we have $\boldsymbol{\nabla} \cdot \boldsymbol{\Omega} = 0$, and by the divergence theorem (7.2) we obtain

$$\oiint_S \boldsymbol{\Omega} \cdot d\mathbf{S} = \iint_A \boldsymbol{\Omega} \cdot d\mathbf{S} + \iint_{S_1} \boldsymbol{\Omega} \cdot d\mathbf{S} + \iint_{S_2} \boldsymbol{\Omega} \cdot d\mathbf{S} = \iiint_R \boldsymbol{\nabla} \cdot \boldsymbol{\Omega} \, dV = 0 \tag{b}$$

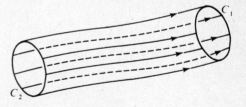

Figure 10-3

Now since $\boldsymbol{\Omega} \cdot \mathbf{n} = 0$ on A,

$$\iint_A \boldsymbol{\Omega} \cdot d\mathbf{S} = \iint_A \boldsymbol{\Omega} \cdot \mathbf{n} \, dS = 0$$

Thus

$$\iint_{S_1} \boldsymbol{\Omega} \cdot d\mathbf{S} = -\iint_{S_2} \boldsymbol{\Omega} \cdot d\mathbf{S}$$

or

$$\iint_{S_1} \boldsymbol{\Omega} \cdot \mathbf{n} \, dS = -\iint_{S_2} \boldsymbol{\Omega} \cdot \mathbf{n} \, dS = \iint_{S_2} \boldsymbol{\Omega} \cdot (-\mathbf{n}) \, dS \tag{c}$$

If $\mathbf{n}$ is the outward normal, it is the positive normal for S_1 and the negative normal for S_2, and conversely. Thus eq. (c) shows that the strength of a vortex tube at any cross section is constant.

We conclude that the strengths of the vortex tubes remain constant in time and that vortices can be neither created nor destroyed.

PROBLEM 10-18 Determine the shape of the surface of an incompressible fluid subject to a gravitational field contained in a cylindrical vessel that rotates about its vertical axis with a constant angular velocity ω.

Solution: If the cylinder is oriented so that the axis of the cylinder is the z-axis, then $\boldsymbol{\omega} = \omega\mathbf{k}$. Euler's equation (10.10) can be written as

$$\frac{d\mathbf{v}}{dt} = -\nabla V - \frac{1}{\rho}\nabla p = -\nabla\psi \tag{a}$$

where $\mathbf{f} = -\nabla V = -\nabla(gz)$ and $\psi = (gz + p/\rho)$ (see Problem 9-22).

Now from (9.12), $\mathbf{v} = \boldsymbol{\omega} \times \mathbf{r}$, and since $\boldsymbol{\omega}$ is a constant vector,

$$\frac{d\mathbf{v}}{dt} = \boldsymbol{\omega} \times \mathbf{v}$$

$$= \boldsymbol{\omega} \times (\boldsymbol{\omega} \times \mathbf{r})$$
$$= \boldsymbol{\omega}(\boldsymbol{\omega} \cdot \mathbf{r}) - \omega^2\mathbf{r}$$
$$= \omega^2 z\mathbf{k} - \omega^2(x\mathbf{i} + y\mathbf{j} + z\mathbf{k})$$
$$= -\omega^2(x\mathbf{i} + y\mathbf{j})$$
$$= -\tfrac{1}{2}\omega^2 \nabla(x^2 + y^2) \tag{b}$$

Substituting eq. (b) into eq. (a), we have

$$\nabla[\psi - \tfrac{1}{2}\omega^2(x^2 + y^2)] = 0$$

or if c is a constant, we obtain

$$\psi - \tfrac{1}{2}\omega^2(x^2 + y^2) = c$$

Now substituting the defined expression of ψ and simplifying, we obtain

$$\frac{p}{\rho} = \frac{1}{2}\omega^2(x^2 + y^2) - gz + c$$

At the free surface p is a constant. Thus the surface is a paraboloid of revolution and is given by

$$z = \frac{1}{2}\frac{\omega^2(x^2 + y^2)}{g} \tag{c}$$

where the origin is taken as the lowest point of the surface.

Equation of Energy

PROBLEM 10-19 Prove **Kelvin's minimum energy theorem:** The irrotational flow of an incompressible fluid in a simply connected region R has less kinetic energy than any other flow having the same normal component of velocity on boundary S.

Solution: Let T_0 represent the kinetic energy of the irrotational flow associated with the velocity potential ϕ. Let T be the kinetic energy of any other flow in which

$$\mathbf{v} = \nabla\phi + \mathbf{v}_1 \tag{a}$$

Then by the boundary condition on S

$$\mathbf{v}_1 \cdot \mathbf{n} = 0 \tag{b}$$

and the equations of continuity yield

$$\nabla \cdot \mathbf{v} = \nabla^2\phi + \nabla \cdot \mathbf{v}_1 = 0, \qquad \nabla^2\phi = 0$$

Hence

$$\nabla \cdot \mathbf{v}_1 = 0 \tag{c}$$

Now from the kinetic energy definition (10.43) the kinetic energy is

$$T = \frac{1}{2}\rho \iiint_R \mathbf{v} \cdot \mathbf{v}\, dV = \frac{1}{2}\rho \iiint_R (\nabla\phi + \mathbf{v}_1) \cdot (\nabla\phi + \mathbf{v}_1)\, dV$$

$$= \frac{1}{2}\rho \iiint_R |\nabla\phi|^2\, dV + \frac{1}{2}\rho \iiint_R v_1^2\, dV + \rho \iiint_R \nabla\phi \cdot \mathbf{v}_1\, dV$$

$$= T_0 + \frac{1}{2}\rho \iiint_R v_1^2\, dV + \rho \iiint_R \nabla\phi \cdot \mathbf{v}_1\, dV \tag{d}$$

Using identity (5.41), we have

$$\nabla\phi \cdot \mathbf{v}_1 = \nabla \cdot (\phi\mathbf{v}_1) - \phi\nabla \cdot \mathbf{v}_1 = \nabla \cdot (\phi\mathbf{v}_1)$$

since from eq. (c) $\nabla \cdot \mathbf{v}_1 = 0$. Thus using the divergence theorem (7.2), we obtain

$$\iiint_R \nabla\phi \cdot \mathbf{v}_1\, dV = \iiint_R \nabla \cdot (\phi\mathbf{v}_1)\, dV$$

$$= \oiint_S \phi\mathbf{v}_1 \cdot \mathbf{n}\, dS$$

$$= 0$$

since from eq. (b), $\mathbf{v}_1 \cdot \mathbf{n} = 0$ on S. Therefore it follows that

$$T = T_0 + \frac{1}{2}\rho \iiint_R v_1^2\, dV \tag{e}$$

Since $\rho > 0$ and the volume integral is nonnegative, $T \geq T_0$ in general and $T = T_0$ only when $\mathbf{v}_1 \equiv \mathbf{0}$, or $\mathbf{v} = \nabla\phi$. This is the required result.

PROBLEM 10-20 For the irrotational flow of an incompressible fluid in region R bounded by surface S, if $d(\rho\, dV)/dt = 0$, where dV is the differential volume, show that

$$\frac{dT}{dt} = \iiint_R \rho\mathbf{v} \cdot \mathbf{f}\, dV - \oiint_S p\mathbf{v} \cdot d\mathbf{S} \tag{a}$$

and give the physical meaning of this result.

Solution: From the definition of kinetic energy in (10.43), we differentiate both sides to obtain

$$\frac{dT}{dt} = \frac{1}{2}\rho \iiint_R \frac{d}{dt}(\mathbf{v} \cdot \mathbf{v})\, dV$$

But $d(\mathbf{v} \cdot \mathbf{v})/dt = 2(\mathbf{v} \cdot d\mathbf{v}/dt)$ (see rule (4.15)), so we have

$$\frac{dT}{dt} = \rho \iiint_R \left(\mathbf{v} \cdot \frac{d\mathbf{v}}{dt}\right) dV \tag{b}$$

By Euler's equation (10.10),

$$\frac{d\mathbf{v}}{dt} = \mathbf{f} - \frac{1}{\rho}\nabla p \tag{c}$$

Substituting eq. (c) into (b) and simplifying, we have

$$\frac{dT}{dt} = \rho \iiint_R \mathbf{v} \cdot \left(\mathbf{f} - \frac{1}{\rho} \nabla p \right) dV$$

$$= \iiint_R \rho \mathbf{v} \cdot \mathbf{f}\, dV - \iiint_R \mathbf{v} \cdot \nabla p\, dV \qquad (d)$$

Proceeding in a manner similar to that of Problem 10-19, we have

$$\mathbf{v} \cdot \nabla p = \nabla \cdot (p\mathbf{v}) - p\nabla \cdot \mathbf{v} = \nabla \cdot (p\mathbf{v})$$

since $\nabla \cdot \mathbf{v} = 0$ for an incompressible fluid. Thus by applying the divergence theorem (7.2), we obtain

$$\frac{dT}{dt} = \iiint_R \rho \mathbf{v} \cdot \mathbf{f}\, dV - \iiint_R \nabla \cdot (p\mathbf{v})\, dV$$

$$= \iiint_R \rho \mathbf{v} \cdot \mathbf{f}\, dV - \oiint_S p\mathbf{v} \cdot d\mathbf{S}$$

To obtain a physical meaning we observe that the left-hand side of eq. (a) is the rate of increase of the kinetic energy of a fluid. The first term of the right-hand side of eq. (a) is an expression for the work of the external forces on the fluid, and the second term is the work of the bounding surface pressure on the fluid.

Supplementary Exercises

PROBLEM 10-21   If a fluid is at rest, show that
$$\nabla + (\rho\mathbf{f}) = \mathbf{0}, \quad \text{and} \quad \mathbf{f} \cdot \nabla \times \mathbf{f} = 0$$
where $\mathbf{f}$ is the external force field acting on the fluid.

PROBLEM 10-22   Derive Bernoulli's equation for the steady flow of an ideal incompressible fluid under the gravitational field.

Answer: $gz + \dfrac{p}{\rho} + \dfrac{1}{2}v^2 = c$, where c is a constant.

PROBLEM 10-23   Show that the function $c(t)$ in eq. (a) of Problem 10-12 can be put equal to zero without loss of generality. [*Hint:* Since $\mathbf{v} = \nabla\phi$, we can add any function of time to ϕ. Replace ϕ by $\phi + \int c(t)dt$.]

PROBLEM 10-24   If the fluid is incompressible and the fluid motion is two-dimensional, show that a **stream function** $\psi(x, y)$ exists such that
$$v_1 = \frac{\partial \psi}{\partial y} \quad \text{and} \quad v_2 = \frac{\partial \psi}{\partial x}$$
where $\mathbf{v} = v_1\mathbf{i} + v_2\mathbf{j}$ is the velocity of the fluid.

PROBLEM 10-25   If the fluid motion is two-dimensional, incompressible, and irrotational, show that the stream function $\psi(x, y)$ of Problem 10-24 satisfies the Laplace equation
$$\frac{\partial^2 \psi}{\partial x^2} + \frac{\partial^2 \psi}{\partial y^2} = 0$$

PROBLEM 10-26   If a fluid is incompressible and has a constant vorticity $\boldsymbol{\Omega}$, show that

$$\nabla^2 \mathbf{v} = \mathbf{0}$$

where $\mathbf{v}$ is the velocity of the fluid. [*Hint:* $\nabla \times (\nabla \times \mathbf{v}) = \nabla \times \boldsymbol{\Omega} = \mathbf{0}$]

PROBLEM 10-27   If the external force field acting on a fluid is conservative and the density of the fluid depends only on the pressure of the fluid, prove the Helmholtz equation

$$\frac{d}{dt}\left(\frac{\boldsymbol{\Omega}}{\rho}\right) = \left(\frac{\boldsymbol{\Omega}}{\rho} \cdot \nabla\right)\mathbf{v}$$

[*Hint:* See Problem 10-15 and use the general continuity equation (10.4).]

PROBLEM 10-28   Show that the vortex lines or vortex tubes can neither begin nor terminate within the fluid; thus they are either closed or reach the boundary.

PROBLEM 10-29   For the irrotational flow of an incompressible fluid in region R bounded by S and a conservative external force field, i.e., $\mathbf{f} = -\nabla V$, show that

$$\frac{dT}{dt} = \iint \rho V \mathbf{v} \cdot d\mathbf{S} + \iint p\mathbf{v} \cdot d\mathbf{S}$$

In other words, the rate of increase of the kinetic energy of a fluid body is equal to the sum of the work of the external forces and the pressures on its surface.

11 APPLICATIONS TO ELECTROMAGNETIC THEORY

THIS CHAPTER IS ABOUT

- ☑ **Equation of Continuity**
- ☑ **Electromagnetic Field**
- ☑ **Equations of Electromagnetic Theory**
- ☑ **Potential Functions of the Electromagnetic Field**
- ☑ **Energy in the Electromagnetic Field and the Poynting Vector**
- ☑ **Static Fields**

11-1. Equation of Continuity

A. Definitions

- An **electromagnetic field** is produced by a distribution of electric current and charge.
- A **current** I is produced by the motion of the charge and is determined by the charge density ρ and velocity $\mathbf{v}$ of the charge.
- The **charge density** ρ is the density of the charge per unit volume.
- The **current density J** is the product of the charge density ρ and the velocity $\mathbf{v}$ of the charge; i.e.,

$$\mathbf{J} = \rho \mathbf{v} \tag{11.1}$$

Thus the current I through a surface S is the rate at which the charge passes through S, or

$$I = \iint_S \rho \mathbf{v} \cdot d\mathbf{S} = \iint_S \mathbf{J} \cdot d\mathbf{S} = \iint_S \mathbf{J} \cdot \mathbf{n}\, dS \tag{11.2}$$

where $\mathbf{n}$ is a unit vector normal to S.

Note: The sign of I depends on the choice of $\mathbf{n}$. If S is a closed surface, $\mathbf{n}$ is taken to be the outward normal vector.

B. Principle of conservation of charge

The **principle of conservation of charge** states that in a region R bounded by a surface S, the rate at which the charge decreases is equal to the rate at which the charge leaves R through the surface S.

C. Equation of continuity

The **equation of continuity** states that for any region R bounded by a surface S,

EQUATION OF CONTINUITY $\qquad \nabla \cdot \mathbf{J} + \dfrac{\partial \rho}{\partial t} = 0 \tag{11.3}$

or, equivalently, upon substituting the current density (11.1),

$$\nabla \cdot (\rho \mathbf{v}) + \frac{\partial \rho}{\partial t} = 0 \tag{11.4}$$

EXAMPLE 11-1: Using the principle of conservation of charge, derive the equation of continuity.

Solution: If $\rho(\mathbf{r}, t)$ is the charge density, then the total charge Q within a region R enclosed by S is given by

$$Q = \iiint_R \rho \, dV \tag{11.5}$$

Now from (11.2)

$$\oiint_S \mathbf{J} \cdot d\mathbf{S} = \oiint \mathbf{J} \cdot \mathbf{n} \, dS \tag{11.6}$$

which is a measure of the rate at which the charge leaves R through S since $\mathbf{n}$ is the outward normal to S.

But the rate at which the charge decreases in R is

$$-\frac{dQ}{dt} = -\frac{d}{dt} \iiint_R \rho \, dV = -\iint_R \frac{\partial \rho}{\partial t} \, dV \tag{11.7}$$

In (11.7) we must use $\partial \rho / \partial t$ inside the integral since $\rho(\mathbf{r}, t)$ is a function of $\mathbf{r}$ and t.

Using the principle of conservation of charge we can equate (11.6) and (11.7) and obtain

$$\oiint_S \mathbf{J} \cdot d\mathbf{S} = -\iiint_R \frac{\partial \rho}{\partial t} \, dV \tag{11.8}$$

From the divergence theorem (7.2) we have

$$\oiint_S \mathbf{J} \cdot d\mathbf{S} = \iiint_R \nabla \cdot \mathbf{J} \, dV \tag{11.9}$$

Thus (11.8) can be rewritten as

$$\iiint_R \left(\nabla \cdot \mathbf{J} + \frac{\partial \rho}{\partial t} \right) dV = 0 \tag{11.10}$$

Since (11.10) holds for an arbitrary region R, the integrand must be zero. Hence

$$\nabla \cdot \mathbf{J} + \frac{\partial \rho}{\partial t} = 0$$

which is the equation of continuity.

Using $\mathbf{J} = \rho \mathbf{v}$ from (11.1), eq. (11.3) can be expressed as

$$\nabla \cdot (\rho \mathbf{v}) + \frac{\partial \rho}{\partial t} = 0$$

By analogy with the corresponding equation of continuity of fluid dynamics given in (10.4), either (11.3) or (11.4) can be called the equation of continuity.

11-2. Electromagnetic Field

A. Lorentz force

The electromagnetic fields produced by charge distributions and their motion (current distribution) are denoted by the vectors $\mathbf{E}(\mathbf{r}, t)$ and $\mathbf{B}(\mathbf{r}, t)$.

A **Lorentz force** is the force $\mathbf{f}$ experienced by a charge q moving with a velocity $\mathbf{v}(\mathbf{r}, t)$ in the electromagnetic field and is given by

LORENTZ FORCE $\qquad\qquad \mathbf{f} = q(\mathbf{E} + \mathbf{v} \times \mathbf{B}) \tag{11.11}$

Equation (11.11) is the defining equation of the vectors **E** and **B**, where **E** is the **electric field intensity** and **B** is the **magnetic induction** or **magnetic flux density**. An important implication of (11.11) is the assumption that the properties of field or space described by **E** and **B** exist whether or not we place q in the field to observe the force. Thus when a test charge is introduced into an actual electric field for measuring purposes, it must be so small that it does not affect the original fields. In this case the Lorentz force equation (11.11) can be expressed as

$$\lim_{q \to 0} \frac{\mathbf{f}}{q} = \mathbf{E} + \mathbf{v} \times \mathbf{B} \tag{11.12}$$

B. Constituent relations

The electromagnetic field is also specified by the vector **D**, called the **electric flux density** or **electric displacement**, and the vector **H**, called the **magnetic field intensity**. In any given medium **D** and **H** are related to **E** and **B** by functional relationships characteristic of the medium. For linear, homogeneous, and isotropic media, the appropriate relations are

$$\mathbf{D} = \varepsilon \mathbf{E} \tag{11.13}$$

$$\mathbf{B} = \mu \mathbf{H} \tag{11.14}$$

where ε and μ are constants called the **permittivity** and **permeability** of the medium, respectively.

The current density **J** and the electric field intensity **E** are related by

$$\mathbf{J} = \sigma \mathbf{E} \tag{11.15}$$

where σ is the **conductivity** of the medium.

Equations (11.13), (11.14), and (11.15) are often referred to as the **constituent relations**.

The values of ε, μ, and σ depend on the system of units. In the mks rationalized system of units used throughout this chapter, the force is measured in newtons, the velocity in meters per second, and the charge in coulombs. The coulomb, designated by C, is the basic electric unit which, along with the meter, the kilogram, and the second, permits the definition of all other electromagnetic units.

11-3. Equations of Electromagnetic Theory

A. Maxwell's equations

Maxwell's equations are the fundamental equations in electromagnetic theory. They are

$$\nabla \times \mathbf{E} + \frac{\partial \mathbf{B}}{\partial t} = \mathbf{0} \tag{11.16}$$

MAXWELL'S EQUATIONS
$$\nabla \times \mathbf{H} - \frac{\partial \mathbf{D}}{\partial t} = \mathbf{J} \tag{11.17}$$

$$\nabla \cdot \mathbf{B} = 0 \tag{11.18}$$

$$\nabla \cdot \mathbf{D} = \rho \tag{11.19}$$

B. Faraday's law

Faraday's induction law states that the electromotive force around any closed contour is equal to the negative of the time rate of change of the magnetic flux linking the contour. In equation form, that is

FARADAY'S INDUCTION LAW
$$\oint_C \mathbf{E} \cdot d\mathbf{r} = -\frac{\partial}{\partial t} \iint_S \mathbf{B} \cdot d\mathbf{S} \tag{11.20}$$

EXAMPLE 11-2: Using the first of Maxwell's equations (11.16), verify Faraday's induction law.

Solution: Integrating (11.16) over a surface S bounded by a closed curve C yields

$$\iint_S \mathbf{\nabla} \times \mathbf{E} \cdot d\mathbf{S} + \iint_S \frac{\partial \mathbf{B}}{\partial t} \cdot d\mathbf{S} = 0 \tag{11.21}$$

Applying Stokes' theorem (7.26) to the first term, we now obtain

$$\oint_C \mathbf{E} \cdot d\mathbf{r} + \iint_S \frac{\partial \mathbf{B}}{\partial t} \cdot d\mathbf{S} = 0$$

If path C is fixed, the partial derivative $\partial/\partial t$ may be brought out from under the sign of integration giving Faraday's induction law:

$$\oint_C \mathbf{E} \cdot d\mathbf{r} = -\frac{\partial}{\partial t} \iint_S \mathbf{B} \cdot d\mathbf{S}$$

The **magnetic flux**, defined as the flux of **B** through S, is represented by

MAGNETIC FLUX
$$\Phi = \iint_S \mathbf{B} \cdot d\mathbf{S} \tag{11.22}$$

Thus the circulation of the electric field intensity **E** around C, sometimes called the **electromotive force** (emf) around C, is

$$\text{emf} = \oint_C \mathbf{E} \cdot d\mathbf{r} \tag{11.23}$$

Note: Both (11.16) and (11.20) verify the experimental fact that a varying magnetic field induces an electric field.

Faraday's experiments showed that the time rate of change of Φ can result from the movement of C in which the electromotive force is induced by **B** or can result from a time variation of **B**. Hence Faraday's law is generally written as

$$\oint_C \mathbf{E} \cdot d\mathbf{r} = -\frac{d}{dt} \iint_S \mathbf{B} \cdot d\mathbf{S} \tag{11.24}$$

or the electromotive force is

$$\text{emf} = -\frac{d}{dt} \Phi \tag{11.25}$$

EXAMPLE 11-3: Express the second of Maxwell's equation (11.17) in an equivalent integral form and give a physical interpretation.

Solution: Integrating (11.17) over a surface S bounded by a closed curve C yields

$$\iint_S \mathbf{\nabla} \times \mathbf{H} \cdot d\mathbf{S} - \iint_S \frac{\partial \mathbf{D}}{\partial t} \cdot d\mathbf{S} = \iint_S \mathbf{J} \cdot d\mathbf{S} \tag{11.26}$$

Applying Stokes' theorem (7.26) to the first term of the left-hand side of (11.26), we obtain

$$\oint_C \mathbf{H} \cdot d\mathbf{r} = I + \iint_S \frac{\partial \mathbf{D}}{\partial t} \cdot d\mathbf{S} \tag{11.27}$$

where I is the total current linking the curve C, as defined in (11.2). Equation (11.27) is the equivalent integral form of (11.17) and indicates that the **magnetomotive force** (mmf) is equal to the sum of the current I linking the curve C and the rate of change of the electric flux linking C. Both (11.17) and (11.27) verify the experimental fact that a magnetic field can be produced not only by a current but also by a time-varying electric field.

Note: Maxwell called the term $\partial \mathbf{D}/\partial t$ the **displacement current**, although nothing is actually displaced in the field. $\mathbf{D}$ is therefore called the **electric displacement** and should be thought of as being parallel to the corresponding term in Maxwell's first equation (11.16). Thus, the mmf is associated with a time-varying flux of $\mathbf{D}$, just as the emf is associated with a time-varying flux of $\mathbf{B}$.

C. Gauss' law

The **electric flux** Φ_e through a surface S is defined by

ELECTRIC FLUX
$$\Phi_e = \iint_S \mathbf{D} \cdot d\mathbf{S} \tag{11.28}$$

Because of (11.28), $\mathbf{D}$ is also called the **electric flux density**.

    Gauss' law for the electric field states that the net outward electric flux through a closed surface S is proportional to the electric charge enclosed by the surface. This is expressed by

GAUSS' LAW FOR THE ELECTRIC FIELD
$$\oiint_S \mathbf{D} \cdot d\mathbf{S} = Q \tag{11.29}$$

    Gauss' law for the magnetic field states that the net outward magnetic flux through a closed surface S is identically zero; that is,

GAUSS' LAW FOR THE MAGNETIC FIELD
$$\oiint_S \mathbf{B} \cdot d\mathbf{S} = 0 \tag{11.30}$$

EXAMPLE 11-4: Using the fourth of Maxwell's equations (11.19), verify Gauss' law for the electric field.

Solution: Since $\nabla \cdot \mathbf{D} = \rho$, an application of Gauss' divergence theorem (7.2) yields

$$\oiint_S \mathbf{D} \cdot d\mathbf{S} = \iiint_R \nabla \cdot \mathbf{D} \, dV = \iiint_R \rho \, dV = Q$$

where Q is the total charge enclosed by S. Thus we get Gauss' law for the electric field:

$$\oiint_S \mathbf{D} \cdot d\mathbf{S} = Q$$

EXAMPLE 11-5: Verify Gauss' law for the magnetic field.

Solution: Since $\nabla \cdot \mathbf{B} = 0$ from the third of Maxwell's equations (11.18), an application of Gauss' divergence theorem (7.2) now yields the desired result

$$\oiint_S \mathbf{B} \cdot d\mathbf{S} = \iiint_R \nabla \cdot \mathbf{B} \, dV = 0$$

11-4. Potential Functions of the Electromagnetic Field

A. Potentials

Using the **vector potential** $\mathbf{A}(\mathbf{r}, t)$ and the **scalar potential** $\phi(r, t)$, the electromagnetic fields $\mathbf{E}$ and $\mathbf{B}$ can be expressed as

$$\mathbf{B} = \mathbf{\nabla} \times \mathbf{A} \tag{11.31}$$

and

$$\mathbf{E} = -\frac{\partial \mathbf{A}}{\partial t} - \mathbf{\nabla}\phi \tag{11.32}$$

The existence of $\mathbf{A}$ completely determines $\mathbf{B}$ by (11.31), but the converse is not true because the curl of the gradient of any scalar vanishes identically. *Note:* The gradient of any scalar function ψ can be added to $\mathbf{A}$ without affecting $\mathbf{B}$ (see Example 5-14).

EXAMPLE 11-6: Verify (11.31) and (11.32).

Solution: Since $\mathbf{B}$ satisfies the third of Maxwell's equations, $\mathbf{\nabla} \cdot \mathbf{B} = 0$, then from Problem 7-22, a vector function $\mathbf{A}$ exists such that

$$\mathbf{B} = \mathbf{\nabla} \times \mathbf{A}$$

Using this, Maxwell's first equation (11.16) becomes

$$\mathbf{\nabla} \times \left(\mathbf{E} + \frac{\partial \mathbf{A}}{\partial t} \right) = 0 \tag{11.33}$$

Thus from the result of Problem 7-24, if ϕ is some scalar function

$$\mathbf{E} + \frac{\partial \mathbf{A}}{\partial t} = -\mathbf{\nabla}\phi$$

or upon transposing,

$$\mathbf{E} = -\frac{\partial \mathbf{A}}{\partial t} - \mathbf{\nabla}\phi$$

B. Gauge transformations

EXAMPLE 11-7: If the vector potential function $\mathbf{A}$ is replaced by

$$\mathbf{A}' = \mathbf{A} + \mathbf{\nabla}\psi \tag{11.34}$$

show that the scalar potential function ϕ should be replaced by

$$\phi' = \phi - \frac{\partial \psi}{\partial t} \tag{11.35}$$

for relationship (11.32) to remain unchanged.

Solution: If $\mathbf{A}$ is replaced by $\mathbf{A}' = \mathbf{A} + \mathbf{\nabla}\psi$, then (11.32) becomes

$$\mathbf{E} = -\frac{\partial}{\partial t}(\mathbf{A}' - \mathbf{\nabla}\psi) - \mathbf{\nabla}\phi = -\frac{\partial \mathbf{A}'}{\partial t} - \mathbf{\nabla}\left(\phi - \frac{\partial \psi}{\partial t} \right) \tag{11.36}$$

But, if we set $\phi' = \phi - \partial\psi/\partial t$, then (11.36) becomes

$$\mathbf{E} = -\frac{\partial \mathbf{A}'}{\partial t} - \mathbf{\nabla}\psi' \tag{11.37}$$

which is identical to (11.32).

The transformations (11.34) and (11.35) are called **gauge transformations**. Thus relations (11.31) and (11.32) are unaltered by the gauge transformations. We therefore say that the field vectors are *invariant* to gauge transformations.

C. Lorentz condition

The **Lorentz condition** for vector and scalar potentials **A** and ϕ is

LORENTZ CONDITION $$\nabla \cdot \mathbf{A} + \mu\varepsilon \frac{\partial \phi}{\partial t} = 0 \qquad (11.38)$$

If potentials **A** and ϕ satisfy this condition, then **A** and ϕ satisfy the following inhomogeneous wave equations:

$$\nabla^2 \mathbf{A} - \mu\varepsilon \frac{\partial^2 \mathbf{A}}{\partial t^2} = -\mu\mathbf{J} \qquad (11.39)$$

$$\nabla^2 \phi - \mu\varepsilon \frac{\partial^2 \phi}{\partial t^2} = -\frac{\rho}{\varepsilon} \qquad (11.40)$$

[For derivation of eqs. (11.39) and (11.40), see Problems 11-4 and 11-5.]

Since the potentials **A** and ϕ may be related by the Lorentz condition (11.38), the electromagnetic field can be represented in terms of a single vector function. This is shown in Problem 11-7.

11-5. Energy in the Electromagnetic Field and the Poynting Vector

A. Poynting vector and energy density

The **Poynting vector P** is defined by

$$\mathbf{P} = \mathbf{E} \times \mathbf{H} \qquad (11.41)$$

The **energy density** $\mathfrak{E}$ of the electromagnetic field is given by

$$\mathfrak{E} = \tfrac{1}{2}(\mathbf{E} \cdot \mathbf{D} + \mathbf{H} \cdot \mathbf{B}) \qquad (11.42)$$

In a homogeneous isotropic medium with constants ε and μ,

$$\nabla \cdot \mathbf{P} + \frac{\partial \mathfrak{E}}{\partial t} = -\mathbf{J} \cdot \mathbf{E} \qquad (11.43)$$

Note: If the conductivity σ of the medium is zero, then $\mathbf{J} = \sigma\mathbf{E} = \mathbf{0}$, and (11.43) reduces to

$$\nabla \cdot \mathbf{P} + \frac{\partial \mathfrak{E}}{\partial t} = 0 \qquad (11.44)$$

which is exactly the same form as the continuity equation (11.3). The "current density **J**" is now **P** and the "charge density ρ" is now $\mathfrak{E}$. Thus we can associate the Poynting vector **P** with the flow of energy density. [For verification of condition (11.43), see Problem 11-8.]

B. Poynting's theorem

Poynting's theorem states that in a homogeneous isotropic space, if R is a region bounded by a closed surface S, then

POYNTING'S THEOREM $$-\frac{\partial}{\partial t} \iiint_R \mathfrak{E}\, dV = \oiint_S \mathbf{P} \cdot d\mathbf{S} + \iiint_R \mathbf{J} \cdot \mathbf{E}\, dV \qquad (11.45)$$

where **P** is the Poynting vector and $\mathfrak{E}$ is the energy density of the electromagnetic field. [For derivation of (11.45), see Problem 11-9.]

EXAMPLE 11-8: Give the physical interpretation of Poynting's theorem (11.45).

Solution: If $\mathfrak{E}$ denotes the energy density of the electromagnetic field, then

$$\iiint_R \mathfrak{E}\,dV \qquad (11.46)$$

is the total energy of the electromagnetic field stored in region R. Hence the left-hand side of (11.45) represents the rate of decrease of the total energy of the electromagnetic field stored in R. The loss of available stored energy must be accounted for by the terms on the right-hand side of (11.45). If the conductivity of the medium is σ, then from relationship (11.15) the second term on the right-hand side of (11.45) becomes

$$\iiint_R \mathbf{J} \cdot \mathbf{E}\,dV = \iiint_R \sigma E^2\,dV = \iiint_R \frac{J^2}{\sigma}\,dV \qquad (11.47)$$

which is the *ohmic loss* or *joule heat loss.*

The first term on the right-hand side of (11.45) is the energy flux flowing out of the region R through the surface S. Thus the decrease of the energy of the electromagnetic field stored in R is partly accounted for by the joule heat loss and the rest flows out of R across the bounding surface S.

From the above interpretation, the Poynting vector $\mathbf{P}$, defined as $\mathbf{P} = \mathbf{E} \times \mathbf{H}$ in (11.41) may be interpreted as the flux of energy of the electromagnetic field, and it gives the amount of field energy passing through a unit area of the surface per unit time.

11-6. Static Fields

A. Maxwell's equations

Static fields are those for which the time derivatives are zero; that is, there is no time variation of the fields.

EXAMPLE 11-9: Derive Maxwell's equations for static fields.

Solution: With the restriction $\partial/\partial t = 0$, Maxwell's equations (11.16) and (11.17) become

$$\mathbf{V} \times \mathbf{E}(\mathbf{r}) = \mathbf{0} \qquad (11.48)$$
$$\mathbf{V} \times \mathbf{H}(\mathbf{r}) = \mathbf{J}(\mathbf{r}) \qquad (11.49)$$

The other equations (11.18) and (11.19) remain the same; that is,

$$\mathbf{V} \cdot \mathbf{B}(\mathbf{r}) = 0 \qquad (11.50)$$
$$\mathbf{V} \cdot \mathbf{D}(\mathbf{r}) = \rho(\mathbf{r}) \qquad (11.51)$$

For static fields the electric and magnetic fields are completely decoupled from each other. The examination of the relations

$$\mathbf{V} \times \mathbf{E} = \mathbf{0} \quad \text{and} \quad \mathbf{V} \cdot \mathbf{D} = \rho$$

and the consequences thereof, constitutes the study of electrostatic fields.
Similarly, the examination of the relations

$$\mathbf{V} \times \mathbf{H} = \mathbf{J} \quad \text{and} \quad \mathbf{V} \cdot \mathbf{B} = 0$$

and the consequences thereof, forms the study of magnetostatic fields.

EXAMPLE 11-10: Show that the electrostatic field **E** is irrotational and therefore conservative, that is, **E** can be represented as the gradient of scalar potential function ϕ in the form

$$\mathbf{E} = -\nabla\phi \qquad (11.52)$$

Solution: Since for an electrostatic field Maxwell's first equation is

$$\nabla \times \mathbf{E} = 0$$

we note that **E** is irrotational and, from Problem 7-24, can be expressed as

$$\mathbf{E} = -\nabla\phi$$

With either form (11.48) or (11.52) and with the result of Problem 7-19 the expression of conservative force (9.23), the line integral of **E** around any closed path is zero and the field is conservative.

Note: Equation (11.52) is also the special case of (11.32) with $\partial\mathbf{A}/\partial t = \mathbf{0}$.

B. Laplace's equation

Poisson's equation for a scalar potential function ϕ of the electrostatic field is

POISSON'S EQUATION $\qquad \nabla^2\phi = -\dfrac{\rho}{\varepsilon} \qquad (11.53)$

where ρ is the charge density and ε is a constant.

Laplace's equation for electrostatic fields is obtained from Poisson's equation by setting the charge density $\rho = 0$; that is,

LAPLACE'S EQUATION $\qquad \nabla^2\phi = 0 \qquad (11.54)$

EXAMPLE 11-11: Derive Poisson's equation for the scalar potential function ϕ of the electrostatic field.

Solution: Since $\mathbf{E} = -\nabla\phi$ from (11.52), from Maxwell's equation (11.51) and relationship (11.13), we have

$$\nabla \cdot \mathbf{D} = \nabla \cdot (\varepsilon\mathbf{E}) = \rho \qquad (11.55)$$

If ε is constant, $\nabla \cdot (\varepsilon\mathbf{E}) = \varepsilon\nabla \cdot \mathbf{E}$, and (11.55) becomes

$$\nabla \cdot \mathbf{E} = \frac{\rho}{\varepsilon} \qquad (11.56)$$

Substituting $\mathbf{E} = -\nabla\phi$ for **E** into this equation, we obtain Poisson's equation $\nabla^2\phi = -\rho/\varepsilon$.

C. Ampere's law

Ampere's circuital law in magnetostatics states that in a magnetostatic field, if I is the total steady current through a surface S bounded by a closed curve C, then

AMPERE'S LAW $\qquad \oint_C \mathbf{H} \cdot d\mathbf{r} = I \qquad (11.57)$

Note: This law corresponds to Gauss's law in electrostatics.

EXAMPLE 11-12: Verify Ampere's circuital law in magnetostatics.

Solution: Integrating Maxwell's equation for a static field (11.49) over a surface S bounded by a closed curve C, from relationship (11.2) we have, for the total steady current I through S,

$$\iint\limits_{S} \mathbf{\nabla} \times \mathbf{H} \cdot d\mathbf{S} = \iint\limits_{S} \mathbf{J} \cdot d\mathbf{S} = I \qquad (11.58)$$

Applying Stokes' theorem (7.26) to this equation yields

$$\oint_{C} \mathbf{H} \cdot d\mathbf{r} = I$$

Note: Ampere's law (11.57) can also be readily obtained from (11.27) by setting $\partial \mathbf{D}/\partial t = \mathbf{0}$.

Ampere's circuital law (11.57) can be used to compute the magnetic field vectors for cases where a high degree of symmetry exists (see Problem 11-15).

SUMMARY

1. The equation of continuity is

$$\mathbf{\nabla} \cdot (\rho \mathbf{v}) + \frac{\partial \rho}{\partial t} = \mathbf{\nabla} \cdot \mathbf{J} + \frac{\partial \rho}{\partial t} = 0$$

where ρ is the electric charge density and $\mathbf{v}$ is the charge velocity and $\mathbf{J} = \rho \mathbf{v}$ is the electric current density.

2. A Lorentz force is the force $\mathbf{f}$ experienced by a charge q moving with a velocity $\mathbf{v}$ in the electromagnetic field and is given by

$$\mathbf{f} = q(\mathbf{E} + \mathbf{v} \times \mathbf{B})$$

where $\mathbf{E}$ is the electric field intensity and $\mathbf{B}$ is the magnetic flux density.

3. Maxwell's equations of electromagnetic theory are

$$\mathbf{\nabla} \times \mathbf{E} + \frac{\partial \mathbf{B}}{\partial t} = \mathbf{0}$$

$$\mathbf{\nabla} \times \mathbf{H} - \frac{\partial \mathbf{D}}{\partial t} = \mathbf{J}$$

$$\mathbf{\nabla} \cdot \mathbf{B} = 0$$

$$\mathbf{\nabla} \cdot \mathbf{D} = \rho$$

where $\mathbf{D} = \varepsilon \mathbf{E}$, $\mathbf{B} = \mu \mathbf{H}$ and ε and μ are the permittivity and permeability of the medium, respectively.

4. Faraday's induction law is given by

$$\oint_{C} \mathbf{E} \cdot d\mathbf{r} = -\frac{\partial}{\partial t} \iint\limits_{S} \mathbf{B} \cdot d\mathbf{S}$$

5. Gauss' law for the electric field is

$$\oiint\limits_{S} \mathbf{D} \cdot d\mathbf{S} = Q$$

where Q is the total electric charge enclosed by a surface S.

6. The vector potential $\mathbf{A}$ and the scalar potential ϕ are defined by the equations

$$\mathbf{B} = \mathbf{\nabla} \times \mathbf{A}$$

$$\mathbf{E} = -\frac{\partial \mathbf{A}}{\partial t} - \mathbf{\nabla}\phi$$

7. Poynting's theorem is given by

$$-\frac{\partial}{\partial t}\iiint_R \mathfrak{E}\,dV = \oiint_S \mathbf{P}\cdot d\mathbf{S} + \iiint_R \mathbf{J}\cdot \mathbf{E}\,dV$$

where $\mathbf{P} = \mathbf{E}\times\mathbf{H}$ is the Poynting vector and $\mathfrak{E} = \frac{1}{2}(\mathbf{E}\cdot\mathbf{D} + \mathbf{H}\cdot\mathbf{B})$ is the energy density of the electromagnetic field.

RAISE YOUR GRADES

Can you explain . . . ?

☑ the relationship between the principle of conservation of charge and the equation of continuity.

☑ how the electromagnetic fields **E** and **B** are defined

☑ what the governing equations are for the electromagnetic fields **E** and **B**

☑ how the potential functions of the electromagnetic field are used to find the electromagnetic fields **E** and **B**

☑ how the energy density and the Poynting vector are defined

SOLVED PROBLEMS

Equations of Electromagnetic Theory

PROBLEM 11-1 Show that the third of Maxwell's equations (11.18) can be derived from the first (11.16).

Solution: Taking the divergence of (11.16), we have

$$\nabla\cdot(\nabla\times\mathbf{E}) + \nabla\cdot\left(\frac{\partial\mathbf{B}}{\partial t}\right) = 0$$

But $\nabla\cdot(\nabla\times\mathbf{E}) = 0$ by identity (5.32) and thus we have

$$\nabla\cdot(\nabla\times\mathbf{E}) = -\nabla\cdot\left(\frac{\partial\mathbf{B}}{\partial t}\right) = 0$$

If all the derivatives of **B** are assumed continuous, interchanging the differentiation of **B** with respect to space and time is allowed and yields

$$\nabla\cdot\left(\frac{\partial\mathbf{B}}{\partial t}\right) = \frac{\partial}{\partial t}(\nabla\cdot\mathbf{B}) = 0$$

or in the time variable

$$\nabla\cdot\mathbf{B} = c$$

where c is a constant. If at any instant in time in its history the value of $\mathbf{B} = \mathbf{0}$, this constant c must be zero. Since it can be assumed that the field originated at some past time, we may without loss of generality set $c = 0$ and obtain

$$\nabla\cdot\mathbf{B} = 0$$

PROBLEM 11-2   Show that the fourth of Maxwell's equations (11.19) can be derived from the second (11.17) with the use of the continuity equation (11.3).

Solution: Taking the divergence of (11.17), we have

$$\mathbf{V} \cdot (\mathbf{V} \times \mathbf{H}) - \mathbf{V} \cdot \left(\frac{\partial \mathbf{D}}{\partial t}\right) = \mathbf{V} \cdot \mathbf{J}$$

But $\mathbf{V} \cdot (\mathbf{V} \times \mathbf{H}) = 0$ by (5.32), and thus we have the variable interchange

$$\mathbf{V} \cdot \left(\frac{\partial \mathbf{D}}{\partial t}\right) = \frac{\partial}{\partial t}(\mathbf{V} \cdot \mathbf{D})$$

if we assume that all derivatives of $\mathbf{D}$ are continuous. So the resulting equation is

$$\mathbf{V} \cdot \mathbf{J} + \frac{\partial}{\partial t}(\mathbf{V} \cdot \mathbf{D}) = 0$$

Now using the continuity equation (11.3), we get $\mathbf{V} \cdot \mathbf{J} = -\partial \rho/\partial t$, and so our previous equation becomes

$$\frac{\partial}{\partial t}(\mathbf{V} \cdot \mathbf{D} - \rho) = 0$$

Thus working in time, if we designate c as constant, we obtain

$$\mathbf{V} \cdot \mathbf{D} - \rho = c$$

Again, the field originated at some time in the past. Because all the changes can be removed from any finite region of space, the constant above must be zero. Then we have the desired result:

$$\mathbf{V} \cdot \mathbf{D} = \rho$$

PROBLEM 11-3   Show that there can be no permanent distribution of free charge in a homogeneous medium having nonzero conductivity.

Solution: In a medium having nonzero conductivity σ, eqs. (11.19) and (11.13) show that

$$\mathbf{V} \cdot \mathbf{D} = \mathbf{V} \cdot (\varepsilon \mathbf{E}) = \rho$$

while the equation of continuity (11.3) and the constituent relation (11.15) yield

$$\mathbf{V} \cdot \mathbf{J} + \frac{\partial \rho}{\partial t} = \mathbf{V} \cdot (\sigma \mathbf{E}) + \frac{\partial \rho}{\partial t} = 0$$

Now using result (5.41) from Example 5-18, we have

$$\mathbf{V} \cdot (\varepsilon \mathbf{E}) = \varepsilon \mathbf{V} \cdot \mathbf{E} + \mathbf{E} \cdot \mathbf{V}\varepsilon = \rho \tag{a}$$

$$\mathbf{V} \cdot (\sigma \mathbf{E}) = \sigma \mathbf{V} \cdot \mathbf{E} + \mathbf{E} \cdot \mathbf{V}\sigma = -\frac{\partial \rho}{\partial t} \tag{b}$$

Eliminating $\mathbf{V} \cdot \mathbf{E}$ between eqs. (a) and (b), we obtain

$$\rho + \frac{\varepsilon}{\sigma}\frac{\partial \rho}{\partial t} = \sigma \mathbf{E} \cdot \left(\frac{\sigma \mathbf{V}\varepsilon - \varepsilon \mathbf{V}\sigma}{\sigma^2}\right) = \mathbf{J} \cdot \mathbf{V}\left(\frac{\varepsilon}{\sigma}\right) \tag{c}$$

This result was obtained using the vector identity of Problem 5-8. If the medium is homogeneous, $\mathbf{V}(\varepsilon/\sigma) = \mathbf{0}$ and eq. (c) reduces to

$$\rho + \tau \frac{\partial \rho}{\partial t} = 0 \tag{d}$$

where

$$\tau = \frac{\varepsilon}{\sigma}$$

is defined as the **relaxation time**. The solution of differential equation (d) is

$$\rho(t) = \rho_0 e^{-t/\tau} = \rho_0 e^{-\sigma t/\varepsilon} \tag{e}$$

where ρ_0 is the value of ρ at $t = 0$. Thus the charge density decays exponentially and is independent of the electric field intensity $\mathbf{E}$ applied. Hence we conclude that there can be no permanent distribution of free charge in a homogeneous medium with nonzero conductivity.

Potential Functions of the Electromagnetic Field

PROBLEM 11-4 Show that in a linear homogeneous isotropic medium in which ε and μ are constants and $\sigma = 0$, the scalar potential ϕ and the vector potential $\mathbf{A}$ satisfy the equations

$$\nabla^2 \mathbf{A} - \mu\varepsilon \frac{\partial^2 \mathbf{A}}{\partial t^2} - \nabla\left(\nabla \cdot \mathbf{A} + \mu\varepsilon \frac{\partial \phi}{\partial t}\right) = -\mu \mathbf{J} \tag{a}$$

$$\nabla^2 \phi + \frac{\partial}{\partial t} \nabla \cdot \mathbf{A} = -\frac{\rho}{\varepsilon} \tag{b}$$

Solution: Using constituent relations (11.13) and (11.14), Maxwell's second equation (11.17) can be written as

$$\nabla \times \mathbf{B} - \mu\varepsilon \frac{\partial \mathbf{E}}{\partial t} = \mu \mathbf{J} \tag{c}$$

Substituting for $\mathbf{B}$ and $\mathbf{E}$ from (11.31) and (11.32), we obtain

$$\nabla \times (\nabla \times \mathbf{A}) + \mu\varepsilon \frac{\partial^2 \mathbf{A}}{\partial t^2} + \mu\varepsilon \nabla\left(\frac{\partial \phi}{\partial t}\right) = \mu \mathbf{J} \tag{d}$$

Substituting the vector identity $\nabla \times (\nabla \times \mathbf{A}) = \nabla(\nabla \cdot \mathbf{A}) - \nabla^2 \mathbf{A}$ (5.44), into eq. (d), we obtain eq. (a) of this problem:

$$\nabla^2 \mathbf{A} - \mu\varepsilon \frac{\partial^2 \mathbf{A}}{\partial t^2} - \nabla\left(\nabla \cdot \mathbf{A} + \mu\varepsilon \frac{\partial \phi}{\partial t}\right) = -\mu \mathbf{J}$$

Next using constituent relation (11.13), Maxwell's fourth equation (11.19) can be written as

$$\nabla \cdot \mathbf{E} = \frac{\rho}{\varepsilon} \tag{e}$$

Substituting for $\mathbf{E}$ from (11.32), we obtain

$$\nabla \cdot \left[-\frac{\partial \mathbf{A}}{\partial t} - \nabla\phi\right] = \frac{\rho}{\varepsilon}$$

or consequently

$$\nabla^2 \phi + \frac{\partial}{\partial t} \nabla \cdot \mathbf{A} = -\frac{\rho}{\varepsilon}$$

PROBLEM 11-5 Verify the inhomogeneous wave equations (11.39) and (11.40), which assume the Lorentz condition.

Solution: If the potentials $\mathbf{A}$ and ϕ satisfy the Lorentz condition (11.38), then

$$\nabla \cdot \mathbf{A} = -\mu\varepsilon \frac{\partial \phi}{\partial t} \tag{a}$$

Hence eq. (a) of Problem 11-4 reduces to the desired result (11.39). Substituting eq. (a) of this problem into eq. (b) of Problem 11-4, we obtain result (11.40).

PROBLEM 11-6 Show that there exist vector and scalar potentials **A** and ϕ that satisfy the Lorentz condition (11.38).

Solution: Suppose that the given potentials **A** and ϕ do not satisfy (11.38); that is,

$$\mathbf{V} \cdot \mathbf{A} + \mu\varepsilon \frac{\partial \phi}{\partial t} \neq 0$$

By making a gauge transformation to **A**′ and ϕ′, that is, letting

$$\mathbf{A}' = \mathbf{A} + \mathbf{V}\psi \quad \text{and} \quad \phi' = \phi - \frac{\partial \psi}{\partial t}$$

we obtain

$$\mathbf{V} \cdot \mathbf{A}' + \mu\varepsilon \frac{\partial \phi'}{\partial t} = \mathbf{V} \cdot \mathbf{A} + \mu\varepsilon \frac{\partial \phi}{\partial t} + \mathbf{V}^2\psi - \mu\varepsilon \frac{\partial^2 \psi}{\partial t^2}$$

Thus it suffices to find a gauge function ψ that satisfies

$$\mathbf{V}^2\psi - \mu\varepsilon \frac{\partial^2 \psi}{\partial t^2} = -\left(\mathbf{V} \cdot \mathbf{A} + \mu\varepsilon \frac{\partial \phi}{\partial t}\right) \tag{a}$$

Under this condition the new potentials **A**′ and ϕ′ will satisfy the Lorentz condition (11.38).

PROBLEM 11-7 The **Hertz vector** $\mathbf{\Pi}$ is a single-valued vector function whose time rate of change is proportional to the potential **A**. Thus in a medium with constants μ and ε we may write the Hertz vector as

HERTZ VECTOR
$$\mathbf{A} = \mu\varepsilon \frac{\partial \mathbf{\Pi}}{\partial t} \tag{a}$$

Show that in a medium with constants μ and ε, the field vectors **E** and **B** can be represented in terms of the Hertz vector $\mathbf{\Pi}$ as

$$\mathbf{E} = -\mu\varepsilon \frac{\partial^2 \mathbf{\Pi}}{\partial t^2} + \mathbf{V}(\mathbf{V} \cdot \mathbf{\Pi}) \tag{b}$$

$$\mathbf{B} = \mu\varepsilon \mathbf{V} \times \frac{\partial \mathbf{\Pi}}{\partial t} \tag{c}$$

Also find the equation that $\mathbf{\Pi}$ must satisfy.

Solution: Substituting the Hertz vector into (11.31) and (11.32) yields

$$\mathbf{B} = \mathbf{V} \times \mathbf{A} = \mathbf{V} \times \left(\mu\varepsilon \frac{\partial \mathbf{\Pi}}{\partial t}\right) = \mu\varepsilon \mathbf{V} \times \frac{\partial \mathbf{\Pi}}{\partial t}$$

$$\mathbf{E} = -\frac{\partial \mathbf{A}}{\partial t} - \mathbf{V}\phi = -\mu\varepsilon \frac{\partial^2 \mathbf{\Pi}}{\partial t^2} - \mathbf{V}\phi \tag{d}$$

Now Maxwell's second equation (11.17) can be written as

$$\mathbf{V} \times \mathbf{B} = \mu\mathbf{J} + \mu\varepsilon \frac{\partial \mathbf{E}}{\partial t} \tag{e}$$

Substituting eqs. (c) and (d) into eq. (e) of this problem, we obtain

$$\mu\varepsilon \frac{\partial}{\partial t} \mathbf{V} \times \mathbf{V} \times \mathbf{\Pi} = \mu\mathbf{J} + \mu\varepsilon \frac{\partial}{\partial t}\left(-\mu\varepsilon \frac{\partial^2 \mathbf{\Pi}}{\partial t^2} - \mathbf{V}\phi\right)$$

or upon simplification

$$\frac{\partial}{\partial t}\left[\mathbf{V} \times (\mathbf{V} \times \mathbf{\Pi}) + \mu\varepsilon \frac{\partial^2 \mathbf{\Pi}}{\partial t^2} + \mathbf{V}\phi\right] = \frac{\mathbf{J}}{\varepsilon} \tag{f}$$

Integrating this with respect to time t, we obtain

$$\nabla \times (\nabla \times \mathbf{\Pi}) + \mu\varepsilon \frac{\partial^2 \mathbf{\Pi}}{\partial t^2} + \nabla\phi = \int \frac{\mathbf{J}}{\varepsilon}\, dt \qquad (\text{g})$$

where the arbitrary constant of integration is set equal to zero because it does not affect the determination of the field.

Using the vector identity (5.44), eq. (g) can be rewritten as

$$\nabla^2 \mathbf{\Pi} - \mu\varepsilon \frac{\partial^2 \mathbf{\Pi}}{\partial t^2} - [\nabla(\nabla \cdot \mathbf{\Pi}) + \nabla\phi] = -\int \frac{\mathbf{J}}{\varepsilon}\, dt \qquad (\text{h})$$

Hence if the condition

$$\nabla(\nabla \cdot \mathbf{\Pi}) + \nabla\phi = 0 \qquad (\text{i})$$

or

$$\nabla\phi = -\nabla(\nabla \cdot \mathbf{\Pi}) \qquad (\text{j})$$

is imposed on ϕ and $\mathbf{\Pi}$, then $\mathbf{\Pi}$ satisfies the inhomogeneous wave equation

$$\nabla^2 \mathbf{\Pi} - \mu\varepsilon \frac{\partial^2 \mathbf{\Pi}}{\partial t^2} = -\int \frac{\mathbf{J}}{\varepsilon}\, dt \qquad (\text{k})$$

Substituting eq. (j) into eq. (d) of this problem yields

$$\mathbf{E} = -\mu\varepsilon \frac{\partial^2 \mathbf{\Pi}}{\partial t^2} + \nabla(\nabla \cdot \mathbf{\Pi})$$

[*Note:* Equation (i) is equivalent to the Lorentz condition (11.38).]

Energy in the Electromagnetic Field and the Poynting Vector

PROBLEM 11-8 Verify the relationship between the energy density and the Poynting vector (11.43).

Solution: Using the vector identity (5.43) and definition (11.41), we have

$$\nabla \cdot \mathbf{P} = \nabla \cdot (\mathbf{E} \times \mathbf{H}) = \mathbf{H} \cdot \nabla \times \mathbf{E} - \mathbf{E} \cdot \nabla \times \mathbf{H} \qquad (\text{a})$$

Substituting Maxwell's equations (11.16) and (11.17) into eq. (a), we obtain

$$\nabla \cdot \mathbf{P} = \mathbf{H} \cdot \left(-\frac{\partial \mathbf{B}}{\partial t}\right) - \mathbf{E} \cdot \left(\mathbf{J} + \frac{\partial \mathbf{D}}{\partial t}\right)$$

or rearranging,

$$\nabla \cdot \mathbf{P} + \mathbf{E} \cdot \frac{\partial \mathbf{D}}{\partial t} + \mathbf{H} \cdot \frac{\partial \mathbf{B}}{\partial t} = -\mathbf{E} \cdot \mathbf{J} \qquad (\text{b})$$

Now, from constituent relations (11.13) and (11.14) we have

$$\mathbf{E} \cdot \frac{\partial \mathbf{D}}{\partial t} = \varepsilon \mathbf{E} \cdot \frac{\partial \mathbf{E}}{\partial t} = \frac{1}{2} \varepsilon \frac{\partial}{\partial t}(\mathbf{E} \cdot \mathbf{E}) = \frac{1}{2} \frac{\partial}{\partial t}(\mathbf{E} \cdot \varepsilon\mathbf{E}) = \frac{\partial}{\partial t}\left(\frac{1}{2}\mathbf{E} \cdot \mathbf{D}\right) \qquad (\text{c})$$

$$\mathbf{H} \cdot \frac{\partial \mathbf{B}}{\partial t} = \mu \mathbf{H} \cdot \frac{\partial \mathbf{H}}{\partial t} = \frac{1}{2} \mu \frac{\partial}{\partial t}(\mathbf{H} \cdot \mathbf{H}) = \frac{1}{2} \frac{\partial}{\partial t}(\mathbf{H} \cdot \mu\mathbf{H}) = \frac{\partial}{\partial t}\left(\frac{1}{2}\mathbf{H} \cdot \mathbf{B}\right) \qquad (\text{d})$$

Thus eq. (b) reduces to

$$\nabla \cdot \mathbf{P} + \frac{\partial}{\partial t}\left[\frac{1}{2}(\mathbf{E} \cdot \mathbf{D} + \mathbf{H} \cdot \mathbf{B})\right] = -\mathbf{E} \cdot \mathbf{J}$$

which, by using the definition of energy density (11.42), becomes

$$\nabla \cdot \mathbf{P} + \frac{\partial \mathfrak{E}}{\partial t} = -\mathbf{E} \cdot \mathbf{J}$$

PROBLEM 11-9   Verify Poynting's theorem (11.45).

Solution: Since (11.43) can be rewritten as

$$-\frac{\partial \mathfrak{E}}{\partial t} = \mathbf{V} \cdot \mathbf{P} + \mathbf{J} \cdot \mathbf{E} \tag{a}$$

integrating this over a region R and using the divergence theorem (7.2), we obtain the desired result

$$-\frac{\partial}{\partial t} \iiint_R \mathfrak{E}\, dV = \iiint_R \mathbf{V} \cdot \mathbf{P}\, dV + \iiint_R \mathbf{J} \cdot \mathbf{E}\, dV$$

$$= \oiint_S \mathbf{P} \cdot d\mathbf{S} + \iiint_R \mathbf{J} \cdot \mathbf{E}\, dV$$

Static Fields

PROBLEM 11-10   Show that for a static field the scalar potential $\phi(r)$ is the negative of the work done by the field on a unit positive charge to bring the charge from a point at infinity to a point designated by $\mathbf{r}$.

Solution: With $q = 1$, the work done by the Lorentz force (11.11) on a unit positive charge is, by (9.22),

$$W = \int_\infty^{\mathbf{r}} \mathbf{f} \cdot d\mathbf{r} = \int_\infty^{\mathbf{r}} (\mathbf{E} + \mathbf{v} \times \mathbf{B}) \cdot d\mathbf{r}$$

$$= \int_\infty^{\mathbf{r}} \mathbf{E} \cdot d\mathbf{r} - \int_\infty^{\mathbf{r}} \mathbf{B} \cdot \mathbf{v} \times d\mathbf{r} \tag{a}$$

where the last expression is obtained with the use of eq. (a) of Problem 1-16.

Since the motion produced by v is along $d\mathbf{r}$, $\mathbf{v} \times d\mathbf{r} = \mathbf{0}$, and the magnetic field does not contribute to the work done on the charge. Thus because $\phi(\infty) \equiv 0$, using the result of Problem 5-4, we have

$$W = \int_\infty^{\mathbf{r}} \mathbf{E} \cdot d\mathbf{r} = \int_\infty^{\mathbf{r}} (-\nabla\phi) \cdot d\mathbf{r} = -\int_\infty^{\mathbf{r}} d\phi = \phi(\infty) - \phi(\mathbf{r}) = -\phi(\mathbf{r}) \tag{b}$$

From eq. (b) the scalar potential of an electrostatic field can also be interpreted as the work required to bring a unit positive charge from a point at infinity to a point within the field; that is,

$$\phi(\mathbf{r}) = -\int_\infty^{\mathbf{r}} \mathbf{E} \cdot d\mathbf{r} \tag{c}$$

PROBLEM 11-11   Find the electric field $\mathbf{E}$ and the potential ϕ of a point charge of magnitude q.

Solution: If a point charge q is located at the origin, Gauss' law for the electric field (11.29) becomes

$$\oiint_S \mathbf{D} \cdot d\mathbf{S} = \oiint_S \mathbf{D} \cdot \mathbf{n}\, dS = q \tag{a}$$

If S is a spherical surface of radius r with its center at the origin, then $\mathbf{n} = \mathbf{e}_r$ and $\mathbf{D} \cdot \mathbf{n} = \mathbf{D} \cdot \mathbf{e}_r = D_r$, where D_r is the radial component of $\mathbf{D}$. Thus eq. (a) becomes

$$\oiint_S D_r\, dS = q \tag{b}$$

From the symmetry of the configuration, $\mathbf{D} = D_r \mathbf{e}_r$, and on a surface of constant radius the value of D_r is a constant. Since the surface area of a sphere of radius r is $4\pi r^2$,

$$\oint_S D_r\, dS = D_r(r)4\pi r^2 = q$$

Hence $D_r(r) = q/(4\pi r^2)$ and

$$\mathbf{D} = D_r \mathbf{e}_r = \frac{q}{4\pi r^2}\, \mathbf{e}_r \qquad (c)$$

Since $\mathbf{D} = \varepsilon\mathbf{E}$, we can write

$$\mathbf{E} = \frac{q}{4\pi\varepsilon r^2}\, \mathbf{e}_r \qquad (d)$$

Now since $\mathbf{E}$ has only a radial component, the spherical coordinates expression (8.53) of $\nabla\phi$, $\mathbf{E} = -\nabla\phi$, reduces to $E_r = -d\phi(r)/dr$, and consequently $d\phi(\mathbf{r})/dr = -q/4\pi\varepsilon r^2$. Integrating both sides, we have

$$\int_a^b \frac{d\phi(r)}{dr}\, dr = -\int_a^b \frac{q\, dr}{4\pi\varepsilon r^2} = -\frac{q}{4\pi\varepsilon}\int_a^b \frac{dr}{r^2}$$

and hence

$$\phi(b) - \phi(a) = \frac{q}{4\pi\varepsilon}\left(\frac{1}{b} - \frac{1}{a}\right) \qquad (e)$$

If in this expression we set $a = \infty$, $b = r$, and $\phi(\infty) = 0$, we obtain

$$\phi(\mathbf{r}) = \frac{q}{4\pi\varepsilon r} \qquad (f)$$

PROBLEM 11-12 Verify Coulomb's law for electrostatic fields: The force $\mathbf{f}$ between two charges q_A and q_B is inversely proportional to the square of the distance r_{AB} between them; in other words, if $\mathbf{e}_r$ is the unit vector directed from q_A to q_B, then

COULOMB'S LAW FOR ELECTROSTATIC FIELDS
$$\mathbf{f} = \frac{q_A q_B}{4\pi\varepsilon r_{AB}^2}\, \mathbf{e}_r \qquad (a)$$

Solution: From the definition of a Lorentz force (11.11) the force $\mathbf{f}$ experienced by a point charge q_B in the electrostatic field is

$$\mathbf{f} = q_B \mathbf{E} \qquad (b)$$

If the field $\mathbf{E}$ is produced by a point charge q_A, we can substitute eq. (d) of Problem 11-11 for $\mathbf{E}$ into eq. (b) above and obtain Coulomb's law for electrostatic fields.

PROBLEM 11-13 An **electric dipole** is formed by two equal and opposite charges separated by an arbitrarily small distance l where the vector $\mathbf{l}$ initiates on the negative charge and terminates on the positive charge. The **electric dipole moment p** is the product of the charge q and the distance vector $\mathbf{l}$, that is,

$$\mathbf{p} = q\mathbf{l}$$

If the electric dipole is located at the origin and the electric dipole moment is in the positive z-direction, show that the potential and the electric field of the dipole in a vacuum (whose permittivity is ε_0) are, respectively,

ELECTRIC DIPOLE MOMENT

$$\phi(\mathbf{r}) = \frac{\mathbf{p} \cdot \mathbf{r}}{4\pi\varepsilon_0 r^3} = -\frac{1}{4\pi\varepsilon_0} \mathbf{p} \cdot \mathbf{V}\left(\frac{1}{r}\right) \qquad (a)$$

$$\mathbf{E}(\mathbf{r}) = \frac{3(\mathbf{p} \cdot \mathbf{r})\mathbf{r} - r^2 \mathbf{p}}{4\pi\varepsilon_0 r^5} \qquad (b)$$

where $\mathbf{r} = r\mathbf{e}_r$ is the vector from the origin to the point of observation and $r \gg l$.

Solution: In Figure 11-1 the potential at a point P due to $\mathbf{p}$ is, from eq. (f) of Problem 11-11,

$$\phi(\mathbf{r}) = \frac{q}{4\pi\varepsilon_0 r_1} + \frac{-q}{4\pi\varepsilon_0 r_2} = \frac{q}{4\pi\varepsilon_0}\left(\frac{1}{r_1} - \frac{1}{r_2}\right)$$

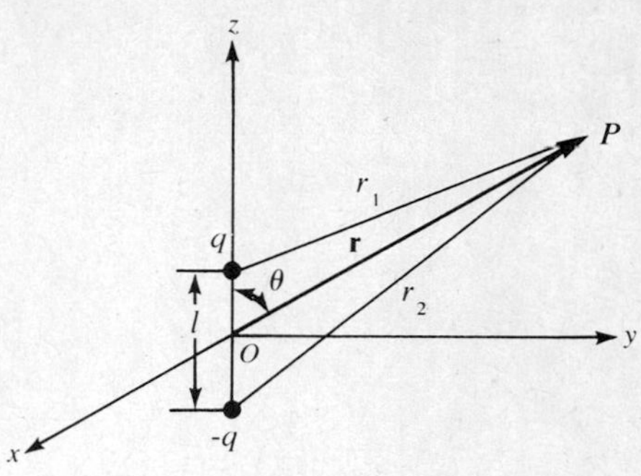

If $r \gg l$, then

$$\frac{1}{r_1} - \frac{1}{r_2} \simeq \frac{1}{r - (l/2)\cos\theta} - \frac{1}{r + (l/2)\cos\theta}$$

$$= \frac{l\cos\theta}{r^2}\frac{1}{1 - (l\cos\theta/2r)^2}$$

$$\simeq \frac{l\cos\theta}{r^2}$$

Substituting this into the value for $\phi(\mathbf{r})$ gives

$$\phi(\mathbf{r}) = \frac{ql\cos\theta}{4\pi\varepsilon_0 r^2} \qquad (c)$$

Since the direction of the electric dipole moment $\mathbf{p}$ is along the z-axis, we observe from Figure 11-1 that

$$\mathbf{p} \cdot \mathbf{r} = ql\mathbf{k} \cdot \mathbf{r} = qlr\cos\theta$$

Therefore from eq. (c)

$$\phi(\mathbf{r}) = \frac{\mathbf{p} \cdot \mathbf{r}}{4\pi\varepsilon_0 r^3} \qquad (d)$$

Also, from the result of Problem 5-7,

$$\mathbf{V}\left(\frac{1}{r}\right) = -\frac{1}{r^3}\mathbf{r} = -\frac{1}{r^2}\mathbf{e}_r$$

Hence we obtain the desired result (a):

$$\phi(\mathbf{r}) = \frac{\mathbf{p} \cdot \mathbf{r}}{4\pi\varepsilon_0 r^3} = -\frac{1}{4\pi\varepsilon_0}\mathbf{p} \cdot \mathbf{V}\left(\frac{1}{r}\right)$$

Using (5.14), we can write the electric field $\mathbf{E}$ as

$$\mathbf{E} = -\nabla\phi = -\frac{1}{4\pi\varepsilon_0}\mathbf{V}\left(\frac{\mathbf{p} \cdot \mathbf{r}}{r^3}\right)$$

$$= -\frac{1}{4\pi\varepsilon_0 r^3}\mathbf{V}(\mathbf{p} \cdot \mathbf{r}) - \frac{(\mathbf{p} \cdot \mathbf{r})}{4\pi\varepsilon_0}\mathbf{V}\left(\frac{1}{r^3}\right) \qquad (e)$$

Now from the result of Problem 5-3 we have

$$\mathbf{V}(\mathbf{p} \cdot \mathbf{r}) = \mathbf{p} \qquad (f)$$

and from the result of Problem 5-7 we have

$$\mathbf{V}\left(\frac{1}{r^3}\right) = -3\frac{\mathbf{r}}{r^5} \qquad (g)$$

Figure 11-1
Dipole.

Substituting eqs. (f) and (g) into eq. (e), we obtain

$$E = \frac{3(\mathbf{p} \cdot \mathbf{r})\mathbf{r} - r^2 \mathbf{p}}{4\pi\varepsilon_0 r^5}$$

PROBLEM 11-14   Show that in a static field produced by steady current the divergence of the current density vector **J** is zero:

$$\mathbf{\nabla} \cdot \mathbf{J} = 0 \qquad\qquad\qquad (a)$$

Solution: The basic equations for the magnetostatic fields are (11.49) and (11.50), respectively:

$$\mathbf{\nabla} \times \mathbf{H} = \mathbf{J}$$
$$\mathbf{\nabla} \cdot \mathbf{B} = 0$$

Since $\mathbf{\nabla} \cdot (\mathbf{\nabla} \times \mathbf{H}) = 0$ from (5.32), taking the divergence of both sides of (11.49), we have

$$\mathbf{\nabla} \cdot (\mathbf{\nabla} \times \mathbf{H}) = \mathbf{\nabla} \cdot \mathbf{J} = 0$$

Note: Equation (a) can also be obtained from the continuity equation (11.3) by setting $\partial\rho/\partial t = 0$.

PROBLEM 11-15   Find the magnetic field **H** of an infinitely long, straight wire carrying a steady current *I*.

Solution: Let a straight wire extend along the *z*-axis from $-\infty$ to ∞. Since there is cylindrical symmetry, we can choose a circular path *C* having a point on the *z*-axis as its center and having radius *a*, as shown in Figure 11-2. Because of symmetry, vector **H** is not only azimuthal but is also in the same direction as *d***r** and its magnitude is constant around the contour *C*. Hence by Ampere's law (11.57),

$$\oint_C \mathbf{H} \cdot d\mathbf{r} = H_\phi(2\pi a) = I$$

Thus

$$\mathbf{H} = H_\phi \mathbf{e}_\phi = \frac{I}{2\pi a}\, \mathbf{e}_\phi$$

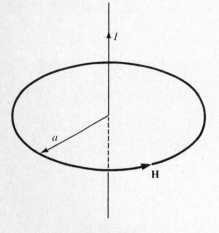

Figure 11-2

PROBLEM 11-16   Show that if we impose the Lorentz condition $\mathbf{\nabla} \cdot \mathbf{A} = 0$ under steady state in a magnetostatic field, the vector potential $\mathbf{A}(\mathbf{r})$ satisfies

$$\mathbf{\nabla}^2 \mathbf{A} = -\mu_0 \mathbf{J} \qquad\qquad\qquad (a)$$

where μ_0 is the permeability of a vacuum.

Solution: Since $\mathbf{B} = \mathbf{\nabla} \times \mathbf{A}$ and $\mathbf{B} = \mu_0 \mathbf{H}$, from Maxwell's equation for a static field we have

$$\mathbf{\nabla} \times (\mathbf{\nabla} \times \mathbf{A}) = \mu_0 \mathbf{J} \qquad\qquad\qquad (b)$$

which reduces to

$$\mathbf{\nabla}(\mathbf{\nabla} \cdot \mathbf{A}) - \mathbf{\nabla}^2 \mathbf{A} = \mu_0 \mathbf{J}$$

Thus if we impose the Lorentz condition (11.38) under steady state $\mathbf{\nabla} \cdot \mathbf{A} = 0$, we obtain

$$\mathbf{\nabla}^2 \mathbf{A} = -\mu_0 \mathbf{J}$$

Note: Equation (a) of this problem can also be obtained from the inhomogeneous wave equation (11.39) using $\partial^2 \mathbf{A}/\partial t^2 = \mathbf{0}$.

Supplementary Exercises

PROBLEM 11-17   Show that in homogeneous isotropic medium **E** and **H** satisfy the inhomogeneous wave equations:

$$\nabla^2 \mathbf{E} - \mu\varepsilon\frac{\partial^2 \mathbf{E}}{\partial t^2} = \mu\frac{\partial \mathbf{J}}{\partial t} + \frac{1}{\varepsilon}\nabla\rho$$

$$\nabla^2 \mathbf{H} - \mu\varepsilon\frac{\partial^2 \mathbf{H}}{\partial t^2} = -\nabla \times \mathbf{J}$$

PROBLEM 11-18   Show that in a homogeneous isotropic medium and source-free region **E** and **H** satisfy

$$\nabla^2 \mathbf{E} - \mu\varepsilon\frac{\partial^2 \mathbf{E}}{\partial t^2} - \mu\sigma\frac{\partial \mathbf{E}}{\partial t} = \mathbf{0}$$

$$\nabla^2 \mathbf{H} - \mu\varepsilon\frac{\partial^2 \mathbf{H}}{\partial t^2} - \mu\sigma\frac{\partial \mathbf{H}}{\partial t} = \mathbf{0}$$

PROBLEM 11-19   A useful gauge for an electromagnetic field where there are no charges is the **Coulomb gauge**, where $\nabla \cdot \mathbf{A} = 0$. Show that in this case the potentials **A** and ϕ satisfy

$$\nabla^2 \mathbf{A} - \mu\varepsilon\frac{\partial^2 \mathbf{A}}{\partial t^2} = -\mu\mathbf{J} + \mu\varepsilon\nabla\left(\frac{\partial \phi}{\partial t}\right)$$

$$\nabla^2 \phi = -\frac{\rho}{\varepsilon}$$

Also show that for the new potentials to satisfy the Coulomb gauge condition, the gauge function ψ must satisfy Laplace's equation $\nabla^2\psi = 0$.

PROBLEM 11-20   Show that the magnetic flux Φ and the vector potential **A** are related by

$$\Phi = \oint_C \mathbf{A} \cdot d\mathbf{r}$$

and hence that the emf in a fixed circuit C is

$$\text{emf} = \frac{d}{dt}\oint_C \mathbf{A} \cdot d\mathbf{r}$$

PROBLEM 11-21   Show that the force per unit volume $\mathbf{f}_V$, referred to as the **Lorentz force density**, on a region of free space (vacuum) containing charges and currents due to an electromagnetic field can be expressed as

$$\mathbf{f}_V = \rho\mathbf{E} + \mathbf{J} \times \mathbf{B}$$

PROBLEM 11-22   Find the electric field **E** and the potential ϕ due to an infinite line charge with charge density per unit length ρ_l.

Answer: $\mathbf{E} = \dfrac{\rho_l}{2\pi\varepsilon\rho}\,\mathbf{e}_\rho,\qquad \phi(\rho_1) - \phi(\rho_2) = \dfrac{\rho_l}{2\pi\varepsilon}\ln\left(\dfrac{\rho_2}{\rho_1}\right)$

PROBLEM 11-23   A spherical volume of radius a centered at the origin contains an electric charge of uniform density ρ_0. Find the electric field $\mathbf{E}(r)$ and the potential $\phi(\mathbf{r})$ due to this charge distribution.

Answer: For $0 \leqslant r \leqslant a$

$$\mathbf{E}(\mathbf{r}) = \frac{\rho_0}{3\varepsilon_0}\,r\mathbf{e}_r,\qquad \phi(\mathbf{r}) = \frac{\rho_0}{3\varepsilon_0}\,a^2 + \frac{\rho_0}{6\varepsilon}(r^2 - a^2)$$

For $r > a$

$$\mathbf{E}(\mathbf{r}) = \frac{a^3\rho_0}{3\varepsilon_0 r^2}\,\mathbf{e}_r,\qquad \phi(\mathbf{r}) = \frac{a^3\rho_0}{3\varepsilon_0 r}$$

PROBLEM 11-24   A coaxial cable consists of a solid inner conductor of circular cross section of radius a surrounded by a hollow conducting cylinder of inner and outer radii b and c, respectively. A total current I flows in the inner conductor and returns through the outer conductor with uniform distribution. Assuming that the axis of the cable is coincident with the z-axis and its length is infinite, find the magnetic field $\mathbf{H}$ inside and between the two conductors.

Answer: For $\rho \leqslant a$

$$\mathbf{H} = \frac{I}{2\pi a^2}\, \rho \mathbf{e}_\phi$$

For $a \leqslant \rho \leqslant b$

$$\mathbf{H} = \frac{I}{2\pi\rho}\, \mathbf{e}_\phi$$

For $b \leqslant \rho \leqslant c$

$$\mathbf{H} = \frac{I}{2\pi\rho}\frac{c^2 - \rho^2}{c^2 - b^2}\, \mathbf{e}_\phi$$

PROBLEM 11-25   Show that the force on an electric dipole in an electric field $\mathbf{E}$ is

$$\mathbf{f} = (\mathbf{p} \cdot \nabla)\mathbf{E}$$

PROBLEM 11-26   Show that the torque on an electric dipole placed in a uniform electric field $\mathbf{E}$ is

$$\tau_e = \mathbf{p} \times \mathbf{E}$$

PROBLEM 11-27   Find the expression for the force exerted on a wire carrying a current I when the wire is immersed in a magnetic field $\mathbf{B}$.

Answer: $\mathbf{f} = I \oint_c d\mathbf{r} \times \mathbf{B}$

PROBLEM 11-28   Show that the electric field of the electric dipole of Figure 11-1 can be expressed as

$$\mathbf{E} = \frac{p}{4\pi\varepsilon_0 r^3}\,(2\cos\theta\mathbf{e}_r + \sin\theta\mathbf{e}_\theta)$$

PROBLEM 11-29   Show that the energy stored in an electrostatic field can be expressed as

$$\frac{1}{2}\iiint_V \mathbf{E} \cdot \mathbf{D}\, dV = \frac{1}{2}\iiint_V \rho\phi\, dV$$

where ρ is the charge density and ϕ is the scalar potential of the electrostatic field.

PROBLEM 11-30   Show that the energy stored in a magnetostatic field can be expressed as

$$\frac{1}{2}\iiint_V \mathbf{B} \cdot \mathbf{H}\, dV = \frac{1}{2}\iiint_V \mathbf{J} \cdot \mathbf{A}\, dV$$

where $\mathbf{J}$ is the current density and $\mathbf{A}$ is the vector potential of the magnetostatic field.

FINAL EXAMINATION

1. Given that $\mathbf{A} = \cos\theta\mathbf{i} - \sin\theta\mathbf{j}$ and $\mathbf{B} = \cos\phi\mathbf{i} + \sin\phi\mathbf{j}$, derive the trigonometric expression for $\cos(\theta + \phi)$ by finding the cosine of the angle between $\mathbf{A}$ and $\mathbf{B}$.

2. Show that
$$(\mathbf{A} \times \mathbf{B}) \cdot (\mathbf{C} \times \mathbf{D}) + (\mathbf{B} \times \mathbf{C}) \cdot (\mathbf{A} \times \mathbf{D}) + (\mathbf{C} \times \mathbf{A}) \cdot (\mathbf{B} \times \mathbf{D}) = 0$$

3. Find $\nabla \cdot (\phi\mathbf{f})$ at the point $(1, 1, -1)$ if $\phi = x^2 - 3yz$ and $\mathbf{f} = xyz^2\mathbf{i} - 2xy^3\mathbf{j} + x^2yz\mathbf{k}$.

4. Show that $\mathbf{f} = 2xy\mathbf{i} + (x^2 + 1)\mathbf{j} + 6z^2\mathbf{k}$ is conservative.

5. Show that if $\mathbf{f} = \nabla\phi$ and $\nabla \cdot \mathbf{f} = 0$ for a region R bounded by surface S, then
$$\iiint\limits_{R} |\mathbf{f}|^2 \, dV = \oiint\limits_{S} \phi\mathbf{f} \cdot d\mathbf{S}$$

Final Examination Solutions

1. Construct **A** and **B** in the plane such that the angle between **A** and **B** is $\theta + \phi$, as shown in Figure E-1:

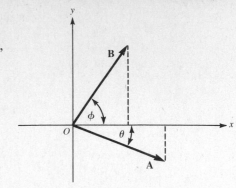

$$A = |\mathbf{A}| = \cos^2\theta + \sin^2\theta = 1$$
$$B = |\mathbf{B}| = \cos^2\phi + \sin^2\phi = 1$$

Thus by the definition of the scalar product we have

$$\cos(\theta + \phi) = \frac{\mathbf{A} \cdot \mathbf{B}}{|\mathbf{A}||\mathbf{B}|} = \cos\theta\cos\phi + (-\sin\theta)\sin\phi$$
$$= \cos\theta\cos\phi - \sin\theta\sin\phi$$

Figure E-1

2. Because the positions of the dot and cross in a triple scalar product can be interchanged, we have

$$(\mathbf{A} \times \mathbf{B}) \cdot (\mathbf{C} \times \mathbf{D}) + (\mathbf{B} \times \mathbf{C}) \cdot (\mathbf{A} \times \mathbf{D}) + (\mathbf{C} \times \mathbf{A}) \cdot (\mathbf{B} \times \mathbf{D})$$
$$= (\mathbf{A} \times \mathbf{B}) \times \mathbf{C} \cdot \mathbf{D} + (\mathbf{B} \times \mathbf{C}) \times \mathbf{A} \cdot \mathbf{D} + (\mathbf{C} \times \mathbf{A}) \times \mathbf{B} \cdot \mathbf{D}$$
$$= [(\mathbf{A} \times \mathbf{B}) \times \mathbf{C} + (\mathbf{B} \times \mathbf{C}) \times \mathbf{A} + (\mathbf{C} \times \mathbf{A}) \times \mathbf{B}] \cdot \mathbf{D}$$
$$= \mathbf{0} \cdot \mathbf{D} = 0$$

The vector on the left is zero since by adding the three identities

$$(\mathbf{A} \times \mathbf{B}) \times \mathbf{C} = (\mathbf{C} \cdot \mathbf{A})\mathbf{B} - (\mathbf{C} \cdot \mathbf{B})\mathbf{A}$$
$$(\mathbf{B} \times \mathbf{C}) \times \mathbf{A} = (\mathbf{A} \cdot \mathbf{B})\mathbf{C} - (\mathbf{C} \cdot \mathbf{A})\mathbf{B}$$
$$(\mathbf{C} \times \mathbf{A}) \times \mathbf{B} = (\mathbf{C} \cdot \mathbf{B})\mathbf{A} - (\mathbf{A} \cdot \mathbf{B})\mathbf{C}$$

we have $(\mathbf{A} \times \mathbf{B}) \times \mathbf{C} + (\mathbf{B} \times \mathbf{C}) \times \mathbf{A} + (\mathbf{C} \times \mathbf{A}) \times \mathbf{B} = \mathbf{0}$.

3. First write $\nabla \cdot (\phi\mathbf{f}) = \nabla\phi \cdot \mathbf{f} + \phi\nabla \cdot \mathbf{f}$. Now

$$\nabla\phi = \left(\frac{\partial}{\partial x}\mathbf{i} + \frac{\partial}{\partial y}\mathbf{j} + \frac{\partial}{\partial z}\mathbf{k}\right)(x^2 - 3yz) = 2x\mathbf{i} - 3z\mathbf{j} - 3y\mathbf{k}$$

$$\nabla \cdot \mathbf{f} = \frac{\partial}{\partial x}(xyz^2) + \frac{\partial}{\partial y}(-2xy^3) + \frac{\partial}{\partial z}(x^2yz) = yz^2 - 6xy^2 + x^2y$$

Thus at $(1, 1, -1)$
$$\nabla\phi = 2\mathbf{i} + 3\mathbf{j} - 3\mathbf{k}$$
$$\mathbf{f} = \mathbf{i} + 2\mathbf{j} - \mathbf{k}$$
$$\phi = 4$$

Also
$$\nabla \cdot \mathbf{f} = -4$$
$$\nabla\phi \cdot \mathbf{f} = (2)(1) + (3)(2) + (-3)(-1) = 11$$
$$\phi\nabla \cdot \mathbf{f} = 4(-4) = -16$$

Hence $\nabla \cdot (\phi\mathbf{f}) = 11 - 16 = -5$ at $(1, 1, -1)$.

4. We have
$$\nabla \times \mathbf{f} = \begin{vmatrix} \mathbf{i} & \mathbf{j} & \mathbf{k} \\ \dfrac{\partial}{\partial x} & \dfrac{\partial}{\partial y} & \dfrac{\partial}{\partial z} \\ 2xy & x^2 + 1 & 6z^2 \end{vmatrix} = \left[\frac{\partial}{\partial x}(x^2 + 1) - \frac{\partial}{\partial y}(2xy)\right]\mathbf{k} = (2x - 2x)\mathbf{k} = \mathbf{0}$$

Thus **f** is conservative.

5. We have $\nabla \cdot (\phi\mathbf{f}) = \nabla\phi \cdot \mathbf{f} + \phi\nabla \cdot \mathbf{f} = \mathbf{f} \cdot \mathbf{f} = |\mathbf{f}|^2$. Hence by the divergence theorem we have

$$\iiint_R \nabla \cdot (\phi\mathbf{f})\, dV = \iiint_R |\mathbf{f}|^2\, dV = \oiint_S \phi\mathbf{f} \cdot d\mathbf{S}$$

APPENDIX A
Summary of Vector Relations

A.1. Vector Algebra

$\mathbf{A} \cdot \mathbf{B} = \mathbf{B} \cdot \mathbf{A}$

$\mathbf{A} \cdot (\mathbf{B} + \mathbf{C}) = \mathbf{A} \cdot \mathbf{B} + \mathbf{A} \cdot \mathbf{C}$

$\mathbf{A} \times \mathbf{B} = -\mathbf{B} \times \mathbf{A}$

$\mathbf{A} \times (\mathbf{B} + \mathbf{C}) = \mathbf{A} \times \mathbf{B} + \mathbf{A} \times \mathbf{C}$

$\mathbf{A} \times \mathbf{A} = \mathbf{0}$

$\mathbf{A} \cdot \mathbf{B} \times \mathbf{C} = \mathbf{A} \times \mathbf{B} \cdot \mathbf{C}$

$\mathbf{A} \cdot \mathbf{B} \times \mathbf{C} = [\mathbf{ABC}]$

$[\mathbf{ABC}] = [\mathbf{BCA}] = [\mathbf{CAB}] = -[\mathbf{ACB}] = -[\mathbf{BAC}] = -[\mathbf{CBA}]$

$\mathbf{A} \times (\mathbf{B} \times \mathbf{C}) = (\mathbf{A} \cdot \mathbf{C})\mathbf{B} - (\mathbf{A} \cdot \mathbf{B})\mathbf{C}$

$(\mathbf{A} \times \mathbf{B}) \times \mathbf{C} = (\mathbf{A} \cdot \mathbf{C})\mathbf{B} - (\mathbf{B} \cdot \mathbf{C})\mathbf{A}$

A.2. Vector Calculus

$\nabla(\phi + \psi) = \nabla\phi + \nabla\psi$

$\nabla(\phi\psi) = \phi\nabla\psi + \psi\nabla\phi$

$\nabla\phi = \nabla\phi(u) = \phi'(u)\nabla u$

$\nabla \cdot (\mathbf{f} + \mathbf{g}) = \nabla \cdot \mathbf{f} + \nabla \cdot \mathbf{g}$

$\mathrm{div}(\mathrm{grad}\,\phi) = \nabla \cdot (\nabla\phi) = \nabla^2\phi$

$\nabla \times (\mathbf{f} + \mathbf{g}) = \nabla \times \mathbf{f} + \nabla \times \mathbf{g}$

$\nabla \times (\nabla\phi) = \mathbf{0}$

$\nabla \cdot (\nabla \times \mathbf{f}) = 0$

$(\mathbf{f} \times \nabla) \cdot \mathbf{g} = \mathbf{f} \cdot (\nabla \times \mathbf{g})$

$\nabla \cdot (\phi\mathbf{f}) = \phi\nabla \cdot \mathbf{f} + \mathbf{f} \cdot (\nabla\phi)$

$\nabla \times (\phi\mathbf{f}) = \phi\nabla \times \mathbf{f} + (\nabla\phi) \times \mathbf{f} = \phi\nabla \times \mathbf{f} - \mathbf{f} \times \nabla\phi$

$\nabla \cdot (\mathbf{f} \times \mathbf{g}) = \mathbf{g} \cdot (\nabla \times \mathbf{f}) - \mathbf{f} \cdot (\nabla \times \mathbf{g})$

$\nabla \times (\mathbf{f} \times \mathbf{g}) = \mathbf{f}(\nabla \cdot \mathbf{g}) - \mathbf{g}(\nabla \cdot \mathbf{f}) + (\mathbf{g} \cdot \nabla)\mathbf{f} - (\mathbf{f} \cdot \nabla)\mathbf{g}$

$\nabla(\mathbf{f} \cdot \mathbf{g}) = \mathbf{f} \times (\nabla \times \mathbf{g}) + \mathbf{g} \times (\nabla \times \mathbf{f}) + (\mathbf{f} \cdot \nabla)\mathbf{g} + (\mathbf{g} \cdot \nabla)\mathbf{f}$

$\mathrm{curl}(\mathrm{curl}\,\mathbf{f}) = \nabla \times (\nabla \times \mathbf{f}) = \nabla(\nabla \cdot \mathbf{f}) - \nabla^2\mathbf{f}$

$\nabla^2\mathbf{f} = \nabla(\nabla \cdot \mathbf{f}) - \nabla \times (\nabla \times \mathbf{f})$

$\nabla\left(\dfrac{\phi}{\psi}\right) = \dfrac{\psi\,\nabla\phi - \phi\,\nabla\psi}{\psi^2}$

$$\iiint_R \nabla \cdot \mathbf{f}\, dV = \oiint_S \mathbf{f} \cdot d\mathbf{S} \qquad \iint_S \nabla \times \mathbf{f} \cdot d\mathbf{S} = \oint_C \mathbf{f} \cdot d\mathbf{r}$$

$$\iiint_R \nabla\phi\, dV = \oiint_S d\mathbf{S}\,\phi \qquad \iint_S d\mathbf{S} \times \nabla\phi = \oint_C \phi\, d\mathbf{r}$$

$$\iiint_R \nabla \times \mathbf{f}\, dV = \oiint_S d\mathbf{S} \times \mathbf{f} \qquad \iint_S (d\mathbf{S} \times \nabla) \times \mathbf{f} = \oint_C d\mathbf{r} \times \mathbf{f}$$

APPENDIX B
Vector Differential Operations in Orthogonal Coordinates

B.1. Rectangular Coordinates (x, y, z)

$$\mathbf{f} = f_1\mathbf{i} + f_2\mathbf{j} + f_3\mathbf{k}$$

$$\nabla\psi = \frac{\partial\psi}{\partial x}\mathbf{i} + \frac{\partial\psi}{\partial y}\mathbf{j} + \frac{\partial\psi}{\partial z}\mathbf{k}$$

$$\nabla\cdot\mathbf{f} = \frac{\partial f_1}{\partial x} + \frac{\partial f_2}{\partial y} + \frac{\partial f_3}{\partial z}$$

$$\nabla\times\mathbf{f} = \begin{vmatrix} \mathbf{i} & \mathbf{j} & \mathbf{k} \\ \dfrac{\partial}{\partial x} & \dfrac{\partial}{\partial y} & \dfrac{\partial}{\partial z} \\ f_1 & f_2 & f_3 \end{vmatrix}$$

$$= \left(\frac{\partial f_3}{\partial y} - \frac{\partial f_2}{\partial z}\right)\mathbf{i} + \left(\frac{\partial f_1}{\partial z} - \frac{\partial f_3}{\partial x}\right)\mathbf{j} + \left(\frac{\partial f_2}{\partial x} - \frac{\partial f_1}{\partial y}\right)\mathbf{k}$$

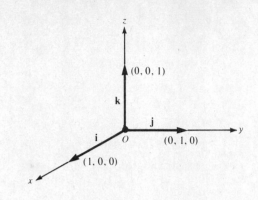

B.2. Cylindrical Coordinates (ρ, ϕ, z)

$$\mathbf{f} = f_\rho\mathbf{e}_\rho + f_\phi\mathbf{e}_\phi + f_3\mathbf{k}$$

$$\nabla\psi = \frac{\partial\psi}{\partial\rho}\mathbf{e}_\rho + \frac{1}{\rho}\frac{\partial\psi}{\partial\phi}\mathbf{e}_\phi + \frac{\partial\psi}{\partial z}\mathbf{k}$$

$$\nabla\cdot\mathbf{f} = \frac{1}{\rho}\frac{\partial}{\partial\rho}(\rho f_\rho) + \frac{1}{\rho}\frac{\partial f_\phi}{\partial\phi} + \frac{\partial f_3}{\partial z}$$

$$\nabla\times\mathbf{f} = \frac{1}{\rho}\begin{vmatrix} \mathbf{e}_\rho & \rho\mathbf{e}_\phi & \mathbf{k} \\ \dfrac{\partial}{\partial\rho} & \dfrac{\partial}{\partial\phi} & \dfrac{\partial}{\partial z} \\ f_\rho & f_\phi & f_3 \end{vmatrix}$$

$$= \frac{1}{\rho}\left(\frac{\partial f_3}{\partial\phi} - \frac{\partial f_\phi}{\partial z}\right)\mathbf{e}_\rho + \left(\frac{\partial f_\rho}{\partial z} - \frac{\partial f_3}{\partial\rho}\right)\mathbf{e}_\phi + \frac{1}{\rho}\left(\frac{\partial f_\phi}{\partial\rho} - \frac{\partial f_\rho}{\partial\phi}\right)\mathbf{k}$$

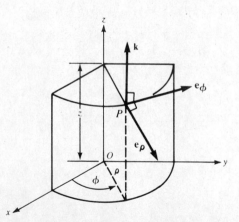

B.3. Spherical Coordinates (r, θ, ϕ)

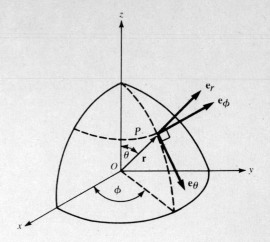

$$\mathbf{f} = f_r\mathbf{e}_r + f_\theta\mathbf{e}_\theta + f_\phi\mathbf{e}_\phi$$

$$\nabla\psi = \frac{\partial\psi}{\partial r}\,\mathbf{e}_r + \frac{1}{r}\frac{\partial\psi}{\partial\theta}\,\mathbf{e}_\theta + \frac{1}{r\sin\theta}\frac{\partial\psi}{\partial\phi}\,\mathbf{e}_\phi$$

$$\nabla\cdot\mathbf{f} = \frac{1}{r^2\sin\theta}\left[\sin\theta\frac{\partial}{\partial r}(r^2 f_r) + r\frac{\partial}{\partial\theta}(\sin\theta f_\theta) + r\frac{\partial f_\phi}{\partial\phi}\right]$$

$$= \frac{1}{r^2}\frac{\partial}{\partial r}(r^2 f_r) + \frac{1}{r\sin\theta}\frac{\partial}{\partial\theta}(\sin\theta f_\theta) + \frac{1}{r\sin\theta}\frac{\partial f_\phi}{\partial\phi}$$

$$\nabla\times\mathbf{f} = \frac{1}{r^2\sin\theta}\begin{vmatrix} \mathbf{e}_r & r\mathbf{e}_\theta & r\sin\theta\mathbf{e}_\phi \\ \dfrac{\partial}{\partial r} & \dfrac{\partial}{\partial\theta} & \dfrac{\partial}{\partial\phi} \\ f_r & rf_\theta & r\sin\theta f_\phi \end{vmatrix}$$

$$= \frac{1}{r\sin\theta}\left[\frac{\partial}{\partial\theta}(\sin\theta f_\phi) - \frac{\partial f_\theta}{\partial\phi}\right]\mathbf{e}_r + \frac{1}{r}\left[\frac{1}{\sin\theta}\frac{\partial f_r}{\partial\phi} - \frac{\partial}{\partial r}(rf_\phi)\right]\mathbf{e}_\theta$$

$$+ \frac{1}{r}\left[\frac{\partial}{\partial r}(rf_\theta) - \frac{\partial f_r}{\partial\theta}\right]\mathbf{e}_\phi$$

INDEX